1st International Conference on Applied Physics and Materials Science 2013

IOP Conference Series: Materials Science and Engineering
Volume 79

Davao City, Philippines
24-26 October 2013

ISBN: 978-1-5108-0747-1
ISSN: 1757-8981

Printed from e-media with permission by:

Curran Associates, Inc.
57 Morehouse Lane
Red Hook, NY 12571

proceedings.com

Some format issues inherent in the e-media version may also appear in this print version.

Printed by Curran Associates, Inc. (2015)

For permission requests, please contact the Institute of Physics
at the address below.

Institute of Physics
Dirac House, Temple Back
Bristol BS1 6BE UK

Phone: 44 1 17 929 7481
Fax: 44 1 17 920 0979

techtracking@iop.org

Additional copies of this publication are available from:

Curran Associates, Inc.
57 Morehouse Lane
Red Hook, NY 12571 USA
Phone: 845-758-0400
Fax: 845-758-2634
Email: curran@proceedings.com
Web: www.proceedings.com

Table of contents

Volume 79

1st International Conference in Applied Physics and Materials Science 24–26 October 2013, Davao City, Philippines

Accepted papers received: 10 March 2015
Published online: 10 June 2015

Preface

Papers

IOP Conf. Series: Materials Science and Engineering **79** (2015) 011001 doi:10.1088/1757-899X/79/1/011001

Preface

We are delighted to come up with thirty two (32) contributed research papers in this proceedings focusing on Materials Science and Applied Physics as an output of the **2013 International Conference in Applied Physics and Materials Science (ICAMS2013)** held on October 22-24, 2013 at the Ateneo de Davao University, Davao City, Philippines. The conference was set to provide a high level of international forum and had brought together leading academic scientists, industry professionals, researchers and scholars from universities, industries and government agencies who have shared their experiences, research results and discussed the practical challenges encountered and the solutions adopted as well as the advances in the fields of Applied Physics and Materials Science. This conference has opened a wide opportunity to establish multidisciplinary collaborations with local and foreign experts.

ICAMS2013, held concurrently with **15th Samahang Pisika ng Visayas at Mindanao (SPVM) National Physics Conference** and **2013 International Meeting for Complex Systems**, was organized by the Samahang Pisika ng Visayas at Mindanao (Physics Society of Visayas and Mindanao) based in MSU-Iligan Institute of Technology, Iligan City, Philippines. The international flavor of converging budding researchers and experts on Materials Science and Applied Physics was the first to be organized in the 19 years of SPVM operation in the Philippines. We highlighted ICAMS2013 gathering by the motivating presence of Dr. Stuart Parkin, a British Physicist, as one of our conference's plenary speakers. Equal measures of gratitude were also due to all other plenary speakers, Dr. Elizabeth Taylor of Institute of Physics (IOP) in London, Dr. Surya Raghu of Advanced Fluidics in Maryland, USA and Prof. Hitoshi Miyata of Niigata University, Japan, Prof. Djulia Onggo of Institut Teknologi Bandung, Indonesia, and Dr. Hironori Katagiri of Nagaoka National College of Technology, Japan.

The warm hospitality of the host university, Ateneo de Davao University, Davao City, Philippines blended with the overwhelming enthusiasm of the conference speakers, participants, and the unwavering support of the conference sponsors and donors and the administration of the MSU-Iligan Institute of Technology, Iligan City, Philippines, all have brought realization to the production of this proceedings.

Prof. Rosario L. Reserva
Prof. Reynaldo M. Vequizo
Prof. Arnold C. Alguno

IOP Conf. Series: Materials Science and Engineering **79** (2015) 011002 doi:10.1088/1757-899X/79/1/011002

2013 International Conference on Applied Science and Materials Science (ICAMS2013)
In conjunction with the 15th SPVM National Physics Conference and 2013 International Scientific Meeting for Complex Systems

Dr. Prof. Stuart Parkin, IBM, CA, USA

IOP Conf. Series: Materials Science and Engineering **79** (2015) 011002 doi:10.1088/1757-899X/79/1/011002

(Front row: Left to right) Dr. Surya Raghu (Maryland, USA), Prof. Jinky B. Bornales (MSU-Iligan Institute of Technology, Philippines), Dr. Beth Taylor (Institute of Physics, London), and Prof. Djulia Onggo (Institut Teknologi Bandung, Indonesia)

IOP Conf. Series: Materials Science and Engineering **79** (2015) 011002 doi:10.1088/1757-899X/79/1/011002

Host University: Ateneo de Davao University, Davao City

Samahang Pisika ng Visayas at Mindanao

Major Sponsor: MSU-Iligan Institute of Technology

Peer review statement

All papers published in this volume of *IOP Conference Series: Materials Science and Engineering* have been peer reviewed through processes administered by the proceedings Editors. Reviews were conducted by expert referees to the professional and scientific standards expected of a proceedings journal published by IOP Publishing.

IOP Conf. Series: Materials Science and Engineering **79** (2015) 012001 doi:10.1088/1757-899X/79/1/012001

Evaluating the B-cell density with various activation functions using White Noise Path Integral Approach

C J G Aban[1], R O Bacolod and M N P Confesor

Department of Physics, Mindanao State University – Iligan Institute of Technology Andres Bonifacio Avenue, Iligan City

E-mail: christinejoy.gravino@gmail.com, marknolan2006@gmail.com

Abstract. A The White Noise Path Integral Approach is used in evaluating the B-cell density or the number of B-cell per unit volume for a basic type of immune system response based on the modeling done by Perelson and Wiegel. From the scaling principles of Perelson [1], the B-cell density is obtained where antigens and antibodies mutates and activation function $f(|S-S_A|)$ is defined describing the interaction between a specific antigen and a B-cell. If the activation function $f(|S-S_A|)$ is held constant, the major form of the B-cell density evaluated using white noise analysis is similar to the form of the B-cell density obtained by Perelson and Wiegel using a differential approach.A piecewise linear functionis also used to describe the activation $f(|S-S_A|)$. If $f(|S-S_A|)$ is zero, the density decreases exponentially. If $f(|S-S_A|)= S-S_A-S_B$, the B-cell density increases exponentially until it reaches a certain maximum value. For$f(|S-S_A|)= 2S_A-S_B-S$, the behavior of B-cell density is oscillating and remains to be in small values.

1. Introduction

Since the proposal of an immune network by Jerne in 1955 [3], various attempts have been made to interpret the behavior of the immune system by studying the collective dynamical properties of interacting cellular or molecular species. Deducing macroscopic properties of the immune system from the properties and interactions among elementary components has a similar purpose to that of statistical mechanics [2]. Thus, results in statistical physics describing a large number of interacting particles can now be applied in immunology. Stocasticity is also a dominating phenomenon in immune system. For instance, T and B cell receptors are assembled by a random combination of germ-line genes during the differentiation of hematopoietic stem cells [3]. Therefore mathematical models of immune response inevitable will be stochastic models.

The white noise theory was develop by Takeyuki Hida in 1975 [4] as a novel approach to infinite dimensional analysis. The white noise analysis is an advance stochastic calculus that has been developed extensively for the last three decades. The basic idea in white noise theory was to take a collection of infinitely many independent random variables ω and treat this as a coordinate system of an infinite dimensional space [5]. White noise theory analysis is often used as a tool in evaluating the Feynman Path Integral, in quantum mechanical system and in problems related to protein folding, (i.e. [6]) stochastic neurodynamics (i.e. [7]), and polymer science [8,9] among others.

[1] To whom any correspondence should be addressed.

This study presents calculation of the B-cell density or the number of B-cells per volume using White Noise analysis. The B-cell density represents the population of B-cells at a certain time after it has encountered foreign pathogens. We use as a starting point the model and design principles for immune system recognition done by Perelson and Wiegel [1].

2. A short review on some design principles for antibody-antigen mutations

During an immune response, many B cells will mutate their antigen-binding receptor. As a result, a clone of identical cells will proliferate into a large number of different points in shape space. The shape space is a conceptual and computational framework to view antibody-antigen affinity and its resultant consequences. It is a linear space of all possible sequences of S_1, S_2, S_3, ... S_n , such that, each antibody is characterized by a specific point in S, which is called the shape [1]. The immune response phenomenon is modeled by the Brownian motion or diffusion in shape space. However, in order to escape immune detection, many antigens will also mutate. There are interesting examples, like the trypanosomes causing African sleeping sickness [11], and the HIV virus [12,14] in which the organisms have the ability to change their surface proteins (antigens). Previous modeling of Perelson *et al.* have the following assumptions for the mutation of antigens [1].

In the modeling of mutations of antigens, it is assumed that the shape $S_A(t)$ of the binding site of a new antigen A at time t is a linear function of time

$$S_A(t) = S_A(0) + v_A t \tag{2.1}$$

where $S_A(0)$ corresponds to the shape of antigen at time t =0. It is assumed that the shape space of the antigens runs through n-dimensional shape space with a constant velocity v_A. The assumption of an antigen moving in a straight-line motion may not be realistic. However, from the point of view of antigen survival, the fastest way to move far from some antibody vicinity is to move in straight line mutations.

Let $b(S_B,t)$ denote the density in shape space of B-cells with shape $S_B$at time t that have been stimulated by the antigen and stimulated B-cells proliferate at per capita rate $c(S,t)$. Assuming that all the stimulated cells divide at the same constant rate c, we have,

$$c(S_B,t) = c\,. \tag{2.2}$$

B-cells death is represented by $-d(S_B,t)b(S_B,t)$ where $-d(S_B,t)$ is the rate of cell death. Within the germinal centers, the location in the body in which most somatic mutation occurs, dividing the B cells appear to be programmed to die [1]. However, they are rescued from dying in an affinity – dependent fashion. Therefore, cells that mutate to be closer in shape space to the antigen are preferably spared from death which gives the following equation

$$d(S_B,t) = d - f(|\,S_B - S_A t\,|) \tag{2.3}$$

where d is a constant and the function f increases when the distance $|S_B$ - $S_A|$ is shape space between the particular B-cell and the antigen decreases. Lastly, stimulated B-cells also mutates in shape space which leads to the term $+D\Delta b$, where D is the diffusion coefficient and Δ, the Laplace operator.

Summing up all the cases, Eq. (2.9) represents the mutation of antigens in shape space

$$\frac{\partial b}{\partial t} = D\Delta b + \{f\,|\,S - S_A\,| - d_0\}b \tag{2.4}$$

where $d_0 = d - c$ and the path integral form of Eq. (2.4) is given by [1]

$$b(S_B,t)=\int \exp\left[-\frac{1}{4D}\int_0^t\left(\frac{dS_B}{d\tau}\right)^2 d\tau+\int_0^t\{f(|S_B-S_A)-d_o\}d\tau\right]d[S_B(\tau)] \tag{2.5}$$

In ref. [1], Eqn. (2.4), the number, $N_B(t)$, of B-cells was computer for shapes located inside a sphere of radius $\varepsilon_l$around the point that represents the shape of the antigen at time t;

$$N_B(t)=\frac{2\pi^{n/2}}{n\Gamma(n/2)}\varepsilon_1{}^n b(S_A(t),t) \tag{2.6}$$

where the asymptotic behavior of $b(S_A(t),t)$ is given by

$$b(S_A(t),t)\equiv(\text{constant})\exp\left[\left\{f(0)-d_o-\frac{v_A{}^2}{4D}\right\}t\right] \tag{2.7}$$

3. The White Noise Analysis

White noise calculus was introduced by Hida [4] in 1978 as a novel approach to infinite dimensional analysis and a tool in evaluating the Feynman Path Integral. The basic idea was to take the collection of infinitely many independent random variables, $\{\omega(t);t\in R\}$, and treat this as the coordinate system of infinite dimensional space. The following table shows the correspondence between the finite-dimensional and infinite-dimensional cases [4].

Table 1. Correspondence: Finite-Dimensional vs. Infinite-Dimensional Cases

FINITE DIMENSIONS	INFINITE DIMENSIONS
Independent variable: x_j	Independent random variable: $\omega(t)$
Coordinate system: $(x_1,\ldots, x_n)$	Coordinate System: $\omega(t);\ t\in \mathbf{R}$
Function: $f(x_1,\ldots, x_n)$	Functional: $\Phi(\omega(t);\ t\in \mathbf{R})$
Space: $\mathbf{R}^n$	Space of tempered distributions: S^*
Lebesgue measure: dx	Gaussian measure: $d\mu(\omega)$

3.1. The T- and S-transform

Each white noise functional is determined by its T-transform defined by [4],

$$T\Phi(\xi)=\int_{S^*}\exp(i<\omega,\xi>)\Phi(\omega)d\mu(\omega) \tag{3.1}$$

which is like a fourier transform analogue in infinite-dimensions. The Gaussian measure $d\mu(\omega)$ has the form

$$d\mu(\omega)=N_\omega\exp\left(-\frac{1}{2}\int\omega(\tau)^2 d\tau\right)d^\infty\omega \tag{3.2}$$

The exponential in $d\mu(\omega)$ is responsible for the Gaussian fall-off [4] and the N_ω is a normalization factor. In order to have correspondence

$$Nd^\infty x\rightarrow N_\omega d^\infty\omega=\exp\left(\frac{1}{2}\int\omega(\tau)^2 d\tau\right)d\mu(\omega) \tag{3.3}$$

S-transform is defined as

$$S\Phi(\xi) = \exp\left(-\frac{1}{2}\|\xi\|^2\right)\int_{S^*} \exp(<\omega,\xi>)\Phi(\omega)d\mu(\omega) \tag{3.4}$$

which can also be written in terms of the T-transform as

$$S\Phi(\xi) = C(\xi)T\Phi(-i\xi) \tag{3.5}$$

and

$$T\Phi(\xi) = C(\xi)S\Phi(i\xi) \tag{3.6}$$

4. B-cell density in immune response: a white noise functional approach

4.1. f(|S-S_A|) is a constant term

The activation function $f(|S_B\text{-}S_A|)$ defines the degree of activation for the B cell population. It is written in the form $f(h)$ where h, which is referred as the external field, is the total stimulation that the B-cell population receives from all the other B cell populations in shape space.

The path of the mutating antibodies in shape space is given as [1],

$$S_B(\tau) = S_A + \sigma(\tau) \tag{4.1}$$

where S_A is the position of the antigen and σ is the difference between the position of the B cell and antigen in shape space. Taking the derivative of S_B with respect to τ, we obtain;

$$\frac{dS_B}{d\tau} = v_A + \frac{d\sigma}{d\tau};\ \left(\frac{dS_B}{d\tau}\right)^2 = \left(v_A + \frac{d\sigma}{d\tau}\right)^2 = \left(\frac{d\sigma}{d\tau}\right)^2 + 2v_A.\frac{d\sigma}{d\tau} + v_A{}^2 \tag{4.2}$$

With Eq. (4.2), the integrand in (2.5) becomes

$$\exp\left\{\left[-\frac{1}{4D}\int_0^t\left\{\left(\frac{d\sigma}{d\tau}\right)^2 + 2v_A.\frac{d\sigma}{d\tau} + v_A{}^2\right\}d\tau\right] + \int_0^t\{f(\sigma) - d_o\}d\tau\right\} \tag{4.3}$$

where $f(\sigma) = f(|S_B\text{-}S_A|)$. We write Eq. (4.3) into the language of white noise through the white noise variable ω and is given by

$$\sigma(\tau) = \sigma_0 - v_A\tau + 2\sqrt{D}\int_0^t\omega(\tau)d\tau \tag{4.4}$$

where the velocity is

$$\frac{d\sigma(\tau)}{d\tau} = -v_A + 2\sqrt{D}\omega(\tau) \tag{4.5}$$

Expressing Eq. (4.3) in terms of Eqs. (4.4) and (4.5), the density equation now becomes

$$b(S_B,t) = \int exp\left\{\int_0^t -\frac{1}{4D}\left[4D\omega(\tau)^2 - 4v_A\sqrt{D}\omega(\tau) + v_A{}^2 - 2v_A{}^2 + 4v_A\sqrt{D}\omega(\tau) + v_A{}^2\right]\right\}d\tau$$

$$\times exp\left\{\int_0^t(f(\sigma) - d_0)d\tau + \frac{1}{2}\int_0^t\omega(\tau)^2d\tau\right\}du(\omega) \tag{4.7}$$

$$b(S_B,t) = \int exp\left\{\int_0^t -\frac{1}{2}\omega(\tau)^2d\tau + \int_0^t(f(\sigma) - d_0)d\tau\right\}du(\omega)$$

To facilitate the detection of antigens, we fixed the endpoints by introducing a Donsker Delta function from position S_B,the position of antibodies, to S_A, the position of antigens, at a final time t and expressing it in its Fourier transform, we have

$$\delta(S_B - S_A) = \frac{1}{2\pi}\int_{-\infty}^{\infty} exp(i\lambda(\sigma_0 - v_At + 2\sqrt{D}\int_0^t\omega(\tau)d\tau))d\tau \tag{4.8}$$

Now, Eq. (4.7) is now written as

$$b(S,t)=\frac{1}{2\pi}\int_{-\infty}^{+\infty}d\lambda\exp(\sigma_0-v_A t+\int_0^t(f(\sigma)-d_0))d\tau\times$$
$$\int\exp\left(\int_0^t-\frac{1}{2}\omega(\tau)^2d\tau\right)\exp\left(i\lambda(2\sqrt{D}\int_o^t\omega(\tau)d\tau\right)d\mu(\omega) \tag{4.9}$$

If we let $I_0=exp\left(\int_0^t-\frac{1}{2}\omega(\tau)^2d\tau\right)$, then the integral with respect to $d\mu(\omega)$ in Eq. (4.9) is equivalent to the T-transform of I_0, thus,

$$TI_0=\int exp\left\{\ i\int_0^t\omega(\tau)\xi d\tau\right\}I_0du(\omega)=exp\left(-\frac{1}{4}\int_0^t\xi^2d\tau\right) \tag{4.10}$$

where $\xi=2\lambda\sqrt{D}$, hence the B cell density is now

$$b(s,t)=\frac{1}{2\pi}\int_{-\infty}^{\infty}d\lambda\exp\left(i\lambda(\sigma_0-v_A t)+\int_0^t(f(\sigma)-d_0)d\tau\right)TI_0d\mu(\omega)$$
$$=\frac{1}{2\pi}\int_{-\infty}^{\infty}d\lambda\exp\left(i\lambda(\sigma_0-v_A t)+\int_0^t(f(\sigma)-d_0)d\tau\right)\times exp\left\{\ -\frac{1}{4}\int_0^t\varepsilon^2d\tau\right\}d\lambda \tag{4.11}$$

Now, provided that t is largeand value ofσ_0 is fixed [1], we have

$$\int_0^t\{f(\sigma)-d_0\}d\tau\approx\{f(0)-d_0\}t \tag{4.12}$$

with Eqs. (4.12), the B-cell density is now given as

$$b(S,t)=\frac{1}{2\pi}\int_{-\infty}^{\infty}exp\{i\lambda(\sigma_0-v_A t)+(f(0)-d_0)t\}\times exp\{-D\lambda^2t\}d\lambda$$
$$=\frac{1}{2\pi}\int_{-\infty}^{\infty}exp\{-D\lambda^2t+i\lambda(\sigma_0-v_A t)+(f(0)-d_0)t\}d\lambda \tag{4.13}$$

To evaluate Eq. (4.13), we use the Gaussian integral given by [10]

$$\int_{-\infty}^{\infty}exp(-ax^2+bx+c)dx=\sqrt{\frac{\pi}{a}}exp\left\{\frac{b^2}{4a}+c\right\} \tag{4.14}$$

where a, b and c are constants. Using Eq. (4.14), we obtain the form of the B-cell density as

$$b(S,t)=\sqrt{\frac{1}{4\pi Dt}}\exp\left\{\frac{-(\sigma_0-v_At)^2}{4Dt}+(f(0)-d_0)t\right\}$$
$$=\sqrt{\frac{1}{4\pi Dt}}\exp\left\{-\frac{{\sigma_0}^2}{4Dt}+\frac{v_A}{2D}(S_{B0}-S_{A0})-\frac{{v_A}^2}{4D}t+(f(0)-d_0)t\right\} \tag{4.15}$$
$$=\mathrm{C}\exp\left[\left\{(f(0)-d_0-\frac{{v_A}^2}{4D})t\right\}\right]$$

where the constant term C is

$$C=\sqrt{\frac{1}{4\pi Dt}}\exp\left\{\frac{v_A}{2D}(S_{B0}-S_{A0})-\frac{{\sigma_0}^2}{4Dt}\right\} \tag{4.16}$$

We note that given in Eq. (4.15), the B-cell density has the same form with that obtained by Perelson and Wiegel using differential approach [2].

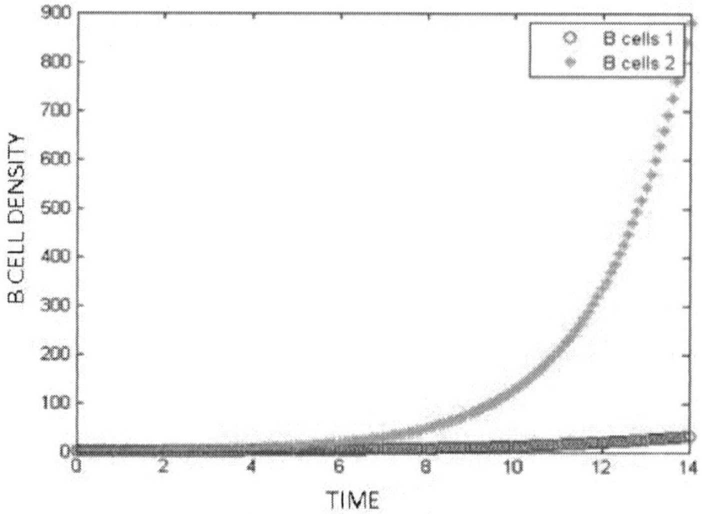

Figure 1. The B cell density when f (0) = 1 for two values of velocities.

In Figure 1, the minimal density of the B cells which settled asymptotically from 0, the plot exponentially increased (B cells 1 and B cells 2). When the velocity is changed from 10 to $v_A = 0.025$, the B cell density still increases exponentially. However, the density for the B cells 1 with $v_A = 10$ has slower increase when compared to the B cell density of B cells 2 with $v_A = 0.025$. A smaller value of the linear velocity of antigens indicates a rapid increase of the B cells since B cells are being activated easily when the antigen is within reach. Antigens that moves slower and too close to antibodies would be quickly recognized and eliminated [8]. Thus, the antigen would move away from regions of high B cell density, and in the extreme case one might envision this as a straight line movement. Fast moving antigens are harder to detect in the shape space by the B cells, where proliferation depends on the activation of the population of B cells. After staying of the mutated antigens in the vicinity of the shape space, the B cells detect these antigens and then activate. Soon after it activate, B cells then proliferate. By the time high-affinity antibodies are generated, the antigens are soon eliminated [8].

4.2. $f(\sigma)$ is a piecewise linear function

Now we consider the case for the B-cell density when we use a piecewise non-negative linear function for the $f(\sigma)$, which is divided into four regions with a stimulatory and suppressive part. Within the stimulatory part, increasing the field will increase the proliferation rate and within the suppressive part, increasing the field decreases proliferation [7].

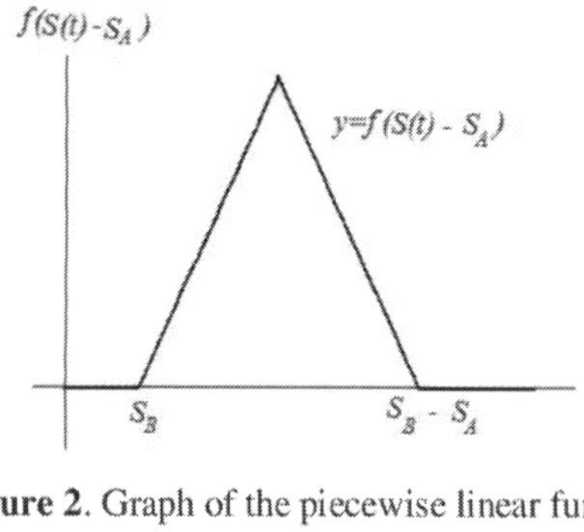

Figure 2. Graph of the piecewise linear function.

The four regions of the piecewise linear functions are given as:

$$f(\sigma)=\begin{Bmatrix} 0 & if\ \ \sigma\le S_B \\ \sigma-S_B & if\ \ S_B<\sigma\le\frac{S_A}{2} \\ S_A-S_B-\sigma & if\ \ \frac{S_A}{2}<\sigma\le S_A-S_B \\ 0 & if\ \ \sigma>S_A-S_B \end{Bmatrix}$$

where $\sigma = S(t) - S_A$. The piecewise linear function incorporates both stimulatory and suppressive part. Within the stimulatory part increasing the field $f(\sigma)$ increases proliferation and within the suppressive part increasing the field decreases proliferation [4].

4.2.1. Piecewise Linear function with $f(\sigma)=0$

The stimulated B-cells proliferate, divide with the same constant rate and mutates with a diffusion coefficient D. The B-cell density is given by

$$b(S,t)=\int_{S_B,o}^{S,t}\exp\left[-\frac{1}{4}\int_0^t\left(\frac{dS}{d\tau}\right)^2 d\tau-\int_0^t\{d_o\}d\tau\right]d[S(\tau)] \tag{4.17}$$

We parametrize the path $S(t)$ of a particle, which starts from an initial point S_B in terms of the Brownian motion $B(t)$ such that

$$S(t)=S_B+2\sqrt{D}\int_0^t\omega(\tau)d\tau \tag{4.18}$$

where the velocity is given by

$$\frac{dS}{dt}=2\sqrt{D}\omega(\tau) \tag{4.19}$$

The integration over all paths ($\lim_{N\to\infty}\prod d[S_j]$ or $d^\infty S$) leads to an integration over the Gaussian white noise measure $d\mu(\omega)=N_\omega\exp\left(-\frac{1}{2}\int\omega(\tau)^2d\tau\right)d\infty\omega$. We multiply $d\mu(\omega)$ by a factor $\exp\left(\frac{1}{2}\int\omega(\tau)^2d\tau\right)$ to obtain the correspondence $d^\infty S\to d^\infty\omega$. The B-cell density expressed in terms of the random white noise variable $\omega(\tau)$ is now written as

$$\begin{aligned} b(S,t)&=\int\exp\left[-\frac{1}{4}\int_0^t\left(2\sqrt{D}\omega\right)^2d\tau-\int_0^t d_o d\tau\right]\times\exp\left(\frac{1}{2}\int_0^t\omega^2d\tau\right)d\mu(\omega) \\ &=\int\exp\left(-\frac{1}{2}\int_0^t\omega^2d\tau-\int_0^t d_0 d\tau\right)d\mu(\omega) \end{aligned} \tag{4.20}$$

To facilitate the detection of antigens, we let the endpoints fixed by introducing a Donsker Delta function $\delta(S-S_A)$;

$$\begin{aligned} b(s,t)&=\int\exp\left[\left(-\frac{1}{2}\int_0^t\omega^2d\tau\right)-\int_0^t d_o d\tau\right]d\mu(\omega)\delta(S-S_A) \\ &=\frac{1}{2\pi}\int_{-\infty}^{\infty}\int\exp\left(i\lambda(S_B-S_A)-\int_0^t d_o d\tau\right)\times\exp\left(i\left(\int_0^t\omega\varepsilon d\tau\right)\right)I_0 d\mu(\omega)d\lambda \end{aligned} \tag{4.21}$$

where

$$\varepsilon=2\lambda\sqrt{D} \tag{4.22}$$

and

$$I_o = \exp\left\{-\frac{1}{2}\int_0^t \omega(\tau)^2 d\tau\right\} \tag{4.23}$$

Simplifying the above expression, we obtain the form of the density as

$$b(s,t) = \frac{1}{2\pi}\int_{-\infty}^{\infty} \exp\left(i\lambda(S_B - S_A) - \int_0^t d_o d\tau\right) TI_0 d\lambda \tag{4.24}$$

Where the T-transform of the Gauss Kernel I_0 is

$$TI_0 = \int \exp\left(i\left(\int_0^t \omega\varepsilon d\tau\right)\right) I_0 d\mu(\omega) = \exp\left(-\frac{1}{4}\int_0^t \varepsilon^2 d\tau\right) \tag{4.25}$$

Using the above transform and substituting it to the density equation, the B-cell density with $f(\sigma)=0$ is,

$$b(s,t) = \frac{1}{2\pi}\sqrt{\frac{\pi}{Dt}}\exp\left(\frac{(S_B - S_A)^2}{4Dt} - d_o t\right) \tag{4.26}$$

In Figure 3, which is the region where $f(\sigma) = 0$, the value of the B-cell density from Equation (4.26) is graphed versus time. We choose $S_A = 100$ and $S_B = 10$ following the values used in reference [7], $D = 100$, $d_0 = 1.0$ and $t = 72$ which is equivalent to three days since the antigen encounter. In appendix B, several values of S_A, S_B, D, d_0 and t are also being plotted and the same pattern is seen for all the values used. It is observed that the initial values of the B-cell density decreases in time until it reaches a stable low value.

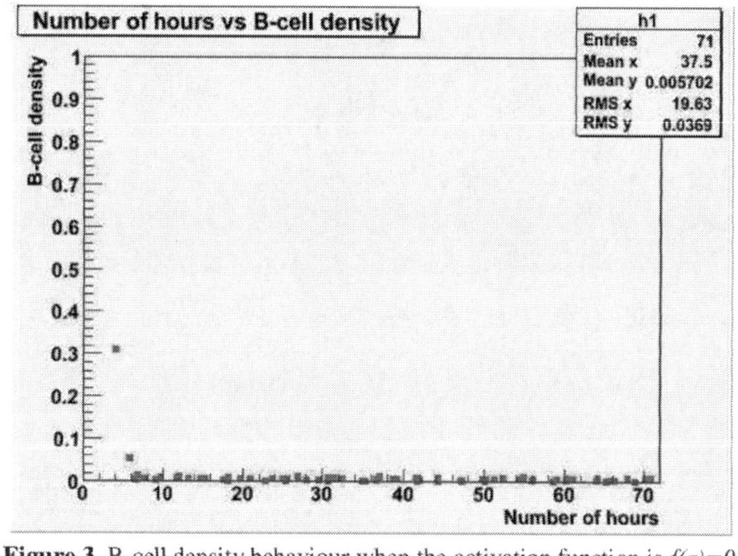

Figure 3. B-cell density behaviour when the activation function is $f(\sigma)=0$.

4.2.2. Piecewise Linear function with $f(\sigma)=S-S_A-S_B$

This region is considered to be a stimulatory region in which increasing the field $f(\sigma)$ will increase the rate of proliferation of the stimulated B-cells. The B-cell density is given by

$$b(s,t) = \int_{S_B,o}^{S,t} \exp\left(-\int_0^t (S_A + S_B + d_0) d\tau\right) \times \exp\left[-\frac{1}{4}\int_0^t \left(\frac{dS}{d\tau}\right)^2 d\tau + \int_0^t S d\tau\right] d[s(\tau)] \tag{4.27}$$

Converting it again into the language of white noise by parametrizing the path S into white noise variable ω:

$$s(\tau) = s_B + 2\sqrt{D}\int_0^t \omega(\tau)d\tau \tag{4.28}$$

where the velocity is given by:

$$\frac{dS}{d\tau} = 2\sqrt{D}\omega(\tau) \tag{4.29}$$

and

$$d[s(\tau)] = N_\omega d^\infty \omega = exp\left(\frac{1}{2}\int_0^t \omega(\tau)^2 d\tau\right)du(\omega) \tag{4.30}$$

the B-cell density equation has the form,

$$b(S,t) = \int \exp\left(-\int_0^t (S_A + d_0)d\tau\right) \times \exp\left(-\frac{1}{2}\int_0^t \omega^2 d\tau + \int_0^t\left(2\sqrt{D}\int_0^t \omega d\tau\right)d\tau\right)d\mu(\omega) \tag{4.31}$$

To fixed the endpoints for the antigen detection, we introduced a Donsker-delta function $\delta(S - S_A)$

$$\begin{aligned} b(s,t) &= \int_{S^*} \exp\left(-\int_0^t (S_A + d_0)d\tau\right) \times \exp\left(-\frac{1}{2}\int_0^t (\omega)^2 d\tau + \int_0^t\left(2\sqrt{D}\int_0^t \omega d\tau\right)d\tau\right)d\mu(\omega) \\ &\quad \times \delta(S - S_A) \\ &= \frac{1}{2\pi}\int_{-\infty}^{\infty}\int \exp\left(i\lambda(S_B - S_A) - \int_0^t (S_A + d_0)d\tau\right) \times \exp\left(-\frac{1}{2}\int_0^t (\omega)^2 d\tau\right) \\ &\quad \times \exp\left(\int_0^t\left(2\sqrt{D}\int_0^t \omega d\tau\right)d\tau + i\lambda 2\sqrt{D}\int_0^t \omega d\tau\right)d\mu(\omega)d\lambda \end{aligned} \tag{4.32}$$

In order to simply the integral, we used the following relationship

$$\begin{aligned} \int_0^t\left(2\sqrt{D}\int_0^t \omega d\tau\right)d\tau &= 2\sqrt{D}\int B(\tau)d\tau \\ &= 2\sqrt{D}\big(B(t) - B(0)\big) = 2\sqrt{D}\left(\int_0^t \omega(t)d\tau - \int_0^t \omega(0)dt\right) \\ &\approx 2\sqrt{D}\int_0^t \omega d\tau \end{aligned} \tag{4.33}$$

Thus, the density would now be written as:

$$\begin{aligned} b(s,t) &= \frac{1}{2\pi}\int_{-\infty}^{\infty}\int \exp\left(i\lambda(S_B - S_A) - \int_0^t (S_A + d_0)d\tau\right) \times \exp\left(-\frac{1}{2}\int_0^t (\omega)^2 d\tau\right) \\ &\quad \times \exp\left(\int_0^t\left(2\sqrt{D}\int_0^t \omega d\tau\right)d\tau + i\lambda 2\sqrt{D}\int_0^t \omega d\tau\right)d\mu(\omega)d\lambda \\ b(s,t) &= \frac{1}{2\pi}\int_{-\infty}^{\infty} \exp\left(i\lambda(S_B - S_A) - \int_0^t (S_A + d_0)d\tau\right) TI_0 d\lambda. \end{aligned} \tag{4.34}$$

The T-transform gives,

$$\begin{aligned} TI_0 &= \int \exp\left(i\left(\int_0^t \omega\varepsilon d\tau\right)\right) I_0 d\mu(\omega) = \exp\left(-\frac{1}{4}\int_0^t \varepsilon^2 d\tau\right) = \exp\left(-\frac{1}{4}\varepsilon^2 t\right) \\ &= \exp\left(-\frac{1}{4}(2\sqrt{D}\lambda - 2i\sqrt{D})^2 t\right) = \exp\left(-\frac{1}{4}(2\sqrt{D}\lambda - 2i\sqrt{D})^2 t\right) \\ &= \exp\left(-Dt\lambda^2 + i2\lambda\sqrt{D} + Dt\right) \end{aligned} \tag{4.35}$$

Thus, we have obtained the value of the density as

$$\begin{aligned} b(s,t) &= \frac{1}{2\pi}\int_{-\infty}^{\infty} \exp\left(i\lambda(S_B - S_A) - \int_0^t (S_A + d_0)d\tau \right) TI_0 d\lambda \\ &= \frac{1}{2\pi}\sqrt{\frac{\pi}{Dt}}\exp\left(\frac{\left((S_B - S_A) + 2\sqrt{Dt}\right)^2}{4Dt} + (D - S_A - d_0)t \right) \end{aligned} \quad (4.36)$$

Figure 4 shows the graph of the B-cell density in equation 4.40 versus time in hours with the field given by $f(\sigma)=S-S_A-S_B$. The values S_A = 100, S_B = 10, D = 100, d_0 = 0.5 and t = 72 is the same as the values used in the first region. It has been observed from the graph that the B-cell density increases from up to 24 hour and then stop. The immune system produces only limited number of cells, it would be difficult for the body to produce many antibodies to pair with all the antigens encountered by the body every day. Thus, for the immune system to detect many possible antigens it must produced high-affinity antibodies from the general antibodies.

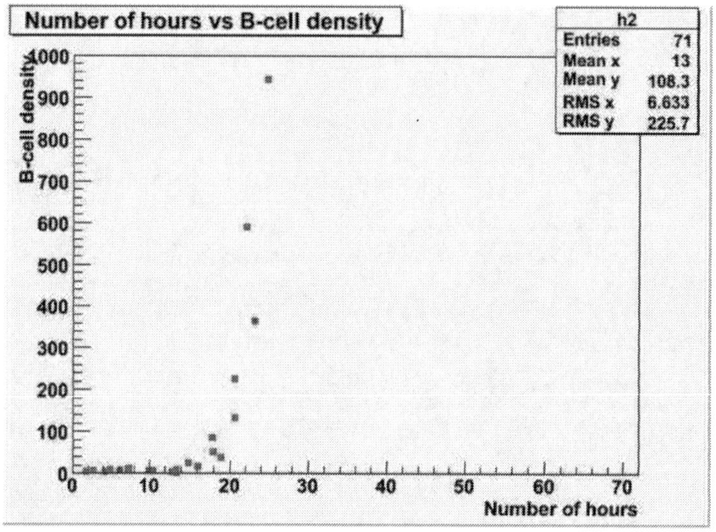

Figure 4. B-cell density behaviour when the activation function is $f(\sigma)=S-S_A-S_B$.

4.2.3. Piecewise Linear function with $f(\sigma)=2S_A-S_B-S$

The third region represents as the suppressive region, in this region, increasing the field $f(\sigma)=2S_A-S_B-S$, will decrease proliferation. At longer times, the B-cell will stop mutating and will start to convert generalist antibodies to high-affinity antibodies. In this region also, the B-cell density will cease to increase its value. Following equation for the B-cell density is given by,

$$b(s,t) = \int_{S_B,o}^{S,t} \exp\left(\int_0^t (2S_A - S_B - d_0)d\tau \right) \times \exp\left[-\frac{1}{4}\int_0^t \left(\frac{dS}{d\tau} \right)^2 d\tau - \int_0^t S d\tau \right] d[s(\tau)] \quad (4.37)$$

Following the procedure done in the previous region, in white noise form is,

$$b(s,t) = \int \exp\left(\int_0^t (2S_A - 2S_B - d_0)d\tau \right) \times \exp\left(-\frac{1}{2}\int_0^t (\omega)^2 d\tau - \int_0^t \left(2\sqrt{D}\int_0^t \omega d\tau \right) d\tau \right) d\mu(\omega) \quad (4.38)$$

Fixing the endpoints by introducing a Donsker-delta function $\delta(S - S_A)$;

$$
\begin{aligned}
b(s,t) &= \int \exp = \int \exp\left(\int_0^t (2S_A - 2S_B - d_0)d\tau\right) \times \exp\left(-\frac{1}{2}\int_0^t (\omega)^2\, d\tau - \int_0^t \left(2\sqrt{D}\int_0^t \omega d\tau\right) d\tau\right) d\mu(\omega) \\
&\quad \times \delta(S - S_A) \\
&= \frac{1}{2\pi}\int_{-\infty}^{\infty} \int \exp\left(i\lambda(S_B - S_A) - \int_0^t (2S_A - 2S_B - d_0)d\tau\right) \\
&\quad \times \exp\left(-\frac{1}{2}\int_0^t (\omega)^2\, d\tau\right) \times \exp\left(-2\sqrt{D}\int_0^t \omega d\tau + i\lambda 2\sqrt{D}\int_0^t \omega d\tau\right) d\mu(\omega) d\lambda \\
&= \frac{1}{2\pi}\int_{-\infty}^{\infty} \exp\left(i\lambda(S_B - S_A) - \int_0^t (2S_A - 2S_B - d_0)d\tau\right) TI_0\, d\lambda
\end{aligned} \tag{4.39}
$$

where the T-transform of I_0 is given by;

$$
\begin{aligned}
TI_0 &= \int \exp\left(i\left(\int_0^t \omega\varepsilon d\tau\right)\right) I_0\, d\mu(\omega) = \exp\left(-\frac{1}{4}\int_0^t \varepsilon^2 d\tau\right) = \exp\left(-\frac{1}{4}\varepsilon^2 t\right) \\
&= \exp\left(-\frac{1}{4}(i2\sqrt{D} + 2\lambda\sqrt{D})^2 t\right) = \exp\left(-\frac{1}{4}(i2\sqrt{D} + 2\lambda\sqrt{D})^2 t\right) \\
&= \exp\left(-Dt\lambda^2 - i2\lambda\sqrt{D} + Dt\right)
\end{aligned} \tag{4.40}
$$

So, the density is now written as;

$$
\begin{aligned}
b(s,t) &= \frac{1}{2\pi}\int_{-\infty}^{\infty} \exp\left(i\lambda(S_B - S_A) + \int_0^t (2S_A - 2S_B - d_0)d\tau\right) TI_0\, d\lambda \\
&= \frac{1}{2\pi}\sqrt{\frac{\pi}{Dt}} \exp\left(\frac{\left((S_B - S_A) - 2\sqrt{Dt}\right)^2}{4Dt} + (2S_B - 2S_A + d_0 - D)t\right)
\end{aligned} \tag{4.41}
$$

In this region as stated in [3], increasing the activation function should decrease the proliferation rate, however, the behavior of Eq.(4.41) shows an oscillating behavior. Verification from other studies are being done however as suggested in [2], a log-bell shape should be used for the activation function.

5. Summary and Recommendations

This study is an approach of using white noise theory in evaluating the B-cell density based from the immune response modeling of Perelson and Wiegel. In the framework of white noise analysis, when the activation is held constant, a somehow similar form is observe but with slight difference in the constant. We also used a piecewise linear function to define the B-cell density and its proliferation behavior. An increase in the B-cell density is observed in the second region since it is believed that in this region, B-cell should proliferate. The piecewise linear function is only a useful mathematical approximation of the B-cell density. However, there are features that are a bit unrealistic, thus, a more appropriate log-bell shape functions should be used in future studies.

6. References

[1] Perelson A and Wiegel F 1999 *Complexity***4**29–37
[2] Perelson S and Weisbuch G 1997 *Reviews of Modern Physics***69**
[3] Szabados T, Tusnády G, Varga L and Bakács T 1998 *Manuscript*
[4] Bernido C and Bernido M V 2002 *Proceedings of the 3rd Jagna Conference and International Workshop*13–9
[5] Retrieved from http://kidshealth.org/parent/general/body_basics/immune.html
[6] Carneiro J, Paixão T, Milutinovic D, Sousa J, LeonaK, Gardner R and Faro J 2005 *Journal of Computational and Applied Mathematics***184**77–100
[7] Bernido C and Bernido M V2006 *Proceedings of the 4th Jagna International Workshop*86–97
[8] Bernido C and Bernido M V2007 *Physics Letters* A1–4

[9] Bernido C, Bernido M V and Bornales J B 2005 *Physics Letters* A **339** 232–6
[10] Retrieved from http://www.ph.ed.ac.uk/~bjp/qt/Gaussian.pdf
[11] Retrieved from http://www.btinternet.com/~ukneqas.parasitologyscheme/Blood_Scheme/Teaching_Information/Trypanosomes/trypanosomes.html
[12] Perelson A 2002 *Nature Reviews* **2** 28–36
[13] Hida T 2003 *Taiwanese Journal of Mathematics* **7** 541–56
[14] PerelsonA and OsterG 1979 *J. Theor. Biol.* **81** 645–70
[15] Noest A J, Takumi K and de Boer R 1997 *Physica D: Nonlinear Phenomena* **105** 285–306

Surface viscosity effects on the motion of self-propelling boat in a channel

M G Aliperio[1] and Mark Nolan Confesor[1,2,3]

[1] Department of Physics, Mindanao State University – Iligan Institute of Technology, Andres Bonifacio Avenue, Tibanga, Iligan City 9200, Philippines

[2] Complex Systems Initiative, Mindanao State University – Iligan Institute of Technology, Andres Bonifacio Avenue, Tibanga, Iligan City 9200, Philippines

E-mail: marknolan2006@gmail.com

Abstract. Self-propelled droplets have been conceived as simple chemical toy models to mimic motile biological samples such as bacteria. The motion of these droplets is believe to be due to the surface tension gradient in the boundary of the droplet. We performed experiments to look at the effect of varying the medium viscosity to the speed of a circular boat that was soaked in Pentanol. We found that the boats undergo oscillatory type of motion inside a channel. Moreover we found the maximum speed of the boat is independent on the viscosity of the medium. On the other a time scale describing the width of the velocity profile of the boat was found to increase with increasing viscosity.

1. Introduction

Motility is an essential need of biological systems that has allowed it to respond to various environmental changes [1]. For instance *Escherichia coli* (E.coli) bacteria via chemotaxis move from regions of low food to regions of higher density of food. Present understanding is that the E.coli uses complex chemical network inside its body that allows it to measure chemical gradients in the environment and as a result control the direction of rotation of the molecular motor of the E.coli. Force generation used by E. coli is through the synchronized rotation of its flagella [2]. We note that E.coli is also sensitive to the presence of temperature (called thermotaxis) and optical field (optotaxis) gradients. Aside from using flagella for motion, other bacteria use crawling mechanism, generally termed as gliding motility, wherein the propulsion is due to active actin polymerization at the cell boundaries. This is for instance used by *Anabaena sp.* a cyanobacteria of family *Nostocaceae* [3]. Understanding cell motility is an important problem with huge potential uses in both biological manipulation and in medical science. For instance understanding cell migration could help elucidate the biochemical and physical dynamics of wound healing [4]. On the physical aspect, cell motility is interesting for a physical understanding of it requires a Non-Equilibrium thermodynamics picture of the system of which the scientific community is still trying to build. An issue of interest on these studies is how the physical and chemical properties affect the dynamics of cell motility. There have been varied approaches to look on this issue. First is the development of 3D scaffolds to look at cell

[3] To whom any correspondence should be addressed.

migration at 3D. Moreover, the scaffolds physical property can be changed. Another approach is to use simple models to mimic cellular motion. It was first reported in 1978 where the spontaneous motion of a surfactant in a two-phase co-existing system is similar to the behavior of an amoeba [5]. Thus the self-motion of a surfactant can be used as simple experimental models for investigating the behavior of micro-organisms.

Separation of a surfactant in a microscopic scale is one manifestation of multiple-phase coexisting systems. The self-motion of a reactive droplet is caused by an imbalance in the interfacial tension of a substrate induced by an inherent chemical effect in the moving droplet. This phenomenon is known as the Marangoni Effect [6]. Spontaneous motion of droplets driven by this effect is one good example of self-motion in physicochemical systems. In this study, a circular boat soaked in Pentanol is observed in water-glycerin aqueous solution. There had been recent studies conducted about the spontaneous motion of a Pentanol droplet driven by the Marangoni effect. Ref. [7] reported the motion of Pentanol with varying droplet size. A small droplet moves irregularly, a middle-sized droplet movesvectorially, and a large droplet splits into smaller droplets. They found out that its volume controls the droplet's type of motion. Also, in similar works in Ref. [8,9], spontaneous motion of Pentanol was found to be dependent on the temperature of the aqueous solution. These previous studies involved a two-phase coexisting system with Pentanol droplet that changes its shape as it moves similar to bacteria, and had not considered varying the viscosity of the aqueous phase.

In this paper, we report the effects of viscosity on the one-dimensional motion of boats soaked in Pentanol with the addition of Glycerin to the aqueous solution. These boats do not change in shape as it moves along the channel and during the diffusion of the pentanol it contain.

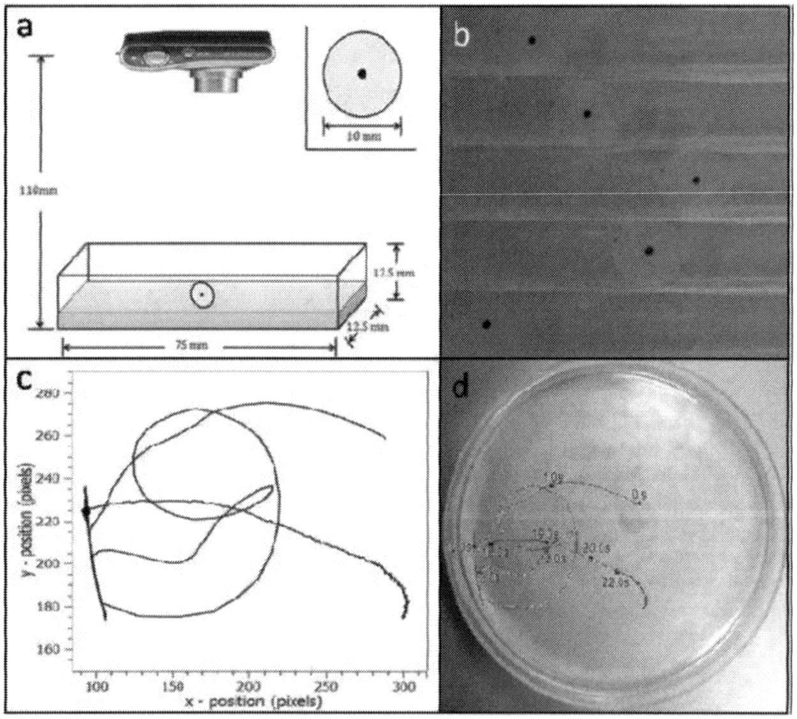

Figure 1. (a) Schematics of the Experimental set-up. (b) Image sequence showing the position of the boat at different time for the case of 3% glycerin concentration. (c) Trajectory of the boat in a petri dish at [Glycerin] = 0% concentration. (d) Image sequence representing the trajectories in (c). The petri dish has a diameter of 85mm.

2. Experimental design and methodology

Experiments are performed using the set-up shown in Figure 1a where 5mL of water-glycerin solution is poured on a rectangular glass channel with 75mm × 12.5mm × 12.5mm dimensions. To determine the effects of viscosity on the mode of motion of camphor boats, the concentration of the solution is varied from 0 vol. % to 21 vol. % glycerin (ρ = 1.261 g/cm3, PHILUSA Corp.). The circular boats, with a diameter of 10mm are made from filter paper. In order to clearly visualize its motion, the center of the camphor boat is inked using a small droplet of black enamel paint (Roosevelt Chemical Inc.). It is then soaked in 100% pentanol for 10 minutes before being added to the water-glycerin solution. The self-motion of the camphor boats on the aqueous solution is monitored using a video camera (Nikon Coolpix L21) at 30 frames per second. The experiments are carried out at room temperature (24°C) and are performed with five trials for each concentration. In the analysis of the movies of the boat self-motion, image sequences are exported using VirtualDub, and its location is determined per 1/30 second using a home-made particle tracking program written in IDL.

3. Results and discussion

Unlike the experiments done in Ref.[7], we used circular boats soaked in Pentanol in all our experiment. Already in our previous work we found that the circular boats undergo rich dynamics for instance circular trajectories as seen in figure 1d [10] when the boats is placed in an aqueous phase. This is very different to the irregular type of motion observe for the case of a plain active Pentanol droplet in water [7]. We take note that in all our other experiments we used a rectangular channel, Figure a, to confine the movement of the boat along one dimension.

Figure 1b shows the self-propelled motion of the boat for the case when the glycerin concentration is 0%. After being placed on the surface of the aqueous solution, the boat accelerates and then undergoes an oscillatory type of motion that is it moves back and forth along the channel. The self-propulsion is due to the surface tension gradient across the boat caused by the unequal distribution of pentanol along the boat diameter [7]. Surface tension and Pentanol concentration has a negative relation, i.e. the higher the concentration of Pentanol, the weaker is the surface tension. We take note that oscillatory type dynamics was also seen for the case of just a plain reactive droplet in channels [11]. In Figure 2a we plot a sample trajectory of a boat in a water medium ([Glycerin] = 0% vol.). The trajectory clearly shows the boat moving from one side of the channel towards the other side and then back and forth although after some time we observe the boat to not be able to reach the other end of the channel before reversing its direction of motion. To investigate the effect of increasing viscosity we made use of a Glycerin/Water solution as medium. We vary the concentration of Glycerin from 0% vol to 21% vol. In figure 2a-f we plot the trajectories of the boat at 0%, 3%, 6%, 9%, 15%, and 21% glycerin concentration respectively. We found that the boat still undergoes oscillatory type of motion irrespective of the concentration of glycerin. At [Glycerin] = 100%vol concentration we found the boat not to move at all. From the particle track, we were able to calculate the particle velocity using the formula where $x[i]$ and $x[i-1]$ are the position of the particle at time interval Δt. In Figure 2g, we plot the velocity for the case of 0% glycerin solution. We superimposed the particle trajectory for comparison. The velocity can be characterized by two parameters; the maximum speed vmax (which corresponds to the height of the peak) and the time width at half maximum, τ_{FWHM} (corresponds to the width of the peak). In figure 3a we plot the τ_{FWHM} values for different Glycerin concentration. It is observed that as the amount of Glycerin on the aqueous solution increases, the time width also increases linearly. On the other hand we found that v_{max} is insensitive to the Glycerin concentration.

To further gain insight to the experimental result we did numerical simulation of the motion of the boat using the model proposed in Ref.[7] where an active droplet motion is given by,

$$\frac{d^2x}{dt^2} = -\mu\frac{dx}{dt} + F(t) + \eta(t) \tag{1}$$

where$x(t)$ is the particle position at time t, μ is the surface tension viscosity, $F(t)$ is the force due to Marangoni effect, and $\eta(t)$ is the thermal noise. The second term of the right-hand side (rhs) of equation(1) corresponds to the friction experienced by the particle as it moves, thus the negative sign.

The third term of the rhs corresponds to the thermal fluctuation due to constant bombardment of the solvent molecules. We numerically evaluate equation (1) to extract features of the motion of the self-propelling particle for varying medium surface viscosity. The force $F(t)$ is modelled as a Heaviside function,

$$F(t) = \begin{cases} f_o, & t > 0.3 \\ 0, & t < 0.3 \end{cases} \tag{2}$$

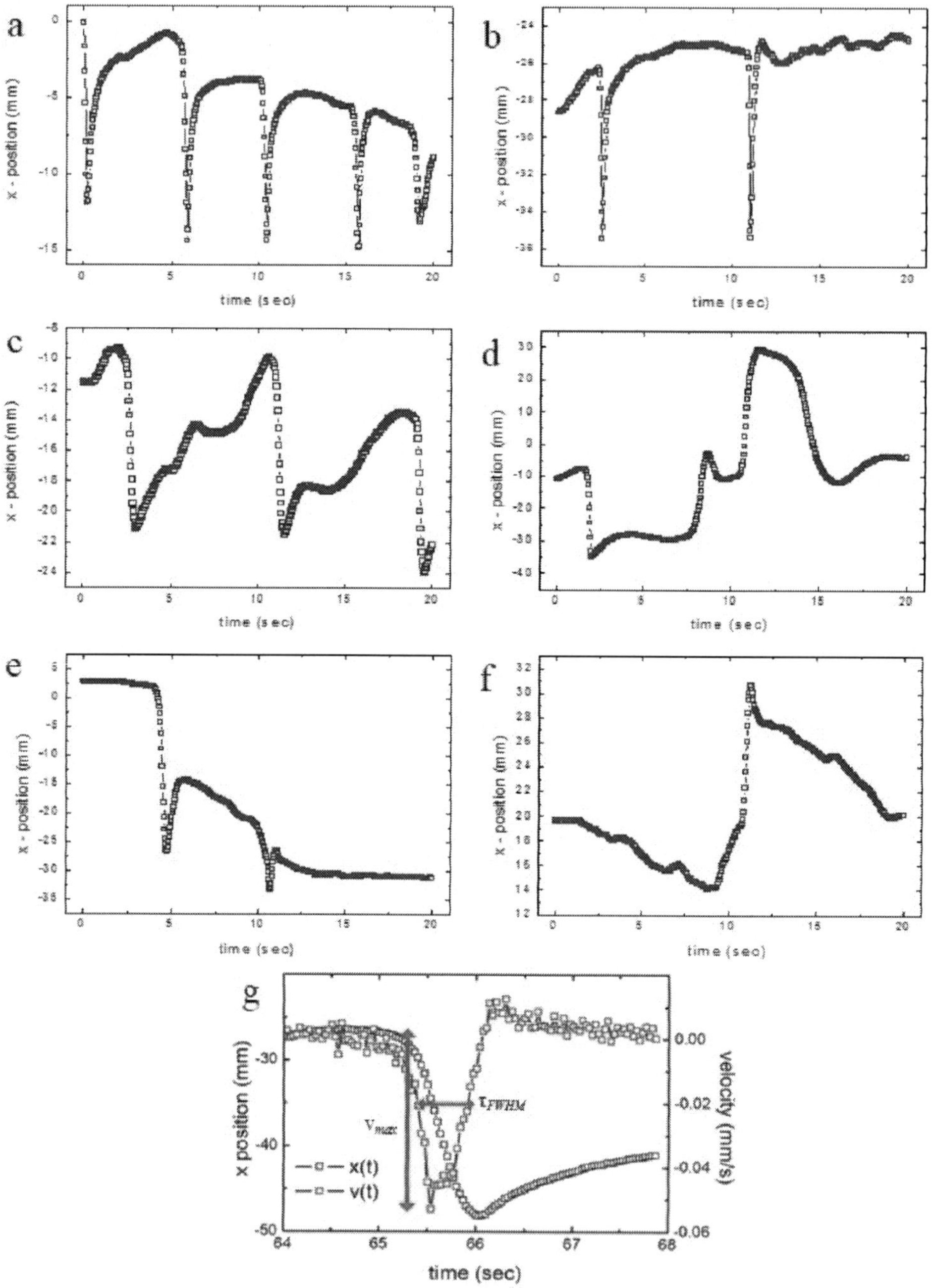

Figure 2. Trajectory plots for (a) 0%, (b) 3%, (c) 6%, (d) 9%, (e) 15%, (f) 21% solution which shows the oscillatory motion of the boat. (g) Position and velocity plot of the boat at 0% concentration. The height and the width of the peak determine v_{max} and τ_{FWHM} values respectively.

and we consider the case of a noise-free scenario. We note that modelling the force by a Heaviside function simplifies the analysis of the motion of the boat. We note that modelling the force by a Heaviside function simplifies the analysis of the motion of the boat. For instance the Pentanol inside the boat could still be diffusing out of the boat during transit and thereby the force may not be considered constant. In Figure 3c we plot the computed boat velocity for varying surface viscosity. Contrary to our experimental result, numerical computations suggest that τ_{FWHM} increases with decreasing viscosity.

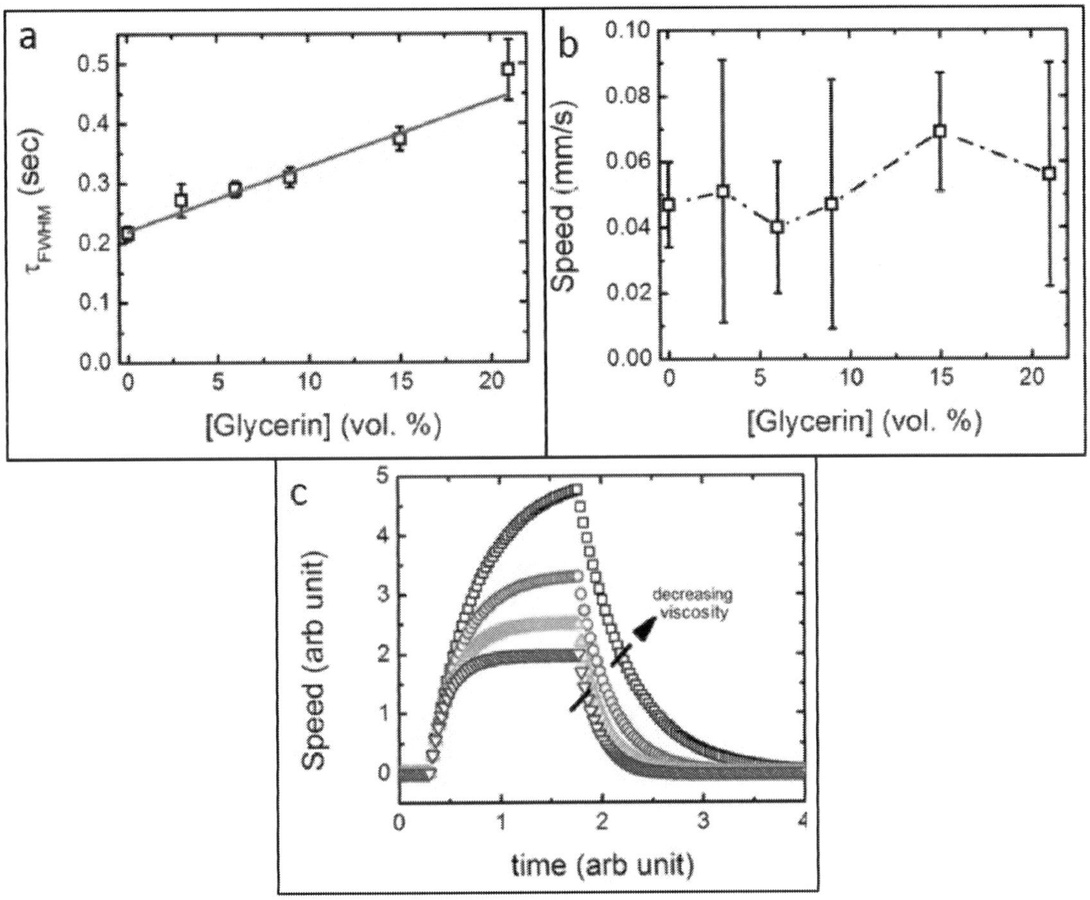

Figure 3. Plot of (a) $\tau_{FWHM}$and (b) speed values for different concentrations of glycerin. (c) Numerical calculation of the speed of the boat via equation (1) for different surface viscosity.

4. Conclusion

In summary, we have found that Pentanol soaked circular boats exhibit oscillatory self-motion when placed on rectangular channel filled with water-glycerin solution. Oscillatory motion of the boat is seen for [Glycerin] $<$ 21% vol. We found that the maximum velocity of the boat was found to be independent of the Glycerin concentration. Moreover the measured τ_{FWHM} was found to increase with increasing viscosity. This results is counter-intuitive, numerical calculation of a simplified model suggest that τ_{FWHM} should increase with decreasing viscosity. Future work will look on this inconsistency.

5. References

[1] Berg H C 1993 *Random Walks in Biology* (NJ USA, Princeton University Press)
[2] Sourjik V 2004 *Trends in Microbiology* **12** 12
[3] Hoiczyk E 2000 *Arch. Microbiol.* **174** 11–17
[4] Apell S P 2012 *Preprint* arXiv:1212.3778
[5] Dupeyrat M and Nakache E 1978 *Bioelectrochem. Bioenerg* .**5** 134
[6] Scriven L E and Sternling C V 1960*Nature* **187** 186
[7] Nagai K 2007*RIMS* **B3** 139–47
[8] Nagai K, Sumino Y, Kitahata H and Yoshikawa K 2006 *Progress of Theoretical PhysicsSuppement* **161** 286–9
[9] Nagai K, Sumino Y, Kitahata H and Yoshikawa K 2005 *Phys. Rev.* E **71** 065301
[10] Aliperio M G and Confesor M N 2014 *Preprint*
[11] Chen Y J, Nagamine Y and Yoshikawa K 2009 *Phys. Rev.* E **80** 016303

Acknowledgement
M.G.E.A. would like to acknowledge the financial support from DOST-SEI ASTHRDP.

Investigation of the Sound Pressure Level (SPL) of earphones during music listening with the use of physical ear canal models

K P Aying[1], R E Otadoy and R Violanda

Theoretical and Computational Sciences and Engineering Group
Department of Physics, University of San Carlos, Nasipit, Talamban, Cebu City, Philippines

E-mail:ayingismylastname@gmail.com, r_otadoy@yahoo.com, renante.violanda@gmail.com

Abstract. This study investigates on the sound pressure level (SPL) of insert-type earphones that are commonly used for music listening of the general populace. Measurements of SPL from earphones of different respondents were measured by plugging the earphone to a physical ear canal model. Durations of the earphone used for music listening were also gathered through short interviews. Results show that 21% of the respondents exceed the standard loudness/duration relation recommended by the World Health Organization (WHO).

1. Introduction

With the advent of affordable and mobile music players, listening to music has become one of the favorite past time of the general population today. With this personalized technology, earphones are massively used and have become a part of everyday activity. While music listening using earphone has become a stress-relieving activity, the effect of excessive loud music listening is often ignored or even unheard of. Though some of the music player issue some warning prior to increasing the volumes very little is known about the actual sound pressure level (SPL) produced inside the auditory canal. Excessive loudness is one of the causes of hearing loss in the world today [1,2]. In fact, the World Health Organization (WHO) already issued a set of standard for the amount of noise exposure level every day for a safe listening [1].

This study measured the sound pressure level used by a representative sample of the general populace in their music listening activity using a physical ear canal model. The duration of the music listening per respondent was also determined through interview. The measured SPL and duration was elucidated in comparison with the issued standard set by WHO.

2. Physical ear canal (PEC) model

The sophistication and complexity necessary to measure SPL in an actual ear canal cannot be address with the current laboratory set-up. Instead of measuring the SPL in an actual ear canal, this study

[1] To whom any correspondence should be addressed.

utilizes a physical ear canal model (PEC) in the measurement. The PEC is a hollow cylindrical tube made of engineering plastic and steel. The PEC has an inner diameter of 8.00 mm which is comparable to the average diameter of the human ear [3-5]. Three different lengths of (29.80±0.05)mm, (29.00±0.05)mm, and (26.50±0.05)mm were utilized to include the range of ear canal length reported by the literatures [4,5]. Figure 1, shows the schematic diagram of the PEC. The earphones to be tested are placed at one end of the PEC while the sound meter is placed on the other end.

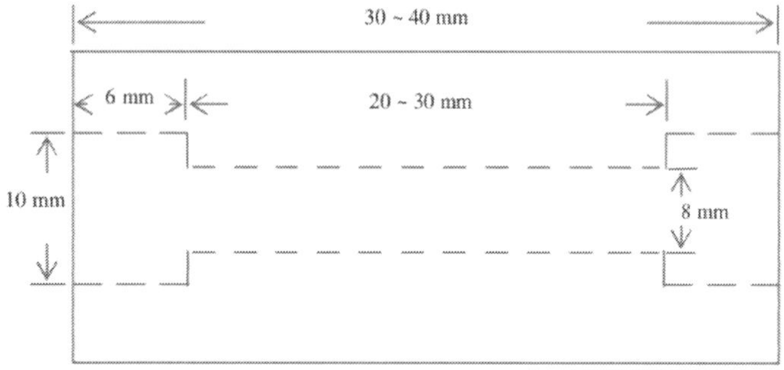

Figure 1. Schematic diagram of a PEC model. The dimensions of the PEC models were obtained from two renowned medical literatures[4,5].

3. Characterizing of PEC models

Prior to using the PEC in the SPL measurements, it is very important to study the acoustical property of the PEC and compare it with the acoustical property of the actual ear canal. Here, only the frequency response property is considered as the actual ear canal acts as a band pass filter of the sound entering the auditory system [6-8]. To determine the frequency response of the PEC, a white noise sound was generated by a LabVIEW computer program using the sound card of the computerand fed it to an earphone which was inserted to one end of the PEC. The sound generated by the earphones was captured by placing a Brüel&Kjær Pre-polarized Free-field ½" Microphone Type 4189 at the other end of the PEC. The microphone was coupled to NI-USB 4431 Data Acquisition module which is connected to the same computer. The power spectrum of the captured sound was then computed using a LabVIEW code that performs a Fast Fourier transformation. A normalization process was then applied to the calculated power spectrum to correct for the distortion of the white noise caused by the earphone itself. Figure 2 shows the experimental setup for the characterization of the PEC.

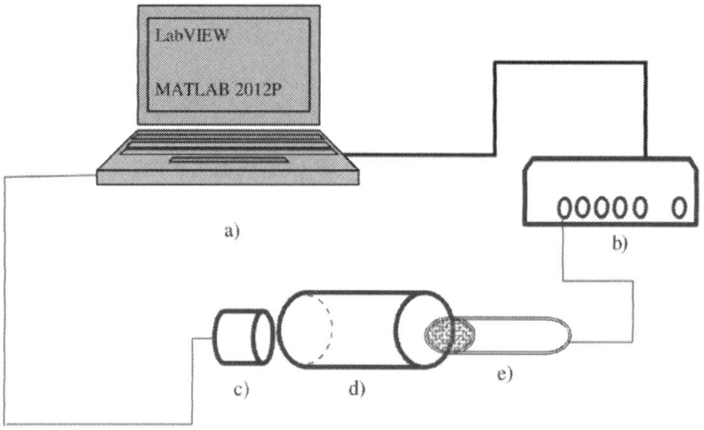

Figure 2. Schematic diagram of experimental setup where a) computer running with LabVIEW program and MATLAB 7.1 software, b) NI DAQ module,c) earphone, d) PEC models, e) Brüel&Kjær Prepolarized Free-field ½" Microphone Type 4189.

4. Measuring of sound pressure level with earphones

Sound pressure levels (SPL) were measured forrandom respondents. The selection of the respondents were done by asking the respondents that were listening to music using earphones that are present in a particular environment such as, university campuses, inside a public transportation vehicle, shopping malls, on the sides of the road and etc. The respondents have ages ranging from 16 to 35 years old with almost the same number of males and females. During each measurement, the respondents were asked to set the volume of their music players to the usual volumes that they used in listening. Their earphones are then inserted to one end of the PEC and a PYLE® PSPL05R Sound Level Meter is inserted in the other end to measure the SPL. Ten SPL values are obtained for each respondent in a span of half a minute. The average SPL was then computed from the ten measured values. The respondents were also asked of the frequency and the average duration of their music listening in a day.

5. Results and discussions

5.1. Characterization of PEC models

The normalized power spectrum of computed from the inside of the engineering plastic PEC is shown in Figure 3 for the different lengths. The vertical axis of the graph shows the SPL and the horizontal axis indicates the frequency of the sound. It can be seen that the graph peaks between 3 kHz and 4 kHz. This is consistent with the frequency response characteristic of the actual ear canal [6-8]. Similar trend were also exhibited by the steel PEC hence the PEC frequency response characteristic is material independent. The frequency response of the PEC is only affected by its length as can be seen in the slight difference of the frequency response for the different length. Ear canal also amplifies sound emitted by the earphones as can be noted in the positive values of the normalized power spectrum. This is due to the confinement of the sound resulting to the sound energies directed towards the microphone or to the ear drum in real situation. With the similarity of the PEC acoustical characteristic to that of actual ear canal, it can be safely assumed that the PEC can represent closely the sound propagation process in the actual ear canal. Hence, PEC can be a useful substitute to an actual ear canal in the SPL measurements inside the ear canal.

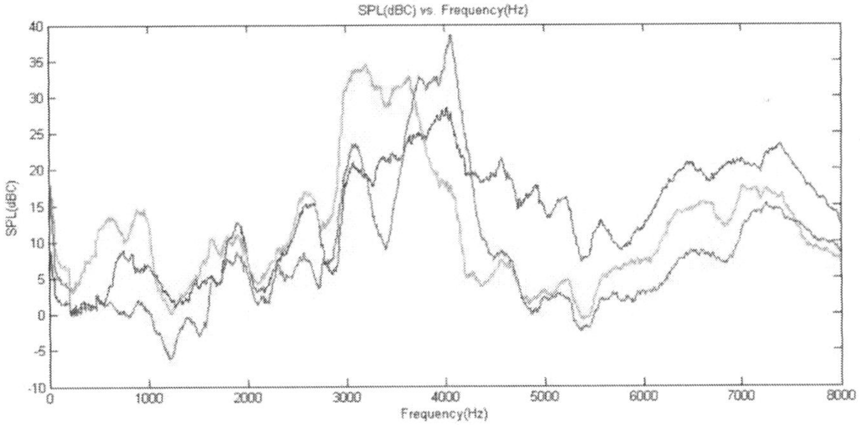

Figure 3. The graph showing the engineering plastic PEC models frequency response. In decreasing order, blue line shows the frequency response of the longest sample, followed by green and then the red.

5.2. Measurement of SPL with earphones

A total of 266 respondents participated in the ear canal SPL measurement. The histogram of the average SPL is shown in Figure 4. The average SPL ranges from about 65 dB up to 105 dB. The most frequent SPL of the sample set is estimated to be around 80-85 dB. This range of SPL is comparable to the noise generated in a busy street. It can also be seen that quite a number of respondents (about 35% of the total sample size) are using excessive volumes (85 dB and above) in their music listening. While it cannot be conclusively stated that these respondents are in the risk of having hearing loss or are already hearing impaired it is clear that a percentage of the populace has the tendency to use excessive loud music and therefore need to be effectively warned.

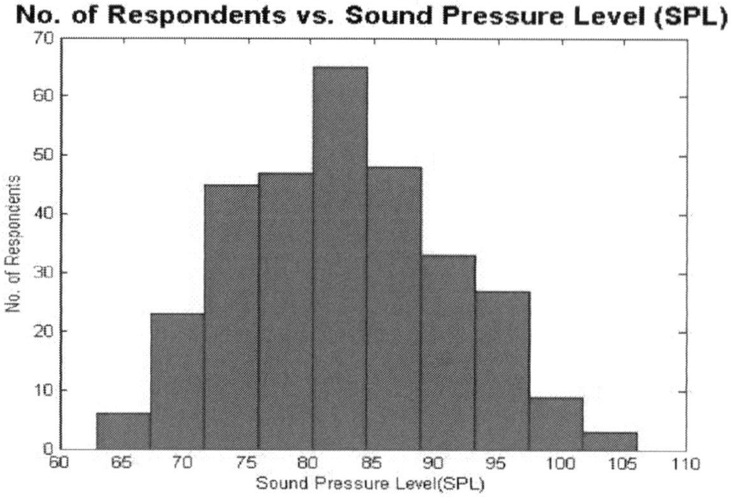

Figure 4. Histogram of the sound pressure level.

Incorporating the listening duration per day to the measured SPL, a scatter plot of the SPL versus duration for all 266 respondents is shown in Figure 5. The dash-line in the graph represents the noise level and recommended daily exposure duration by the WHO. The plot shows that about 21% of the respondents, points in the right of the dash-line, exceed the exposure duration set by WHO. These respondents have a risked of acquiring noise induced hearing loss (NIHL).

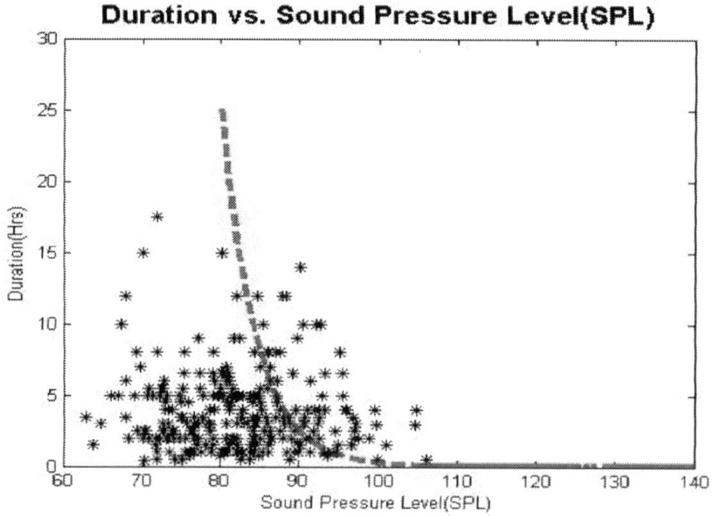

Figure 5. The red broken line shows the standard set by the WHO for the amount of time exposure for a specific loudness level. The asterisks are the data for the 266 respondents of the average time of exposure vs. the sound pressure level in music listening.

6. Conclusion

We have modeled an ear canal as hollow cylinders with dimensions representative to the actual dimension of an ear canal. The physical ear canal model has a peak frequency response in the range of about 3 kHz up to 4 kHz. The acoustical properties of the physical ear canal model are comparable to the acoustical properties of a real ear canal basing from literature. This allows the model to be a substitute for the real ear canal for sound pressure level measurement.

Approximation of the sound pressure level inside the ear canal during music listening of the sample set of the populace was carried. To minimize the complexity of the experiment, a physical ear canal model was utilized instead of the actual ear canal of the respondents. We have found out that the most of frequent sound pressure level of the respondents is within 80 dB to 85 dB. Thirty five percent of the respondents listen to music with an average of more than 85 dB SPL. In comparison to the WHO standard for noise exposure level, about 21% of the respondents exceed the WHO criteria. In conclusion, we have successfully approximated the sound pressure level during music listening. The result of this study could be used in formulating effective warnings to minimize noise induced hearing loss of the listener.

As a recommendation, it would be important to include more respondents to have a much better statistics. It is also interesting to correlate the sound pressure level with that of background noise level to identify potential behavior of the listener.

References

[1] U.S. Department of Health and Human Services 1998 *Criteria for a Recommended Standard: Occupational Noise Exposure, Revised Criteria 1998* (Ohio: Public Health Service Centers for Disease Control and Prevention National Institute for Occupational Safety and Health)

[2] Levey S, Fligor B, Ginocchi C and Kagimbi L 2012 *Contemporary Issues in Communication and Disorders* **39** 76–83

[3] Hiipakka M T2008 *Measurement Apparatus and Modelling Techniques of Ear Canal Acoustics*Master Thesis, Helsinki University of Technology, Helkinsi, Finland

[4] Moore K L and Dalley A F 1980 *Clinically Oriented Anatomy* (Lippincott Williams & Wilkins) pp 962–76

[5] Shier D, Butler J and Lewis R*Hole's Human Anatomy & Physiology* (New York: McGraw-Hill) pp 454–62

[6] Pielemeier W H 1946 *Journal of the Acoustical Society of America* **18** 248

[7] Mehrgardt S and Mellert V 1977 *Journal of the Acoustical Society of America* **61** 1567

[8] Shaw E A G 1974 *Journal of the Acoustical Society of America* **56** 1848–61

Correlation between the residual stress in 3C-SiC/Si epifilm and the quality of epitaxial graphene formed thereon

R Bantaculo[1,2,3], **H Fukidome**[2] **and M Suemitsu**[2]

[1]Department of Physics, MSU-Iligan Institute of Technology, Bonifacio Avenue,Tibanga, 9200 Iligan City, Philippines

[2]Research Institute of Electrical Communication, Tohoku University, Katahira 2-1-1, Aoba-ku, Sendai 980-8577, Japan

E-mail: rolando.bantaculo@g.msuiit.edu.ph

Abstract. The quality of epitaxial graphene on silicon (GOS) is negatively correlated with the residual stress of the 3C-SiC films grown on the Si substrates. This has been systematically demonstrated by use of a series of 3C-SiC films formed on Si(110) substrates with varied residual stress. The residual stress of the3C-SiC film and the grain size of graphene were estimated with Raman-scattering spectroscopy while the crystallinity of the3C-SiC filmwas evaluated by x-ray diffraction. The more it reduces the residual stress the better the GOS quality it becomes. In particular, use of the rotated epitaxial film of 3C-SiC formed on Si(110) substrate, which gives the lowest residual stress, is found to produce a graphene with one of the best quality ever obtained in GOS. The revealed GOS quality improvement opens new opportunities for the production of high-performance GOS-based devices.

1. Introduction

Graphene, a single atomic layer of hexagonally arranged carbon atoms, is attracting much attention owing to its very interesting physical properties, such as the high electron mobility reaching up to ~200,000 $cm^2/V\cdot s$, anomalously quantized Hall effect [1-5], and spin polarization of the current [6]. These physical properties make graphene a promising material for future electronic devices.

Graphene can be formed by many ways. One can form graphene by micromechanical [1] and chemical [7] exfoliation of graphite as well as by CVD on metal substrates [8]. Among others, however, the industrially most practical way of forming graphene should be theepitaxial growth of graphene on surfaces of SiC crystals, which can be conducted on both bulk [9] and epitaxial films [10-11]. By simply annealing the SiC surfaces to desorb Si atoms therefrom, epitaxial graphene forms. The epitaxial graphene on SiC thin films, or further theepitaxial graphene on silicon (GOS), is especially intriguing in that it can enjoy various device technologies accumulated in Si devices in making graphene-based devices [11-12]. Field-effect transistors have already been demonstrated by using GOS [13].

The challenge for GOS production is its inclusion of defects [11], which is related to imperfections already present in the 3C-SiC films formed on the Si substrates such as stacking faults, misfit dislocations, antiphase boundary domains, surface roughness and cracks. For instance, Fukidome *et*

[3] To whom any correspondence should be addressed.

al.[14] have demonstrated that the quality of graphene is determined, at least partly, by the surface roughness of the SiC film used in the GOS [14] process. These imperfections in the SiC films used in GOS are somehow related to the large lattice mismatch (~20%) present between the 3C-SiC film and the Si substrate [11,15]. This large lattice mismatch induces a significant tensile stress accumulated in the SiC film. Little is known, however, about the relationship between the residual stress in the 3C-SiC film and the quality of GOS. We have prepared in this study four kinds of GOS samples with varying residual stresses in the 3C-SiC film.

We used Si(110) wafers as the starting substrates. The merit of using the Si(110) substrate is that we can widely change the residual stress in the 3C-SiC thin films on this surface. In particular, presence of the rotated epitaxy of 3C-SiC(111) on Si(110) [15-16] can greatly minimize the residual stress. In this rotated epitaxy, the accumulated stress in the film is minimized by forming a well-lattice-matched stacking of $[\bar{1}\,\bar{1}\,0]$Si//$[\bar{1}\,\bar{1}\,0]$3C-SiC and [001]Si//$[\bar{1}\,\bar{1}\,2]$3C-SiC [15,17] orientation. In fact, it was experimentally demonstrated that the lattice constant anisotropy $(a_{//}-a_{\perp})/a_{//}$ between the lateral and the growth directions is decreased by a factor of four from that of 3CSiC(111)/Si(111) in this rotated epitaxy [16]. We have conducted in this study x-ray diffraction (XRD) and Raman-scattering measurements on the four samples mentioned above, and have obtained correlations between the residual stress in the 3C-SiC film, crystallinity of 3C-SiC evaluated by the XRD-FWHM, and the domain size of the GOS film evaluated by the Raman analysis.

2. Experimental methods

An N-type Si(110) wafer with a resistivity of 1~20 Ω·cm., cut into 7x25 mm^2 substrates, was used. In prior to growth, substrateswere wet cleaned employing the RCA technique [18-19]. The growth was conducted by a gas-source molecular beam epitaxy (GSMBE) using monomethylsilane (MMS, CH_3SiH_3) as a single source. After introduction to the GSMBE chamber, the sample was degassed at 600 °Cand flash annealed at 1100°C for several times to clean the surface. MMS (Trichemical, 99.999%) was then introduced into the chamber. In the first step, a buffer-layer was formed at 600 °C [20] which was followed by the second step: formation of the 3C-SiC thin film [11-12]. After the 3C-SiC growth, graphene was formed on the SiC surface by annealing the sample at 1250 °C *in vacuo* [21-22]. Four GOS samples were prepared, whose preparation details are summarized in Table 1.

Table 1. Summary of the multi-step growth method for growing graphene with varying stresses and crystal qualities.

Sample Name	Step 1 Buffer layer	Step 2 SiC Growth	Step 3 SiC Growth	Step 4 Graphene Growth
Sample (a)	$P_{MMS}=4\times10^{-3}$ Pa	$P_{MMS}=4\times10^{-3}$ Pa	$P_{MMS}=4\times10^{-2}$ Pa	$P_b=2\text{~}3\times10^{-7}$ Pa
Thickness:	T=600°C	T=1050°C	T=1050°C	T=1250°C
~78 nm	t_{depo} : 5 min	t_{depo} : 40 min	t_{depo} : 60 min	t_{anneal}: 30 min
Sample (b)	$P_{MMS}=5\times10^{-3}$ Pa	$P_{MMS}=4\times10^{-3}$ Pa	$P_{MMS}=2.5\times10^{-2}$ Pa	$P_b=2\text{~}3\times10^{-7}$ Pa
Thickness:	T=600°C	T=1000°C	T=1000°C	T=1250°C
~108 nm	t_{depo} : 15 min	t_{depo}: 220 min	t_{depo}: 110 min	t_{anneal}: 30 min
Sample (c)	$P_{MMS}=1.25\times10^{-1}$ Pa	$P_{MMS}=5\times10^{-3}$ Pa	$P_{MMS}=5\times10^{-2}$ Pa	$P_b=2\text{~}3\times10^{-7}$ Pa
Thickness:	T=600°C	T=1000°C	T=1050°C	T=1250°C
~111 nm	t_{depo}: 5 min	t_{depo}: 88 min	t_{depo}: 65 m in	t_{anneal} : 30 min
Sample (d)	$P_{MMS}=1.25\times10^{-1}$ Pa	$P_{MMS}=5\times10^{-3}$ Pa	$P_{MMS}=5\times10^{-2}$ Pa	$P_b=2\text{~}3\times10^{-7}$ Pa
Thickness:	T=600°C	T=1000°C	T=1000°C	T=1250°C
~83 nm	t_{depo}: 15 min	t_{depo} : 35 min	t_{depo}: 65 min	t_{anneal}: 30 min

The thickness of the 3C-SiC thin film was determined by *ex situ* spectroscopic ellipsometry (SE) (SpecEL-2000, Mikropack GmbH, Germany). The crystal orientation and the crystallinity of the 3C-

SiC films were characterized by x-ray diffraction (XRD) using a diffractometer (RINT2000, Rigaku Corp., Japan) with a monochromatic CuKα x-ray source operated at 40 kV and 30 mA. The stress in the 3C-SiC thin film and the graphene quality were characterized using a Raman-scattering microspectroscopy apparatus (RM1000, Renishaw plc, UK) with an Ar^+ laser (wavelength: 514 nm, power:10 mW being used as the excitation source.

3. Results &discussions

The XRD patterns of the 3C-SiC films on Si are shown in Fig. 1. The peak at 47.34° corresponds tothe Si(220) planesoriginating from the substrate. Thepeak at 35.75° isattributed to the diffraction due to the rotated 3C-SiC(111) planes with its second-order 3C-SiC(222) peak appearing at 75.6°. The peak at 60.05° corresponds to the non-rotated 3C-SiC(220) planes. The sample (a) is thus a rotated 3C-SiC(111) thin film on Si(110), while sample (b)-(d) are non-rotated SiC(110) thin films on Si(110) but with different residual stresses as we will see shortly. The impact of the growth parameters on the rotation/non-rotation nature of the 3C-SiC/Si hetero-epitaxy is itself an interesting topic, which is currently investigated separately. The full widths at half maximum (FWHM) values of the 3C-SiC peaks of the samples are listed in Table 2. As we can see, the qualities of the 3C-SiC thin films, judged from the FWHM values, are in the order: (a) > (d) > (c) > (b).

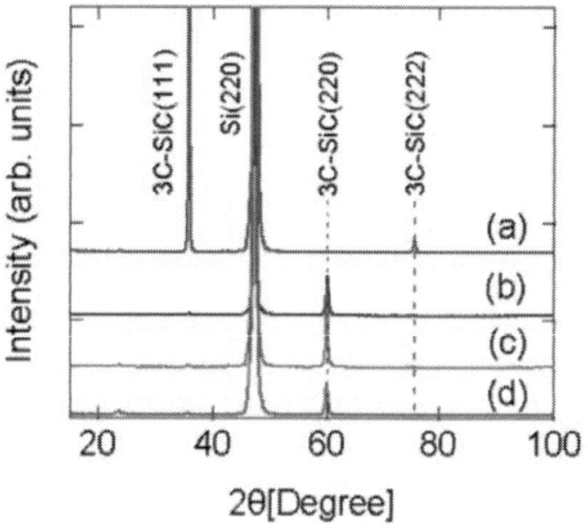

Figure 1. XRD patterns of graphene on 3C-SiC/Si films.
Patterns (a)-(d)corresponds to samples (a)-(d), respectively.

Table 2. FWHM values of the 3C-SiC films.

Sample Name	3C-SiC FWHM
Sample (a)	0.25°±0.012°
Sample (b)	0.5°±0.012°
Sample (c)	0.45°±0.012°
Sample (d)	0.4°±0.012°

The residual stress accumulated in the 3C-SiC film can be estimated from the peak frequency of the transverse-optical (TO) phonon mode of 3C-SiC since the TO peak is quite sensitive to the stress in the crystal [23]. Figure 2 shows the Raman spectra of the samples in the low-frequency regions where the TO mode appears. In the spectra, two peaks can be seen, one at ~790 cm^{-1} [24-25] and the other at820 cm^{-1} [27]. The former is attributed to the TO mode of 3C-SiC because this peak is absent in the spectrum of the Si substrate (top). The TO peaks of 3C-SiC are redshifted from that (796 cm^{-1}) of bulk 3C-SiC crystals [26] and varies with the process condition. This indicates that the residual stress in the 3C-SiC film varies with the process condition as well. Quantitatively, the residual stress can be estimated using the empirical formula obtained by Oleo *et al.* [28], where the measured TO phonon energy ω_{TO} in cm^{-1} is described in terms of the stress P (GPa) as

$$\omega_{TO} = (796.2 \pm 0.3) + (3.88 \pm 0.08)P - (2.2 \pm 0.4) \times 10^{-2} P^2 . \quad (1)$$

The results obtained using Eq. (1) are summarized in Table 3. The minus signs of the pressure indicate that the stresses in the present 3C-SiC epilayers are of tensile ones. This result is consistent with the smaller lattice constant of 3C-SiC as compared to Si (a_{SiC}=0.436 nm, a_{Si}=0.543 nm) as well as the ~8 % larger thermal expansion coefficient of 3C-SiC [26].

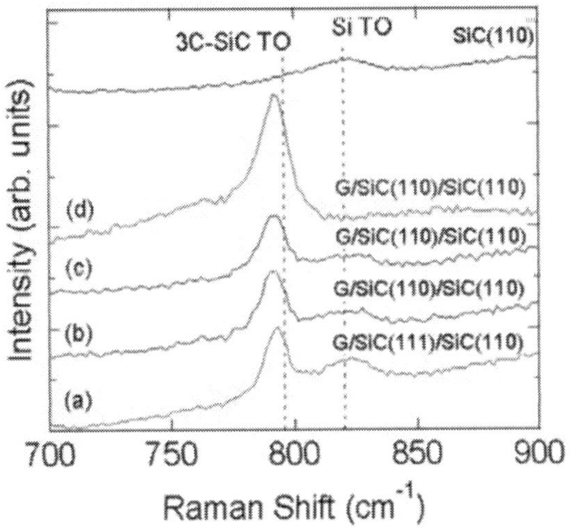

Figure 2. Raman scattering spectra of GOS. The onefromthe Si(110) substrate is also shown as a reference. Spectra (a)-(d) correspond to samples (a)-(d), respectively.

Table 3. Peak positions of the TO-phonon mode of 3C-SiC and stress *P* values estimated with Eq. (1).

Samples	TO peak (cm^{-1})	P (GPa)
Sample (a)	792.41±0.78	-0.824±0.05
Sample (b)	791.84±0.16	-1.261±0.05
Sample (c)	791.72±1.63	-1.188±0.05
Sample (d)	791.34±0.53	-1.160±0.05

Figure 3 shows the XRD FWHM of the 3C-SiC thin films plotted against the estimated residual stress. The XRD FWHM increases with the stress. Namely, the 3C-SiC film crystallinity is negatively correlated with the 3C-SiC residual stress. This is consistent with the previous experimentalobservations that the residual stress causes crystal imperfections such as cracks, misfit dislocations, micropipes, and voids [21,27-28].

It is quite natural and has been partially demonstrated [14] that the quality of GOS is strongly dependent on the quality of the 3C-SiC film inserted between the Si substrate and graphene. Figure 4 shows the Raman-scattering spectra from the samples. In all the spectra, three fundamental vibrational modes of graphene appear at ~1600 cm^{-1} (G band), ~2700 cm^{-1} (G' band), and ~1360 cm^{-1} (D band) [11,14,31]. The G band is the only band coming from a single-resonant Raman process, and is associated with a doubly degenerate (in-plane TO (iTO) and longitudinal optical (LO)) phonon mode (E_{2g} symmetry) at the center of the Brillouin zone (Γ point) [32-34]. The G' (also known as 2D) band is due to an intervalley double resonant (DR) process involving excitation/de-excitation of an electron with wavevector k in the vicinity of K point and two iTO phonons with wave vectors $q \approx 2k$ [33-36]. Here, both k and q are measured from the Dirac K point. The D band involves an electron transition at the wave vector k in the vicinity of K point and one iTO phonon with the wavevector $q \approx 2k$ [33-38]. It arises from an intervalley process, which occurs around two inequivalent K points (K and K') at the corners of the Brillouin zone [33-38]. Since energy and momentum are not conserved within the simple electron-phonon scheme, the D band process is related to microscopic defects. The G' band, on the other hand, is related to the intrinsic electronic structure of graphene [33-36,39], and is not related to any defects.

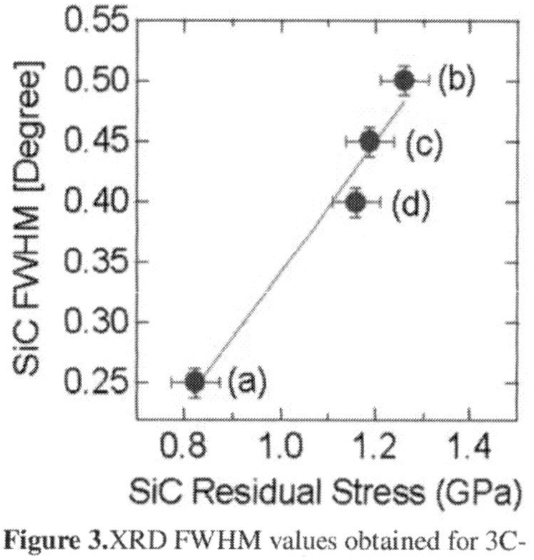

Figure 3.XRD FWHM values obtained for 3C-SiC vs. the 3C-SiC residual stress obtained by Raman scattering spectroscopy. The solid line is guide to the eyes.

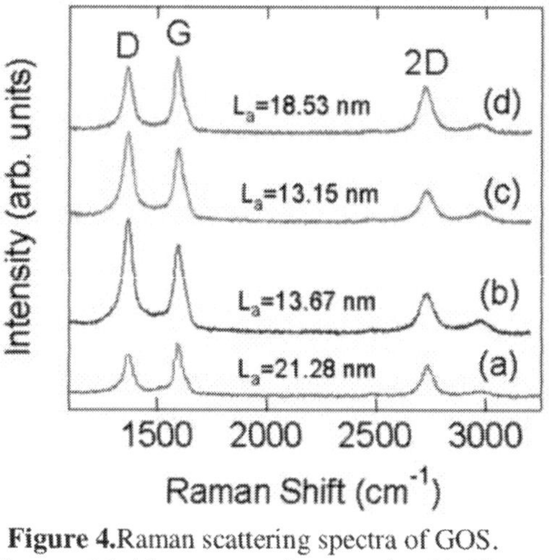

Figure 4. Raman scattering spectra of GOS. Spectra (a)-(d) correspond to samples (a)-(d), respectively.

A plot of the grain size (L_a) of graphene versus the residual stress is shown in Figure 5. The L_a is empirically estimated from the ratio of the Raman scattering intensity of the D and G bands [40] as;

$$L_a = \frac{560}{E_{laser}^4}\left(\frac{I_D}{I_G}\right)^{-1}, \tag{2}$$

where E_{laser} is the excitation energy in eV used in the Raman scattering experiment, and I_D and I_G the intensities of the D and G bands, respectively. The result in Fig. 5 shows that L_a is negatively correlated with the residual stress in 3C-SiC film. This indicates that the quality (grain size) of GOS is negatively affected by the residual stress of the 3C-SiC film at least on the Si(110) surface. It is

IOP Conf. Series: Materials Science and Engineering **79** (2015) 012004 doi:10.1088/1757-899X/79/1/012004

understood that the residual stress of 3C-SiC somehow influences the quality of the 3C-SiC film, which serves as the template for the epitaxy of GOS. In particular, the use of the stress-relaxed, rotated epitaxy of 3C-SiC(111)/Si(110) turns out to be quite effective in obtaining a high quality GOS.

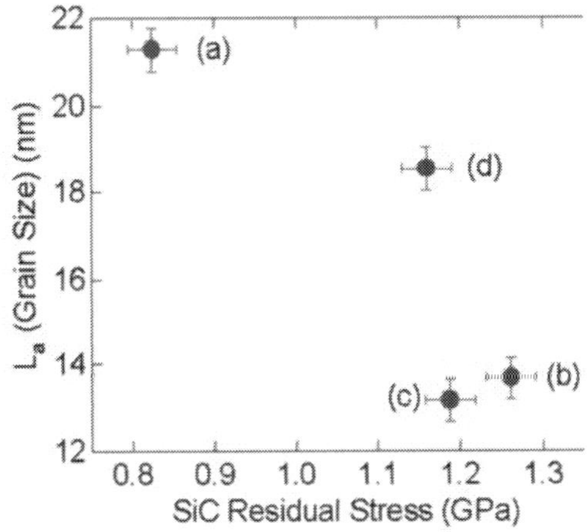

Figure 5. The grain size La of GOS vs. the 3C-SiCresidual stress both obtained by the Raman scattering spectroscopy.

4. Conclusions

By conducting XRD and Raman-scattering measurements on a series of samples with varied residual stress in the 3C-SiC films, a systematic study has been conducted to obtain relations amongthe residual stress, the XRD-FWHM crystallinity of the 3C-SiC film, and the grain size of GOS. As far as the GOS on Si(110) substrates is concerned, clearcorrelations were found to exist among these three parameters. In particular, use of the rotated epitaxy, 3C-SiC(111)/Si(110), was found to be quite effective in obtaining stress-relaxed, high-quality 3C-SiC films and thus high-quality GOS films. This finding will serve to the production of high-performance GOS-based devices.

5. References

[1] Novoselov K S, Geim A K, Morozov S V, Jiang D, Zhang Y, Dubonos S V, Grigorieva I V and Firsov A A2004 *Science* **306** 666
[2] Novoselov K S, Geim A K, Morozov S V, Jiang D, Katselson M I, Dubonos S V, Grigorieva I V and Firsov A A2005 *Nature* **38** 197
[3] Zhang Y B, Tan YW, Stormer H L and Kim P 2005*Nature* **438** 201
[4] GeimA K and NovoselovK S2007 *Nature Mater.***6** 183
[5] Bolotin K I, Sikes K J, Hone J, Stormer H L and Kim P 2008 *Phys. Rev. Lett.* **101** 096802
[6] Son Y W, Cohen M L and Louie S G2006 *Nature* **444** 347
[7] Stankovich S *et al.*2006 *Nature* **442** 282
[8] Sutter P W, Flege J and Sutter E A 2008*Nature Mater.*7406
[9] Abe S, Handa H, Takahashi R, Imaizumi K, Fukidome H and Suemitsu M 2010 *Nanoscale Res.Lett.* **5** 1888
[10] Ni Z H, Chen W, Fan X F, Kuo J L, Yuo T, Wee A T S and Shen X 2008*Phys. Rev.* B **77**115416
[11] Suemitsu M, MiyamotoY, Handa H and Konno A 2009 *e-J. Surf. Sci. Nanotech.* **7** 311
[12] Miyamoto Y*et al.*2009 *e-J. Surf. Sci. Nanotech.* **7** 107
[13] Olac-vaw R *et al.* 2010 *Jap. J. Appl. Phys.* **49** 06GG01
[14] Fukidome H, Miyamoto Y, Handa H, Saito E and Suemitsu M 2010 *Jap. J. Appl. Phys* **49**

01AH03
[15] Nishiguchi T, Nakamura M, Nishio K, Isshiki T and Nishino S 2004*Appl. Phys. Lett.* **84** 3082
[16] Konno A, Narita Y, Ito T, Yasui K, Nakazawa H, Endoh T and Suemitsu M 2006*ECS Trans.* **3** 449
[17] Ito T, Kanno T, Akiyama T, Nakamura K, Konno A and Suemitsu M 2008 *Appl. Phys. Express* **1** 111201
[18] Kern W and PuotienD A 1970*RCA Rev.***31**187
[19] KernW1984 *Semicond. Int.*794
[20] Nakazawa H and Suemitsu M 2001 *Appl. Phys. Lett.* **79** 755
[21] Riedl C, Starke U, Bernhardt J, Franke M and Heinz K 2007*Phys. Rev.* B **76** 245406
[22] Rollings E *et al.* 2006 *J. Phys. Chem. Solids* **67** 2172
[23] Lysenko V, Barbier D and Champagnon B 2001 *Appl. Phys. Lett.***79** 2366
[24] Nakashima S and Harima H 1997 *Phys. Status Solidi* A **162** 39
[25] Nakashima S and Harima H 2004 *Silicon Carbide* ed. W J Choyke *et al.*(Berlin: Springer) p 585
[26] Zhu J, Liu S and Liang J 2000 *Thin Solid Films* **368** 307
[27] Temple P A and Hathaway C E 1973*Phys. Rev.* B **7** 3685
[28] Olego D, Cardona M and Vog l P 1982 *Phys. Rev.* B **25** 3878
[29] Steckl A J and Li J P, *IEEE Trans. Electron Devices* **39** 64
[30] Scholz R, Gösele U, Niemann E and Leidich D 1995 *Appl. Phys. Lett* .**67** 1453
[31] Suemitsu M and Fukidome H 2010*J. Phys. D: Appl. Phys.* **43** 374012
[32] Nakamura J, Konogi H and Osaka T 1995*Phys. Rev.* B **51**5433
[33] Saito R, Dresselhaus G and Dresselhaus M S 1998 *Physical Properties of Carbon Nanotubes*(London: Imperial College Press) p 183
[34] Dresselhaus M S, Dresselhaus G, Saito R and Jorio A 2005 *Phys. Rep.***409**47
[35] Pimenta M A, Dresselhaus G, Dresselhaus M S, Cancado L G, Jorio A and Saito R 2007*Phys. Chem. Chem. Phys.***9** 1276
[36] Malard L M, Pimenta M A, Dresselhaus G and Dresselhaus M S 2009*Phys. Rep.***473** 51
[37] Thomsen C and Reich S 2000 *Phys. Rev. Lett.***85** 5214
[38] Saito R, Jorio A, Souza A G, Dresselhaus G, Dresselhaus M S and Pimenta M A 2002*Phys.Rev. Lett.* **88** 027401
[39] Ferrari A C 2007*Solid State Commun.***143** 47
[40] Tuinstra F and Koening J L 1970*J. Chem. Phys.* **53** 1126

Acknowledgments
This work has been partly carried out at the Evaluation Division of Fundamental Technology Center, Research Institute of Electrical Communication, Tohoku University and supported by JSPS KAKENHI (21360017), Grant-in-aid for Young Scientist (B) (20760485), Ministry of Education, Culture, Sports, Science, and Technology (MEXT) of Japan, and the Tohoku University Electro-Related Departments Global COE Program.

IOP Conf. Series: Materials Science and Engineering **79** (2015) 012005 doi:10.1088/1757-899X/79/1/012005

10-bit segmented current steering DAC in 90nm CMOS technology

R Bringas Jr[1], F Dy and OJ Gerasta

Mindanao State University-Iligan Institute of Technology
Iligan City, Philippines

E-mail: franzemyll_dy@yahoo.com, kitchensink956@gmail.com, olgajoy.labajo@g.msuiit.edu.ph

Abstract. This special project presents a 10-Bit 1Gs/s 1.2V/3.3V Digital-to-Analog Converter using1 Poly 9 Metal SAED 90-nm CMOS Technology intended for mixed-signal and power IC applications. To achieve maximum performance with minimum area, the DAC has been implemented in 6+4 Segmentation. The simulation results show a static performance of ±0.56 LSB INL and ±0.79 LSB DNL with a total layout chip area of 0.683 mm^2.The segmented architecture is implemented using two sub DAC's, which are the LSB and MSB section with certain number bits. The DAC is designed using 4-BitBinary Weighted DAC for the LSB section and 6-BitThermometer-coded DAC for the MSB section. The thermometer-coded architecture provides the most optimized results in terms of linearity through reducing the clock feed-through effect especially in hot switching between multiple transistors. The binary-weighted architecture gives better linearity output in higher frequencies with better saturation in current sources.

1. Introduction

Data conversion gives the link between the analog world and the digital system and is performed by means of sampling circuits, analog-to-digital converters (ADCs), and digital-to-analog converters (DACs). Such applications like present and future communication systems including high-speed modems and broadband wired and wireless communication subsystems require increasingly higher performance D/A converters. The key issues of these applications are low power consumption, higher conversion rate, high resolution and small die area.

Particularly all the high speed D/A converters are based on the current steering architecture due to its capability of driving resistive loads without buffering. This architecture also provides good static characteristics with reduced power dissipation and area. But the main disadvantage with this architecture is the rapid increase in harmonic distortion caused by the glitches generated from the current switches when the signal frequency is increased.

The current steering architecture can be implemented by either binary or unary architecture. In this paper, an attempt has been made to optimize the segmentation between the unary and binary implementation in this architecture to achieve a good linearity with reduced power and area.

[1] To whom any correspondence should be addressed.

Furthermore, the so-calledsegmented current-steering architectureis employed to achieve fast, constant output impedance, and to compensate glitch energy, nonlinearity and overall layout area.

In this project, a design for a 10-bit, 1GS/s segmented current-steering DAC is developed, implemented with a behavioral MATLAB Simulink model and Synopsys Custom design tool. It attempts to show a rigorous analysis of the performance of Segmented DAC architecture.

2. Design architecture

Figure 1 shows the functional block diagram of the proposed project. The DAC is divided into two sub-DACs, which consist of the four least significant bits (LSBs) using binary array current cells and the six most significant bits (MSBs) with thermometer-coded unary array.

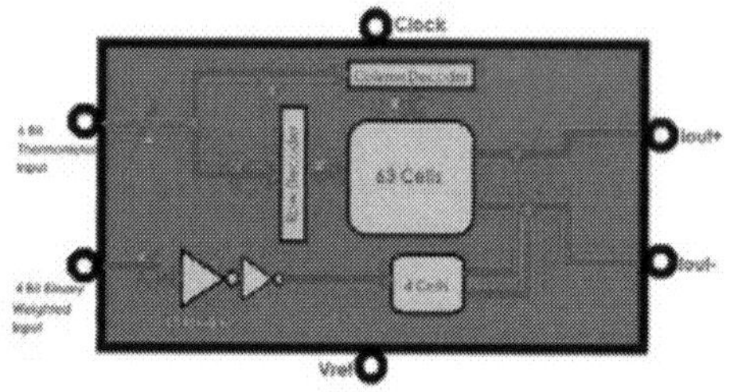

Figure 1. Functional block diagram of 10-bit segmented current-steering DAC.

Now, the segmented current-steering DAC undergoes through a swatch array which consists of latches, drivers and differential switches. It then undergoes through resistive networks that have been known as a method of providing local interaction between cells with minimum requirement in terms of space and interconnection.

In order to enhance the design of the DAC, modifications are made which include the application of a segmented topology with row and column matrix type array to improve INL and the reduction of the number of cells. The DAC consists of a column and a row decoder: 63 cells for the thermometer-coded DAC, 16 cells for the binary-weighted DAC, switches, latches, level shifters, a bias and clock block.

2.1. Segmented DAC architecture

The design will make use of a commonly used segmented DAC architecture for high resolution which is a hybrid between the binary-weighted and the thermometer-coded DAC.In this N -bit architecture, the M more significant bits are encoded into a thermometer code and the M– N less significant bits are binary weighted. There is a need of an encoder for the MSBs, which takes an M -bit input signal and generates one-bit output signals. When increasing M, the size of the encoder grows exponentially and a larger amount of switches and interconnection is needed.

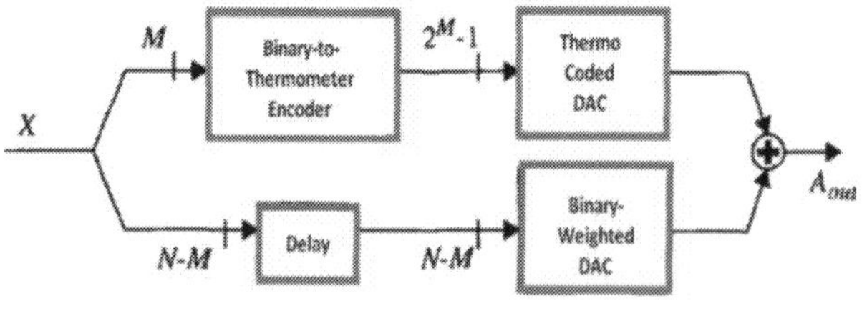

Figure 2. An N-bit Segmented DAC where M MSBs are thermometer coded.

2.2. Percentage of segmentation

A theoretical analysis of method of proper segmentation is presented. A fully binary-weighted design is referred to as 0% segmented, whereas a fully thermometer-coded design is referred to as 100% segmented. This analysis discusses the minimization of the chip area, under the constraints of 10-bit DC performance, while simultaneously trying to optimize the frequency domain performance.

In this present application, 60% segmentation is chosen which implies 6 MSBs are implemented by thermometer coded DAC and 4 LSBs are implemented with binary weighted DAC.

2.3. Binary weighted sub-DAC

The general architecture of a binary weighted current-steering DAC is shown in Figure 3. The switches are controlled by the input bits, where N is the number of bits. B0 is the LSB and the corresponding current source has the DC value ILSB. The source controlled by bit bi, i.e., the i–the LSB current source, is preferably formed by connecting 2¬i LSB current sources (unit current sources) in parallel, hence the MSB current source has the DC value IMSB = 2N-1*ILSB.

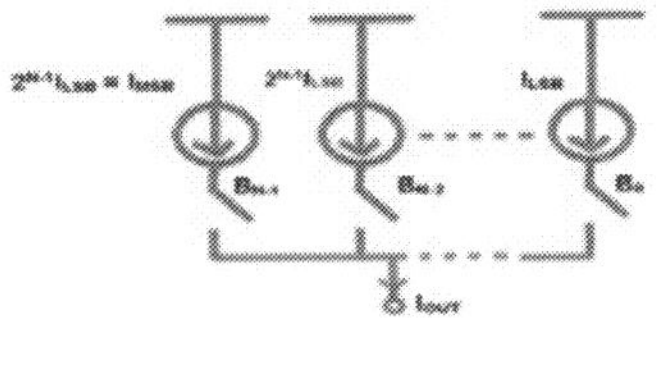

Figure 3. An N-bit current-steering binary weighted DAC.

2.4. Thermometer-coded subDAC

In the thermometer coded DAC, the reference elements are all equally large and the matching of the individual elements becomes simpler than for the binary case. The total sum of all weights is 2N-1. The transfer function of the thermometer coded converter is monotonic and the DNL and INL is improved compared to the binary version. The requirements on element matching are also relaxed. In fact, if the matching is within a 50% margin, the converter is still monotonic.

In this design 6 MSBs have been implemented by unary architecture, so a thermometer decoder with 6 binary inputs and 63 thermometer coded output is required. For a more systematic design the MSB current sources are implemented in unit current matrix structure, and the thermometer decoder is divided into two parts, a row and a column decoder with each part having 3 inputs and 8 outputs, also called a 3-bit thermometer decoder. Figure 4 shows the logic behind the decoder.

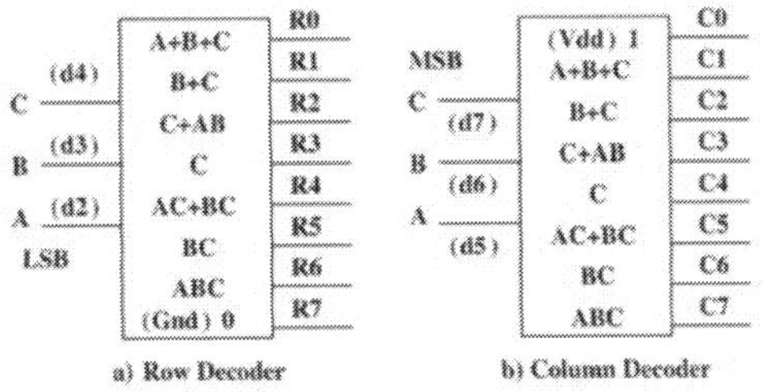

Figure 4. Thermometer decoder (a)row, (b)column.

3. Matlab behavioural simulation

To understand the behavior of the D/A block, a behavioral model is developed for segmented current-steering DAC using MATLAB SIMULINK® environment. Simulation results confirm the accuracy of the model.

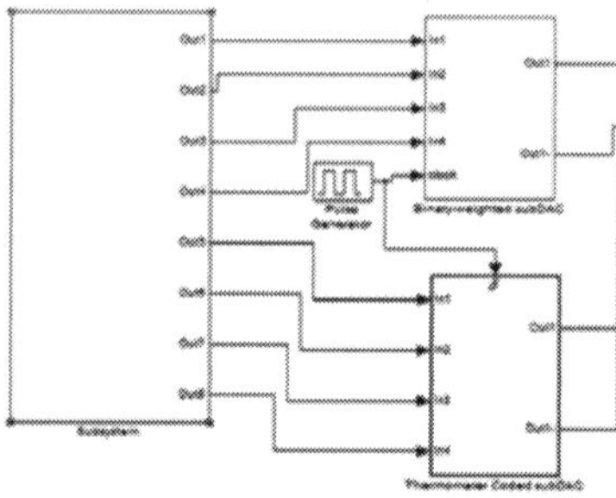

Figure 5. Behavioral model of the 8-bit segmented DAC.

The figure 6 below shows the linear ramp response of both the binary-weighted and the thermometer-coded sub-DACs showing step resolution linearity. Figure 7 shows the reconstructed sinewave output.

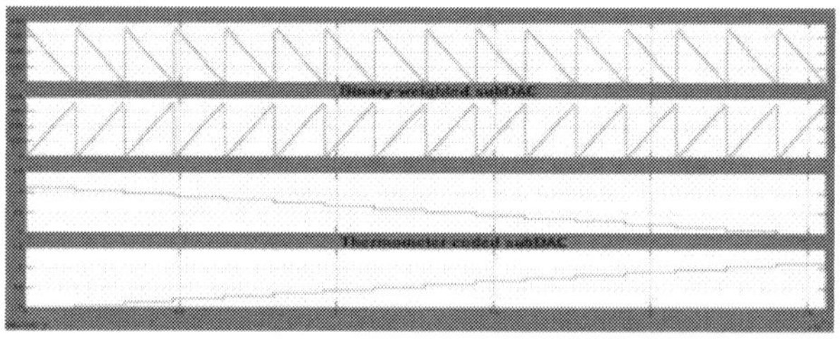

Figure 6. Linear ramp output of two sub-DAC's.

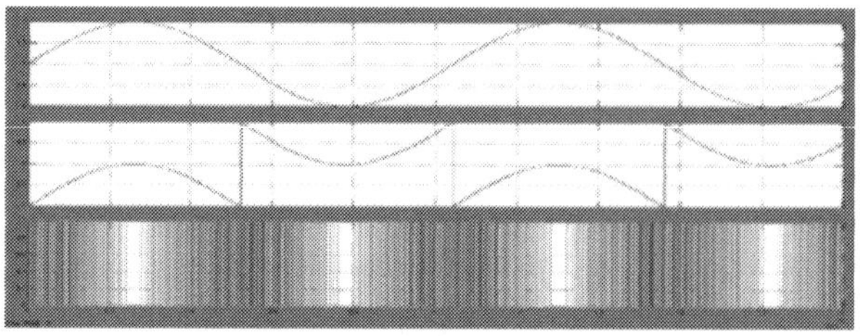

Figure 7. Reconstructed output of the DAC.

4. Cell based implementation using synopsys

4.1. Current source and switch

The basic unit of the DAC is the current cell which is composed of a cascade PMOS which comprises the analog part and a differential switch that uses inversion-mode PMOS varactors to reduce the effect of channel charge, which comprises the digital part. It is very important to test the switch first for its compatibility with the current.

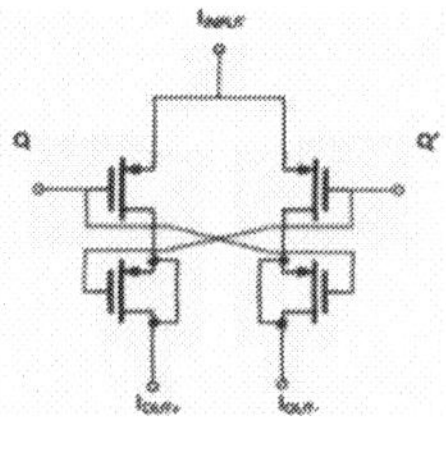

Figure 8. Inversion-mode PMOS varactors.

4.2. Voltage bias

Furthermore, all bias voltages and testbench voltages driving the system must be properly defined. The supply voltage for the analog and digital blocks are 3.3 V and 1.2 V, respectively.

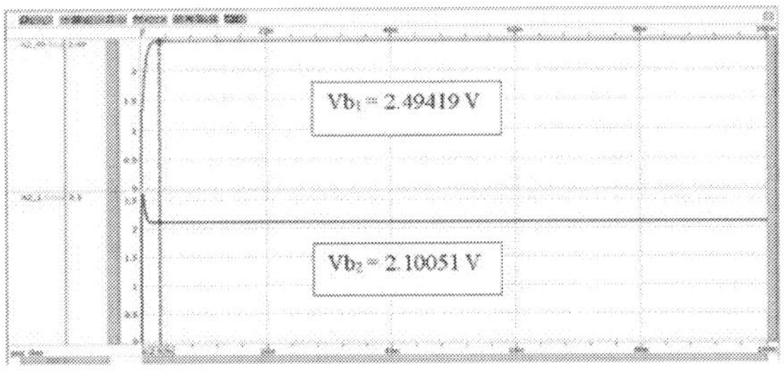

Figure 9. Voltage bias simulation.

4.3. D-latch

D-latches also play a vital role in this project since they are responsible for the synchronization or the timing of the differential switches.

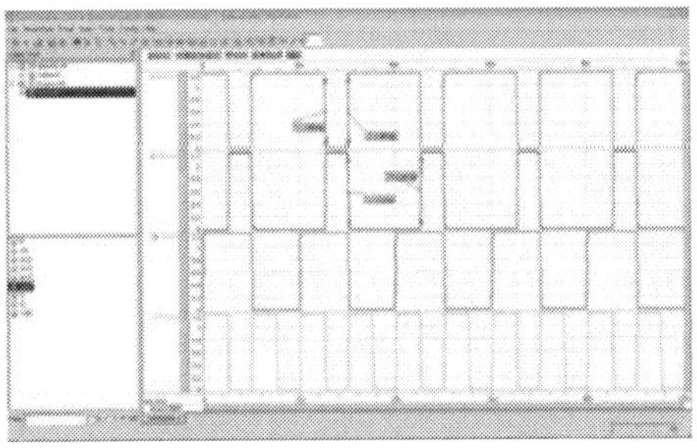

Figure 10. D-latch simulation.

The figure above is the simulated D-latch which is used for the whole system. It has delays that vary from 85-150 ps.

4.4. Thermometer decoder

The thermometer decoder, which is divided into row and column, plays an essential part in the thermometer coded sub-DAC. The logic behind the thermometer decoder is different from that of an ordinary decoder. Since the digital inputs pass through this block, it is very important to design this properly.

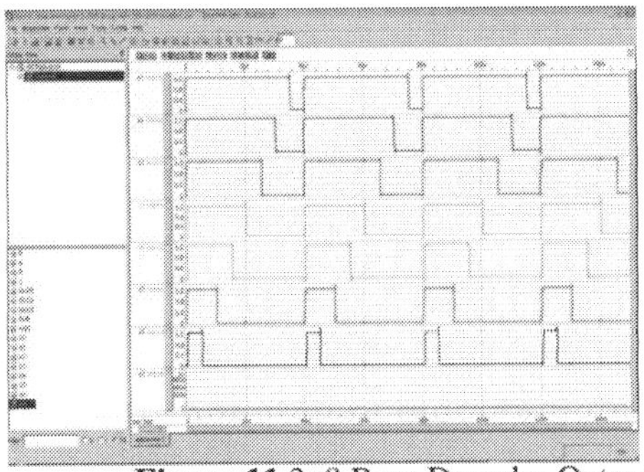

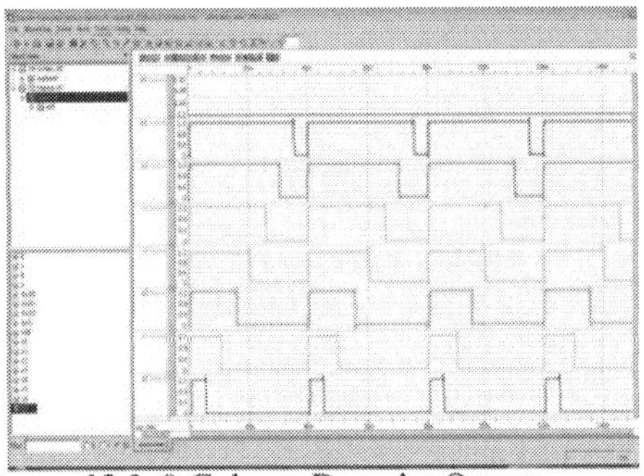

Figure 11.3x8 Row Decoder Output. **Figure 12.**3x8 Column Decoder Output.

4.5. Four-bit binary-weighted sub DAC

Given a pulse input for every bit, the output of the DAC will show a staircase waveform that must have equal time steps. For a 4-LSB binary-weighted DAC shown in Figure 13, the number of steps must be 2^4-1 = 15. For a desired Vref of 1 V, the step size of a 4-bit binary weighted (LSB) must be $1/(2^{10}-1)$=0.9775 mV. Therefore, the voltage swing must be 0.9775 mV multiplied by the number of time steps. Hence, the voltage swing is 0.9775 mV * 15 = 14.66 mV.

4.6. Six-bit thermometer-coded sub DAC

Given a pulse input for every bit, the output of the DAC will show a staircase waveform that must have equal time steps. For a 6-MSB thermometer coded DAC, the number of steps must be 2^6-1 = 63.

Figure 14 gives the staircase waveform for the 6-bit thermometer coded DAC. As observed, it is monotonic with almost negligible glitches and with a voltage swing of approximately 1.04 V. However, for a 6-bit thermometer-coded (MSB) with desired Vref of 1 V, the step size must be $1/(2^{10}-1)*2^n$ where n is the number of bits for the binary-weighted (LSB). Thus, the step size must be 16/1023 = 15.64 mV. Thus, the voltage swing must be 15.64 mV * 63 time steps = 0.985 V.

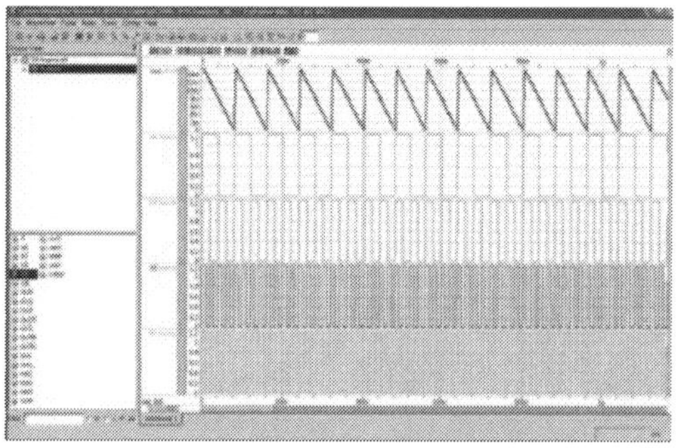

Figure 13. Binary-weighted DAC output with digital inputs.

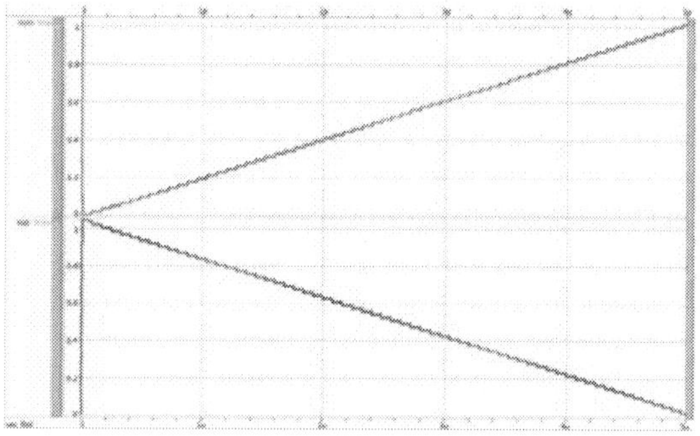

Figure 14. Thermometer-coded DAC output.

4.7. Overall 10-bit segmented current-cteering DAC output

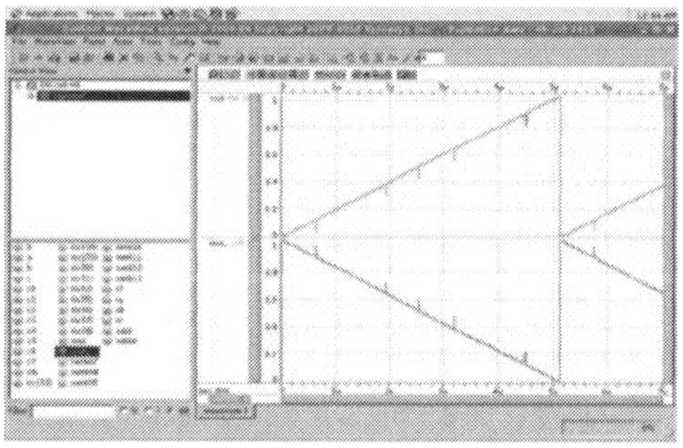

Figure 15.10-Bit segmented DAC ramp output.

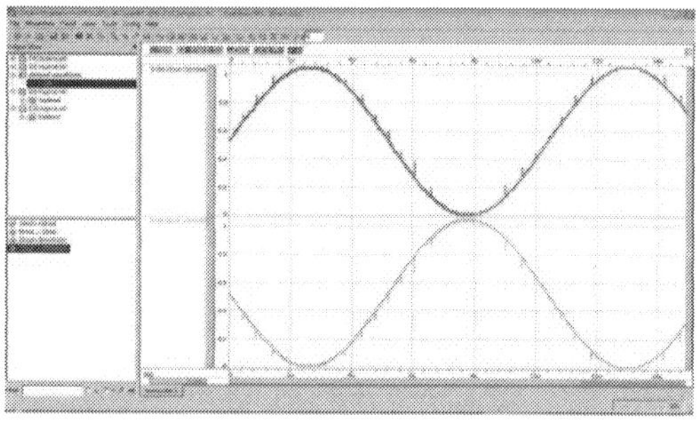

Figure 16. 10-Bit segmented DAC 96 kHz sine wave reconstructed output.

Figure 15 shows the total and combined waveform of the 2 sub-DACs. This is now the final output of the 10-bit segmented current-steering DAC with a voltage swing of approximately 1.04 V + 15.2 mV = 1.0552 V.

Figure 16 shows the total reconstructed output with its LSB and MSB components. The LSB output is the same as that of the behavioral model output. The same also goes for the MSB component. The post-simulation result turns out to be spiky with a maximum glitch of approximately 1.5 mV. This is due mainly by the latch where it is responsible for the switching synchronization and yet it obtained large amount of delay. Another contributing factor of this result is the routing of the MOS, wherein more parasitic capacitances are created by often using 2 consecutive metal types (i.e., metal 1 and metal 2 or metal 2 and metal 3, etc.). Furthermore, the filter design used is only a first order RC filter.

5. Layout floor plan

The layout size is illustrated in the figure below. Including the dummy MOS, the chip layout area is $0.7179*0.95187 = 0.683\ mm^2$. Excluding the dummy MOS, the total layout area would be just 0.605 mm^2.

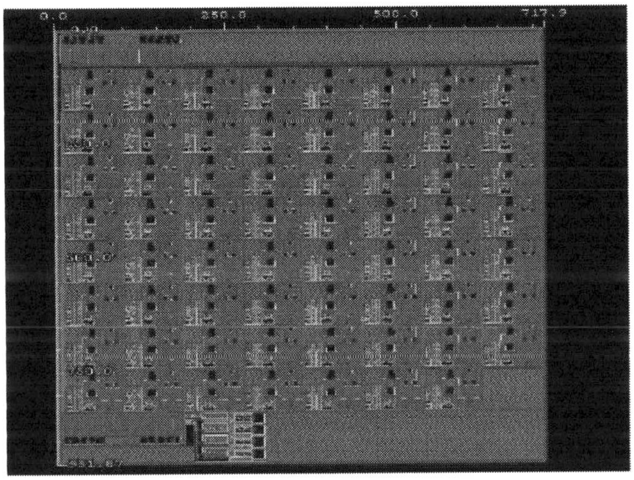

Figure 17. Design lay-out.

6. References

[1] Marques A, Bastos J, Steyaert M and Sansen W 1998*1998 IEEE International Conference on Electronics, Circuits and Systems*123–6

[2] Haider S *et al.* 2005*18th International Conference on VLSI Design*319–22

[3] Sarkar S and Banerjee S 2009 *IEEE Computer Society Annual Symposium on VLSI* 268–73

[4] Kester W 2005*The Data Conversion Handbook* (MA, USA: Newnes) pp 161–2

Acknowledgement

This work was supported by Department of Science and Technology – ERDT Fund, Philippines.

IOP Conf. Series: Materials Science and Engineering **79** (2015) 012006 doi:10.1088/1757-899X/79/1/012006

Vertical and rotational motion of mahogany seed

A V C Camposano, N C Virtudes, R E S Otadoy and R Violanda[1]

Theoretical and Computational Sciences and Engineering Group
Department of Physics, University of San Carlos, Nasipit, Talamban, Cebu City, Philippines

E-mail:renante.violanda@gmail.com

Abstract. Starting with a set of basic assumptions and with the application of well-known Newtonian physics, a theoretical model has been established for the flight of the mahogany winged seed. Using a high-speed camera, we successfully confirmed that the mahogany winged seed attains a vertical and rotational terminal velocity. From our model the mahogany seed has a terminal speed of 1.45 m/s. The experimental value of the terminal velocity on the average is 1.47 m/s (only about 1% error). The experimental value of the angular velocity was found to be around 54.33 rad/s, about 14% error compared to the predicted terminal velocity of 47.5 rad/s. The high predictable nature of a mahogany's terminal velocity can facilitate the biologist's study of mahogany mass seed dispersal.

1. Introduction

It is an observed fact that not all falling objects travel straight downward under the influence of gravity. Leaves, paper and seeds all follow complicated trajectories as they fall. This certain phenomenon is greatly utilized by nature; one good example can be seen in trees in their seed dispersal.

Seed dispersal is an important process in plant population dynamics. There are a lot of tree species among them are Triplariscaracasana, Swetenia mahogany, and Acer macrophyllum that have developed rather special way of dispersing seeds in the environment. One of the trees that exhibit this seed dispersal behavior, Swetenia mahogany, can be found around the University of San Carlos Talamban campus. The seed of these trees will be the subject of this paper.

The objective of this paper is to model the aerodynamics of mahogany winged seeds to discover whether there is a systematic limit to the vertical and rotational terminal velocity. Studying seed's terminal velocity is important. In the work of Green [1] he pointed out that modeling seed dispersal by wind for tropical species of seeds requires that one understand the magnitude of the variation in terminal velocity; that is, the first step in understanding the dispersal of the seeds of these trees is to acquire knowledge on how aerodynamic forces and the mahogany morphology act to regulate its rate of descent.

[1] To whom any correspondence should be addressed.

2. Theory

2.1 Physical model of mahogany

In order to have any successful model of the flight of mahogany seed a lot of sophisticated formalism should be considered such as Kutta-Zhukovsky formalism or Navier-Stoke theorem. Mahogany seed complex structure is apparently the reason that suggests the use of advanced aerodynamic theories. In the study of C. L. Ladera and P. A. Pineda [2] on the flight of samara, a three-winged seed that disperses comparable to mahogany, they have shown, using reasonable assumptions that their simplified model agreed with their experimental results.

In this paper we use the same assumptions, but modified the differential equations to cater the different wing structure of the mahogany seed. The assumptions are: when a mahogany seed descends to the ground its axis of rotation is seen to remain essentially vertical coincident with its main symmetry axis all along, that is, the mahogany wings behave as a rigid body which rotates with respect to the vertical axis, another simplification is that the camber of arching of the wings can be neglected [3]. Also, it is assumed that the speed of the wind is constant. The differential equations that govern the translation and rotational motion of the mahogany seed are discussed next.

2.2. Model of the vertical and rotational motion of mahogany seed

Essentially, as the mahogany leaf falls, it begins to rotate on its own around its center of mass, which is the seed, by the action of the air sweeping through its wing. Figure 1 refers to the diagram of the forces acting on the mahogany seed. The downward force is the weight mg. While falling to the ground a force pushes upward on the mahogany seed called the aerodynamic vector force L. We can write the vertical forces in differential form as,

$$m\frac{d^2y}{dt^2} = mg - L_v \quad , \qquad (1)$$

where$\ddot{y}$ is the acceleration while $\boldsymbol{L_v}$ is the magnitude of the vertical components of the lift force respectively.

To describe the kinematics of the falling mahogany seed, equation 1 was solved. Knowledge on the physical quantities involved in the vertical component of lift was determined beforehand. Experiments in the wind tunnel show that the magnitude of lift may be, within a good precision, considered proportional to the square air-speed v of the wing [1, 2].

$$L = K_L v^2 , \qquad (2)$$

where$\boldsymbol{K_L}$ is the lift coefficient depending on many factors.

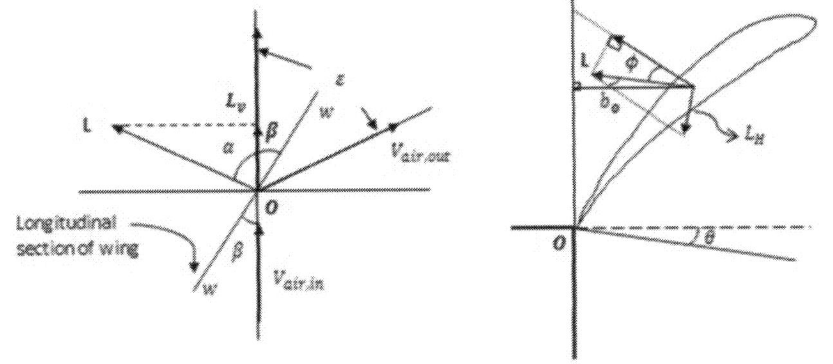

Figure 1. a) The mahogany seed is modeled as a flat slab ww', with an inclination angle $\boldsymbol{\beta}$ with respect to the z-axis. The lift force L is inclined the angle $\boldsymbol{\alpha}$ with respect to the z-axis. $\boldsymbol{V_{air}}$is the velocity of the wind and $\boldsymbol{\varepsilon}$ is the angle of the deflected wind b) Horizontal component of the lift force L, a small angle $\boldsymbol{\phi}$ is the tilt of the wing. The angle $\boldsymbol{\theta}$ is the angular displacement

Let $\boldsymbol{\rho}$ be the air density, l the length of the mahogany wing and $\boldsymbol{\alpha, \beta, \theta,}$ ***and*** $\boldsymbol{\varepsilon}$ are the angles defined in the model depicted in Figure 1a and 1b. In the wing's reference frame, the angles are defined as follows; $\boldsymbol{\beta}$ is the angle with respect to the vertical axis. In the full dynamical analysis of fluids the angle $\boldsymbol{\varepsilon}$ from the deflected wind to the vertical axis also affects the magnitude of the lift. This further can be explained by the momentum theory [3].

The factors in lift of coefficient can be determined with the proportionality that follows; the lift is proportional to the density $\boldsymbol{\rho}$ of air and it is also proportional to the cross-sectional area of the wings [2, 3].

The wings of the mahogany seeds are actually like an "actuator disk", a circular region which retards the passage of air over its entire surface. In the work of Green [1] the area they considered was the area of the disk swept by the wing of samara. For a given angle of inclination $\boldsymbol{\beta}$ the relevant area is not the cross-section area of the wing but the area $\boldsymbol{\pi(l \sin \beta)^2}$, which is exactly the same area that would be involved if rotating helicopter wings is involved. Considering the change in momentum of the air passing over the wing, the lift is also proportional to the $\mathbf{sin}\,\boldsymbol{\varepsilon}$, $\boldsymbol{\varepsilon}$ being the deflection angle. The coefficient of the lift $\boldsymbol{K_v}$, can now be written as,

$$K_L = \rho\pi(l \sin\beta)^2 \sin\varepsilon. \tag{3}$$

The vertical component of the lift force is given by the cosine of equation 2 in which we can write equation now as,

$$m\frac{d^2y}{dt^2} = mg - \rho\pi(l\sin\beta)^2(\sin\varepsilon)(\cos\alpha)v^2. \tag{4}$$

For the rotational motion of the falling mahogany, Newton 2nd law is given by

$$\tau = I\frac{d^2\theta}{dt^2}, \tag{5}$$

I being the moment of inertia, $\boldsymbol{\theta}$ is the relevant coordinate (refer to figure 1b) and $\boldsymbol{d^2\theta/dt^2}$ is the angular acceleration.

The rotational aerodynamic torque $\boldsymbol{\tau_{rot}}$ and the drag torque $\boldsymbol{\tau_{drag}}$ are two opposing torques that determines the rotation. The wing of mahogany seed shows a tilt $\boldsymbol{\phi}$ considered as the angle of attack of the air flow as it rotates. The tilt produces small horizontal forces that generate the rotational torque [2]. Thus, the horizontal component of the lift causes a rotational aerodynamic torque as a result of the interaction of the air with the wings as the mahogany seed descends. From the center of the wing to the axis of rotation it was considered the mean value of the lever arm $\boldsymbol{b_o}$. The rotational torque is given by,

$$\tau_{rot} = b_o L_H, \tag{6}$$

The horizontal component $\boldsymbol{L_H}$ can be shown to be given as,

$$L_H = \pi\rho(l\sin\beta)^2(\sin\varepsilon)(\sin\phi)v^2. \tag{7}$$

The drag force results to an opposing torque to the rotational torque as a result of air drag on the upper surfaces of the wings as the mahogany rotate [3]. The differential of the drag on an infinitesimal length of the surface of the mahogany wing was considered which was integrated over all the length of the wing to give us the drag torque as,

$$\tau_{drag} = \frac{1}{8}C_d\rho(al^4)(\sin\beta)^3\omega^2, \tag{8}$$

where a is the width of the wing, $\boldsymbol{\omega}$ is the rotational speed of the wing and $\boldsymbol{C_d}$ is the drag coefficient. The net torque of the mahogany with the use of Newton 2nd law of rotational motion is given as,

$$\tau_{rot} - \tau_{drag} = I\frac{d^2\theta}{dt^2}, \tag{9}$$

The rotational differential equation of the physical model is given by replacing the equation above with the rotational torque and drag torque and by dividing it with the moment of inertia. The result is,

$$\frac{d^2\theta}{dt^2} = Cv^2 - D\left(\frac{d\theta}{dt}\right)^2, \tag{10}$$

where the coefficient C and D are given by the following,

$$C = \frac{\pi b_o \rho (l \sin\beta)^2 (\sin\varepsilon)(\sin\phi)}{I} \tag{11}$$

and

$$D = \frac{C_d \rho (al^4)(\sin\beta)^3}{8I}. \tag{12}$$

For the vertical motion, the solution of equation 4 was obtained using Mathematica™. The equation of motion and velocity are,

$$y = y_o + \frac{1}{A}\log\left(\cosh\left(t\sqrt{AB}\right)\right) \tag{13}$$

$$v = \left(\sqrt{A/B}\right)\tanh\left(t\sqrt{AB}\right) \quad , \tag{14}$$

where $\boldsymbol{A} = \left(\boldsymbol{l^2/m}\right)\boldsymbol{C}$, B=9.8, and $\boldsymbol{C} = \boldsymbol{\pi\rho(\sin\beta)^2\sin\varepsilon\cos\alpha}$, $\boldsymbol{y_o}$ is the initial position.

The non-linear differential equation for rotational motion given in equation 10 was solved using numerical method. Runge and Kutta algorithm in Scilab™ was utilized.

The constants A, B, C and D were calculated using the field collected sample measurement of mahogany for each required parameters. The average mass was $(\boldsymbol{5.17 \pm 0.01}) \times \boldsymbol{10^{-4}kg}$, average wing length of $\boldsymbol{8.73\ cm}$, air density of $\boldsymbol{1.05kg/m^3}$, wing width $\boldsymbol{a = 24\ mm}$, the inclination $\boldsymbol{\beta = 48}$°, mean of the torque arm $\boldsymbol{b_o = 30\ mm}$, moment of inertia $\boldsymbol{I = (1.14 \pm 0.01) \times 10^{-6}kg \cdot m^2}$, tilt $\boldsymbol{\phi = 2}$°, assumed lift force angle $\boldsymbol{\alpha = 42}$°, assumed deflection angle $\boldsymbol{\varepsilon = 90}$° and drag coefficient of $\boldsymbol{C_d = 0.2}$ [2].

The solutions of the differential equation of our physical model using the required parameters were graphed and are given in figure 2.

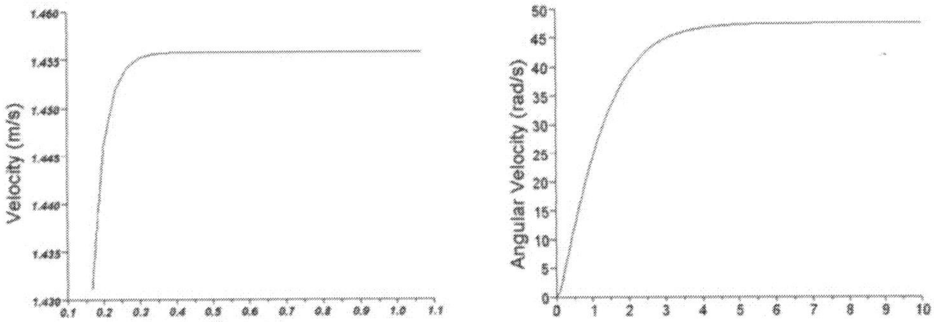

Figure 2. Left) Speed descent of a modelled mahogany seed, as it falls to the ground. The predicted terminal is close to 1.45 m/s. This terminal velocity was attained about 0.55 s after it falls. Right) Shows the predicted angular speed of the single-winged mahogany seed as a function of time. The predicted terminal angular speed is around 48.7 rad/s attained about 5 s after it falls.

3. Methodology

The experimental set-up as shown in figure 3 are composed of a high speed camera that is connected to a pre-made LabVIEW™ program that can change the threshold of the camera and crop the range of interest in which the camera would capture, then eventually export the position-time data to Microsoft Excel so it could be conveniently analyzed. Mahogany samples were then collected then spray painted with white so it would stand out on the black background that was hanged on a wall that is to be captured by the camera. The selection of the range was then done and the proper thresholding was done on the LabVIEW™ program, after which the sample was then dropped at a certain measured height simultaneously as the camera starts recording the position-time data of the sample with a capture rate of 30 frames per second. The data generated by this set-up is used to describe the vertical component of the mahogany's flight.

The set-up shown in figure 3B was used to be able to describe the rotational motion of the mahogany's flight utilizing the same concept as the one on figure 3A but only having the cameras line of sight at worm's view tothe descent of the mahogany sample. One can then determine the average period of the spinning motion of the leaf from the images captured.

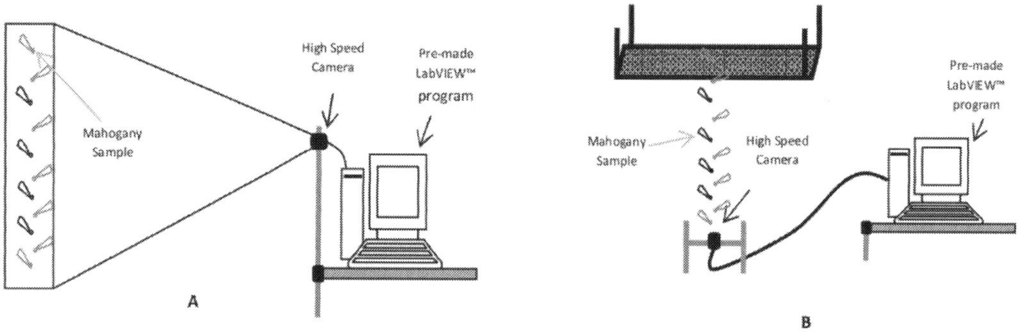

Figure 3. A) Shows the experimental set-up used to measure the terminal velocity of mahogany seed. The high-speed camera was used to gather the position-time data. B) Shows the set-up to calculate the angular frequency of mahogany seed. From the image collected the period of mahogany rotational motion was crudely measured by taking note of the time it takes to complete one revolution.

4. Results and discussion

The position of the mahogany winged seed was plotted against its time of descent, as shown in figure 4. The theoretical points (broken line) approached a final uniform regime from a short non-linear transient.

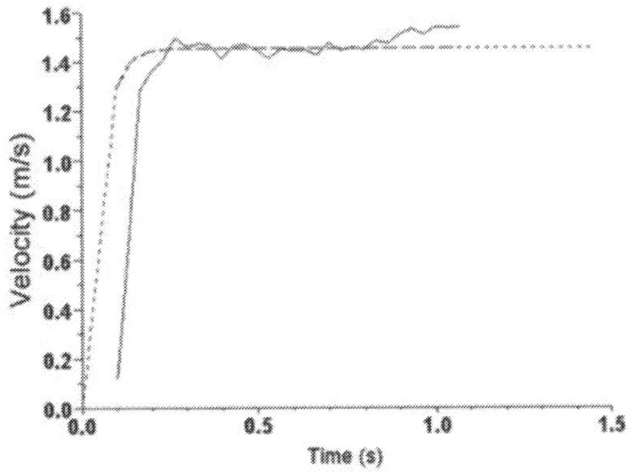

Figure 4. The theoretical graph (broken line) was super imposed with the experimental velocity (solid line) time graph of a sample of mass 0.47 g and length of 79 mm. An oscillating part is observed in the terminal regime of the motion.

Our data provides unifying properties of the mahogany winged seed's descent. From our model the mahogany seed reaches ground with a mean speed close to 1.45 m/s, the actual value depending on its physical parameters (mass, wing, inclination and the like). Hence, the samples showed a little variation on the value of each terminal velocity. The value of the experimental terminal velocity on the average was found to be around 1.47 m/s. This result agreed with our theoretical model shown in figure 2(Left) (only about 1% error). In figure 4, 'small oscillating" departure are observed with respect to the theoretical curve. This can be explained from the small wobbling of mahogany axis as it falls. We have assumed that the mahogany falls to the ground with its axis exactly along a vertical however a real mahogany seed can wobble a bit so a refined model must account to these small wobbling of its axis as it falls.

With the known initial condition $\omega(0)=0$, Runge Kutta method was used to obtain the solution of the nonlinear equation which gives us a predicted terminal angular speed of 47.5 rad/s. As shown in figure 2(Right) this terminal angular speed is reached about 5 s after the mahogany falls. From our crude method of determining the angular frequency, we have calculated a mean angular speed of $\omega_t = 54.33$ rad/s (this is about 14% error) from the data in Table 1. This is a result of the poor quality of the experimental technique used to determine the angular frequency and its large variation as shown in table 1.

Table 1. Rotational terminal velocity with its respective mahogany seed's mass and length.

Mahogany Sample Number	(Mass±0.0001)kg	(Length±0.001)m	Angular velocity (rad/s)
20	0.0053	0.087	61.83
26	0.0059	0.100	41.40
28	0.0063	0.094	43.39
31	0.0051	0.042	62.83
39	0.0035	0.050	62.20

IOP Conf. Series: Materials Science and Engineering **79** (2015) 012006 doi:10.1088/1757-899X/79/1/012006

5. Conclusion

We have confirmed that the vertical motion of the mahogany winged seed attained a terminal velocity. This is a confirmation of our model simulation which predicted a terminal velocity of around 1.5 m/s. We also have modeled the rotational motion of the mahogany seed as it falls and predicted a terminal angular speed of about 48 rad/s. The experiment with a sample of mahogany gives a mean angular velocity of about 54.3 rad/s. This is about 14% error compared to the predicted value and is due to the poor quality of the experimental technique used to determine the angular frequency. The vertical terminal velocity has been experimentally obtained, with good agreement with our theoretical value. For the rotational motion of mahogany a better experimental technique should be used to measure the angular frequency. In addition, a refined model must take into account the small wobbling of its axis at it falls.

References

[1] Green DS1980 *American Journal of Botany* **67**(8) 1218–20
[2] Ladera C L and Pineda P A2009 *Latin American Journal of Physics Education* **3**(3) 557–65
[3] Waltham C 1998 *The Physics Teacher* **36** 457–62
[4] Fowles G R and CassidayG L2004 *Analytical Mechanics* (California: Thomson Brooks/Cole) Chapter 4 pp 144–73
[5] Bramwell A R and Done G 2001 *Bramwell's Helicopter Dynamic* (Woburn, Ma: Butterworth-Heinemann) p 44

Euclidean distance and Kolmogorov-Smirnov analyses of multi-day auditory event-related potentials: a longitudinal stability study

M V Durato[1,4], A M Albano[2], P E Rapp[3] and S A Nawang[1]

[1] Physics Department, Mindanao State University–Iligan Institute of Technology, A. Bonifacio Avenue, Iligan City 9200, Philippines

[2] Department of Physics, Bryn Mawr College, 101 N Merion Ave, Bryn Mawr, Pennsylvania 19010, USA

[3] Department of Military and Emergency Medicine, Uniformed Services University of the Health Sciences, 4301 Jones Bridge Road, Bethesda, Maryland 20814, USA

E-mail: mvdurato@gmail.com

Abstract. The validity of ERPs as indices of stable neurophysiological traits is partially dependent on their stability over time. Previous studies on ERP stability, however, have reported diverse stability estimates despite using the same component scoring methods. This present study explores a novel approach in investigating the longitudinal stability of average ERPs—that is, by treating the ERP waveform as a time series and then applying Euclidean Distance and Kolmogorov-Smirnov analyses to evaluate the similarity or dissimilarity between the ERP time series of different sessions or run pairs. Nonlinear dynamical analysis show that in the absence of a change in medical condition, the average ERPs of healthy human adults are highly longitudinally stable—as evaluated by both the Euclidean distance and the Kolmogorov-Smirnov test.

1. Introduction

Among numerous adaptive dynamical systems known to humanity, many consider the human brain as the most complex due to its sophisticated neural networks[1] comprising over 100 billion interacting nerve cells or neurons that process and transmit information through electrical signals[2]. Tracking this electrical activity is important in studying brain function, in medical diagnosis, and in monitoring the effectiveness of therapies used to treat neurological or psychiatric disorders[3]. To date, recording *event-related potentials* (ERPs) from the human scalp is still the best non-invasive and inexpensive method[4] for tracking the rapid changes in the brain's electrical activity in preparation for, or in response to sensory, affective, or cognitive stimuli[3,5].

Since the first published ERP experiment in 1964, ERPs have remained a useful diagnostic tool in both neuroscience and psychiatry. But if ERPs are to be legitimately used in clinical trials and in clinical diagnosis or screening, it is imperative to first establish that they exhibit good measurement properties, especially test-retest stability[6–8]. Sufficient knowledge regarding ERP stability helps

[4] To whom any correspondence should be addressed.

evaluate how much of the variance embodied in a given ERP waveform is indicative of the task-pertinent brain mechanisms that the ERP must encapsulate[8].

1.1. Previous studies on ERP stability
Over the last two decades, however, published ERP stability studies have reported quite diverse results. Table 1 provides a summary of some earlier findings on the test-retest stability of major auditory ERP components commonly encountered in psychophysiology and neuroscience research. As depicted, stabilities vary from as low as .07 to as high as .80 despite using the same component scoring methods (e.g. *peak latency*, *base-to-peak amplitude*) and the same stability measures—specifically, the Pearson product moment test-retest correlation coefficient, r to index inter-subject stability, and the intraclass test-retest correlation coefficient, ICC to index score agreement.

Table 1. Representative summary of earlier findings on ERP stability for electrode site Cz.

Peak	Study	No. of subjects	Latency		Amplitude	
			r	ICC	r	ICC
N1	[9]	19	0.48	0.27	0.09	0.09
N1	[10]	10		0.53		0.47
N1	[11]	46		0.40; 0.40		0.75; 0.75
N1	[12]	29; 30	0.35; 0.42		0.63; 0.50	
P2	[9]	19	0.51	0.50	0.23	0.23
P2	[11]	46		0.45; 0.15		0.80; 0.75
P2	[12]	29; 30	-0.07; 0.07		0.63; 0.52	
P3	[9]	19	0.74	0.71	0.62	0.61
P3	[10]	10	0.12 - 0.46	0.47	0.54 - 0.68	0.54
P3	[11]	46		-0.07; 0.45		0.75; 0.80
P3	[12]	29; 30	0.54; 0.14		0.58; 0.66	

To some extent, this considerable variation in stability estimates may have been caused by methodological differences between studies in: (i) the technique or algorithm used to detect, extract, and quantify specific ERP components[13,14]; (ii) the choice of electrode channels to analyze[14,15]; and (iii) the number of subjects that constitute the averaged results[7]. Given these conflicting estimates, there is apparently a lack of evidence in current ERP literature to verify whether or not ERP waveforms exhibit substantial consistency over time.

1.2. Rationale, objective, and research question
This present study explores a novel approach in investigating the test-retest stability of average ERPs—that is, by treating the ERP waveform as a time series and then applying two nonlinear dynamical measures, namely, Euclidean Distance (ED) and Kolmogorov-Smirnov Test (KST) to evaluate the similarity or dissimilarity between the ERP time series of different sessions. Since the entire ERP waveform will be subjected to analysis, this time series approach overcomes the inherent difficulty in isolating individual ERP components and in quantifying their respective feature.

ED and KST are established proximity metrics that have been successfully used in numerous diverse fields of science and engineering research, including the nonlinear analysis of multi-channel EEG recordings [16].Nevertheless, despite ERPs being derived from EEG data, there is no published study as yet that employs any of these two nonlinear metrics to investigate the behavior of ERPs.In this present study,EDand KST are used to quantify the degree of longitudinal stability of multi-day auditory ERPs derived from multi-channel scalp EEG recordings of healthy human adultsby comparing pairs of trials across five different experiment sessions. The predominant question encompassing all aspects of this study is: *In the absence of a change in medical condition, will the average ERPs of healthy human adults be longitudinally stable—as evaluated by a specific measure—or will they exhibit a high degree of day-to-day variability?*

2. Methods

The ERP dataused in this study are from the laboratory of Prof. Paul Rapp of the Department of Military and Emergency Medicine, Uniformed Services University of the Health Sciences, Bethesda, Maryland. Four male adults participated in five different experiment sessions separated by days.All participants were in good health, as determined from their medical records at the university clinics. More specifically, they had no history of neurological or psychiatric disorders or of drug abuse, had normal hearing, and were not under any medication during the entire course of the experiment.

2.1. ERP stimuli

ERPs were evaluated using two auditory stimulus categories, both presented binaurally to the subjects via headphones, as the subjects were comfortably seated in a dimly lit, electrically and acoustically shielded room. In both categories, the time series began 250 milliseconds prior to stimulus presentation. In the first category (also referred to as Task 1), a single click is delivered at $t = 0$ milliseconds—that is, after 250 baseline samples have been collected. In the second category (also referred to as Task 2), a click is delivered at $t = 0$ milliseconds and then followed by another click at $t = 500$ milliseconds. The inter-stimulus interval in both stimulus categories varied randomly between three and five seconds.

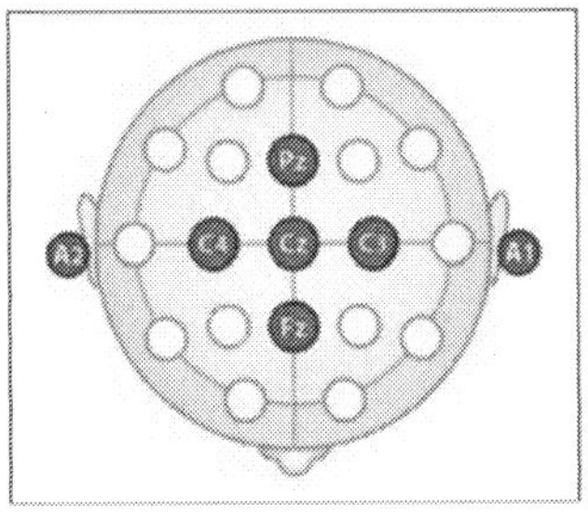

Figure 1. Scalp locations (marked red) of the five electrode channels used in the experiment. In the International 10-20 system, electrode pairs are typically placed in 19 standard positions distributed over the head, with letters to indicate position (F = frontal, C = central, P = parietal) and numbers to identify hemisphere location (odd #s = left, even #s = right).

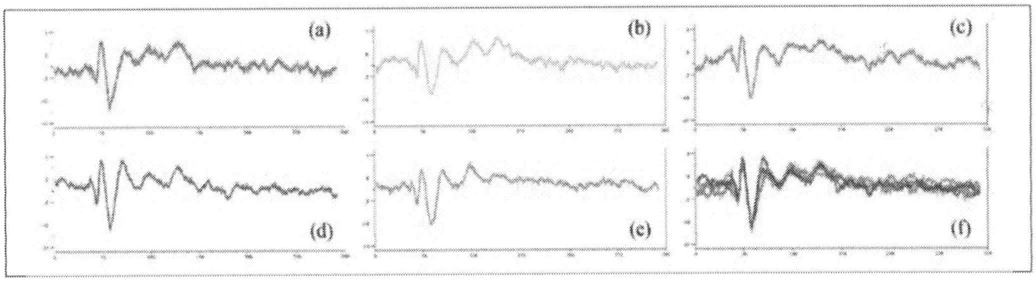

Figure 2. ERP waveforms of subject 1 (for task 1, electrode site Fz). The x-axis is in milliseconds while the y-axis is in microvolts. Each of the plots for runs 1 (a), 2 (b), 3 (c), 4 (d), and 5 (e) consists of seven trials. Thus, the superimposed plots of runs 1 to 5 in (f) comprise a total of 35 trials.

2.2. EEG acquisition

EEG signals were simultaneously recorded and analyzed from multiple scalp electrodes—specifically, from three mid-sagittal electrode channels (Fz, Cz, Pz) and two mid-lateral electrode channels (C3, C4). These five electrode channels whose scalp locations are shown in figure 1 were consistently employed across all trials per session and across all five sessions. Aside from these five electrodes measured in reference to linked earlobes, other electrodes were also used to monitor the vertical and horizontal electro-oculogram (EOG). All electrode impedances were kept below 5 $k\Omega$.

The signals were sampled every millisecond and subsequently digitized using a 12-bit digitizer. For each trial, the baseline was derived from the average of the first 250 samples. A single trial was rejected if it violated any of these two cases: (1) the amplitude of the horizontal and vertical EOG was $100\ \mu V$ greater or less than the baseline; or (2) the amplitude of any of the five EEG channels was $50\ \mu V$ greater or less than the baseline. The first 80 single trials that satisfy the above requirements were then combined to produce an average signal. Each experiment session consisted of seven average signals, denoted as trials 1,2,…,7, thereby totaling 35 trials for all five sessions (see figure 2).

2.3. Pairwise analysis of ERP waveforms

This study employs two nonlinear measures of similarity, namely EDand KST to detect how closely or distantly related the ERP waveforms of the five different experiment sessions are to each other, despite the apparent complexity in their shapes and behavior.This similarity characterization is accomplished for each subject, task, and electrode channel through rigorous, comprehensive pairwise analysis involving a total of 10 different test-retest pairsderived from all possible combinations of the five experiment sessions. Furthermore, for each test-retest pair, e.g. Run 1 vs. Run 2, each of the seven trials of Run 1 will be paired to all seven trials of Run 2. Therefore, for every test-retest pair associated with a particular task, electrode channel, and subject, there are a total of 49 trial pairsto be analyzed using EDand KST.

Table 2 and table 3 list the 10 test-retest pairs and the 49 trial pairs, respectively, along with their corresponding indices. These indices will be used to refer to any particular test-retest pair or trial pair throughout the rest of this paper. More importantly, results of the analysis will be plotted using these indices as x-coordinates and the ED and KST values as y-coordinates.

Table 2. Test-retest pairs.

N	**Test-retest pair**
1	Run 1 vs. Run 2
2	Run 1 vs. Run 3
3	Run 1 vs. Run 4
4	Run 1 vs. Run 5
5	Run 2 vs. Run 3
6	Run 2 vs. Run 4
7	Run 2 vs. Run 5
8	Run 3 vs. Run 4
9	Run 3 vs. Run 5
10	Run 4 vs. Run 5

Table 3. Indexed trial pairs for every test-retest pair.

m	**Trial pair**
1, 2, …, 7	Trial 1 vs. Trial x, where $x = 1, 2, \dots, 7$
8, 7, …, 14	Trial 2 vs. Trial x, where $x = 1, 2, \dots, 7$
15, 16, …, 21	Trial 3 vs. Trial x, where $x = 1, 2, \dots, 7$
22, 23, …, 28	Trial 4 vs. Trial x, where $x = 1, 2, \dots, 7$
29, 30, …, 35	Trial 5 vs. Trial x, where $x = 1, 2, \dots, 7$
36, 37, …, 42	Trial 6 vs. Trial x, where $x = 1, 2, \dots, 7$
43, 44, …, 49	Trial 7 vs. Trial x, where $x = 1, 2, \dots, 7$

2.4. Euclidean distance

Given two time series, $X = \{x_1, x_2, \dots, x_n\}$ and $Y = \{y_1, y_2, \dots, y_n\}$, the Euclidean distance, d_E (also referred to as *Pythagorean distance*) between X and Y is given by:

$$d_E = \sqrt{(x_1 - y_1)^2 + (x_2 - y_2)^2 + \cdots + (x_n - y_n)^2} = \left(\sum_{i=1}^{n} (x_i - y_i)^2\right)^{1/2}. \quad (1)$$

This simple measure, which reflects the proximity in space of two time series [17] is one of the most commonly used similarity criteria aimed at assessing how close or how distant the time patterns of two series are [18]. In this study, ERPs are normalized to zero mean and unit variance before calculating EDs in order to minimize the effects of differences in amplification and of amplifier drift.

2.5. Kolmogorov-Smirnov test

The Kolmogorov-Smirnov test is a well-known nonparametric test used to distinguish between distributions [19]. More specifically, it is used to ascertain whether two distributions disagree (two-sample KST), or whether a probability distribution disagrees from a reference distribution(one-sample KST). This test is applicable to continuous, unbinned, one-dimensional data and is sensitive to changes in both location and shape of the empirical cumulative distribution function of two data sets or time series [20, 21]. Particularly, the KST Scilab implementation [22, 23] used in this study outputs the probability that the two ERP waveforms being compared come from the same distribution, i.e., the probability that corresponding time series of the ERPs are statistically similar.

2.6. Schreiber-Schmitz surrogates

The method of surrogate data is normally employed in nonlinear time series analysis to establish scale and context for the measured topological properties of the experimental time series [24]. First introduced by Theiler et al. in 1992, this method has developed into a valuable tool in establishing whether or not the structures and irregularities exhibited by a particular time series are indeed indicative of nonlinear dynamics. If not, then the time series may be considered basically stochastic, whose structures can be due to random inputs to the system or fluctuations in some parameters. In this case, the time series may not be worth analyzing further [25]. There are currently several methods of calculating surrogate data sets.This studyuses the *Schreiber-Schmitz iterative surrogate* which preserves both the power spectrum and the distribution of the original data set or time series [22, 26].

3. Results and discussion

3.1. Euclidean distance analysis

Preliminary ED analysis of the ERP data of subjects 1 to 4 reveals that for any particular subject, task, and electrode site/channel, the average EDs of the 10 test-retest pairs are consistently and significantly lower than the mean EDs of 1000 pairs of Schreiber-Schmitz surrogates derived from 10 sets of randomly chosen trial pairs. This indicates that the ERPs of the five different experiment sessions are more similar to each other than their corresponding surrogates are.

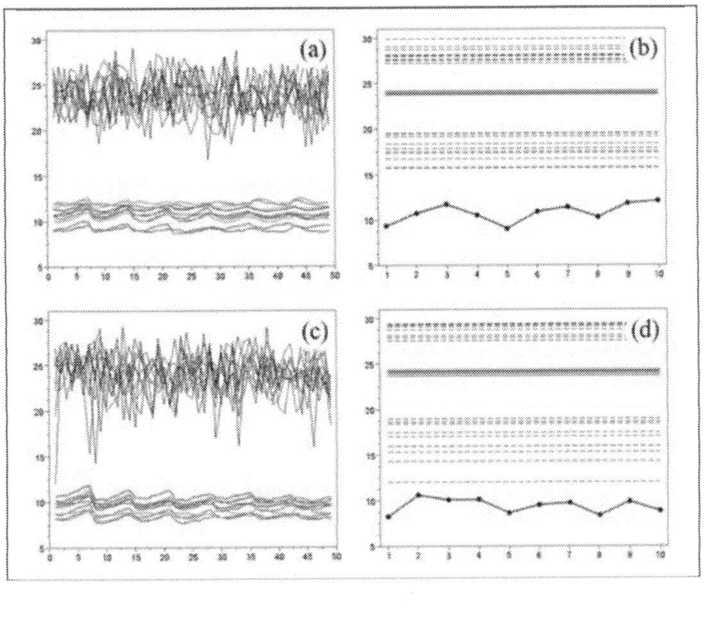

Figure 3. ED values (left column) and average ED values (right column) of subject 1, task 1, electrode channels Fz ((a), (b)), Cz ((c), (d)), Pz ((e), (f)), C3 ((g), (h)), and C4 ((i), (j)).

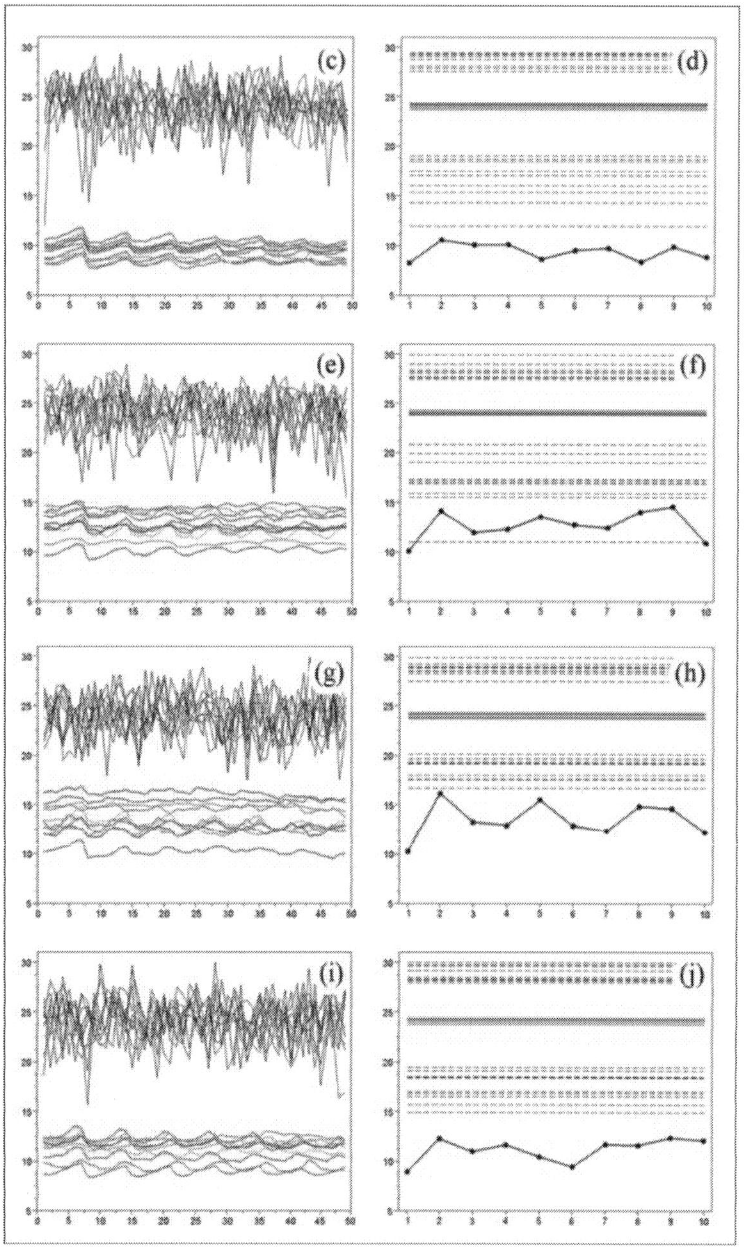

Figure 3*cont.* The x-axis of the ED plots correspond to the indexed trial pairs in table 3, while the x-axis of the average ED plots correspond to the indexed test-retest pairs in table 2. The solid black lines in the upper portion of the ED plots correspond to the EDs of 49 pairs of Schreiber-Schmitz (S-S) surrogates of 10 sets of randomly-chosen trial pairs. For the average ED plots, the solid red lines are the mean EDs of 100 pairs of S-S surrogates of 10 sets of randomly-chosen trial pairs. The dashed black lines are the maximum and the minimum values of the S-S surrogate EDs.

Figure 3 displays the resulting graphs of the ED analysis for subject 1, task 1. The same analysis was performed for task 2 of subject 1, and also for tasks 1 and 2 of subjects 2 to 4. Due to limited space, however, only the graphs for subject 1, task 1 are shown. As evident in the ED plots (left column of figure 3), the Euclidean distances between trials of the same experiment run or session (denoted by a single-colored line) are approximately the same, suggesting good within-session consistency and reliability. On the other hand, the Euclidean distances of different runs show some considerable spread, as manifested in the average ED plots (right column of figure 3). Nevertheless, this spread is almost always less than that of the surrogate EDs—that is, in terms of both the computed range and standard deviation (stdev) of the ED values for both data and surrogates. This behavior is better exemplified in tables 3 and 4, which depict the data EDs and surrogate EDs corresponding to the graph in figure 3(b) of subject 1, task 1, electrode channel Fz.

Table 3. Average data EDs for subject 1, task 1, electrode channel Fz.

n	Mean
1	9.34
2	10.75
3	11.69
4	10.53
5	9.06
6	10.91
7	11.41
8	10.30
9	11.81
10	12.08

Table 4. Surrogate ED statistics for subject 1, task 1, electrode channel Fz. Surrogate analysis involves 10 random trial pairs (denoted by s), each of which consists of 100 pairs of Schreiber-Schmitz surrogates.

s	Mean	Max	Min	Range	Stdev
1	23.72	27.96	18.36	9.61	2.13
2	23.86	28.02	15.72	12.30	2.10
3	24.05	27.56	16.79	10.77	2.08
4	23.94	29.01	15.82	13.20	2.81
5	23.73	27.23	19.10	8.14	1.94
6	23.85	28.71	17.83	10.89	2.15
7	23.85	27.61	19.54	8.07	2.01
8	23.82	28.15	17.47	10.68	2.40
9	24.15	29.88	19.38	10.50	2.25
10	24.06	28.13	17.34	10.79	2.14

The range and the standard deviation of the data EDs of the 10 test-retest pairs in table 3 are 3.02 and 1.02, respectively. These values are significantly lower than the computed average range (10.49) and average standard deviation (3.02) ofthe total 1000 pairs of Schreiber-Schmitz surrogates involved in the analysis.A similar trend is consistently manifested in the results of the range and standard deviation comparisonssummarized in figures 4 and 5, in the form of bar graphs.

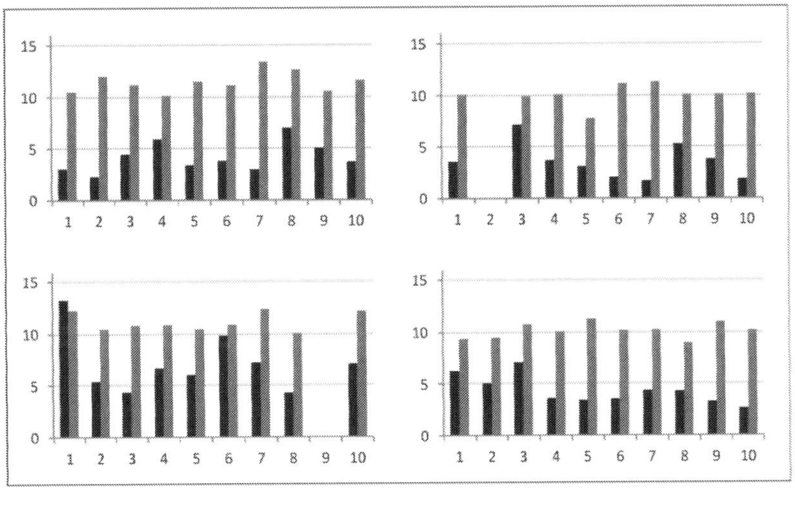

Figure 4. Range comparison of the data EDs and surrogate EDs of subjects 1 (a), 2 (b), 3 (c), and 4 (d). For each graph, the y-axis is the range of the EDs, while the x-axis corresponds to: (1) Fz, task 1; (2) Cz, task 1; (3) Pz, task 1; (4) C3, task 1; (5) C4, task 1; (6) Fz, task 2; (7) Cz, task 2; (8) Pz, task 2; (9) C3, task 2; and (10) C4, task 2.

It is evident from figures 4 and 5 that the variation in the EDs between successive experimental runs separated by days is almost always less than that between surrogates, $S1(i)$ and $S2(j)$, $i,j = 1,2,\ldots,1000$ for which all the $S1(i)$ have identical distributions and spectra and all the $S2(j)$ have identical distributions and spectra, distinct from those of the $S1(i)$. More specifically, 97.5% of the range of the data EDs are less than the range of the surrogate EDs, while 87.5% of the standard deviation of the data EDs are less than that of the surrogate EDs. These results suggest that, using Euclidean distance as a criterion, the auditory ERPs of subjects 1 to 4 are longitudinally stable.

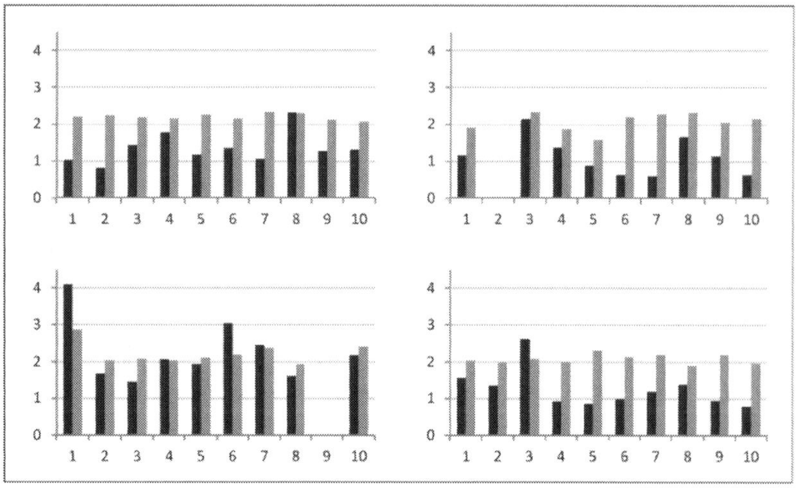

Figure 5. Standard deviation comparison of the data and surrogate EDs of subjects 1 (a), 2 (b), 3 (c), and 4 (d). For each graph, the y-axis is the standard deviation of the EDs, while the x-axis has identical labels as in figure 4.

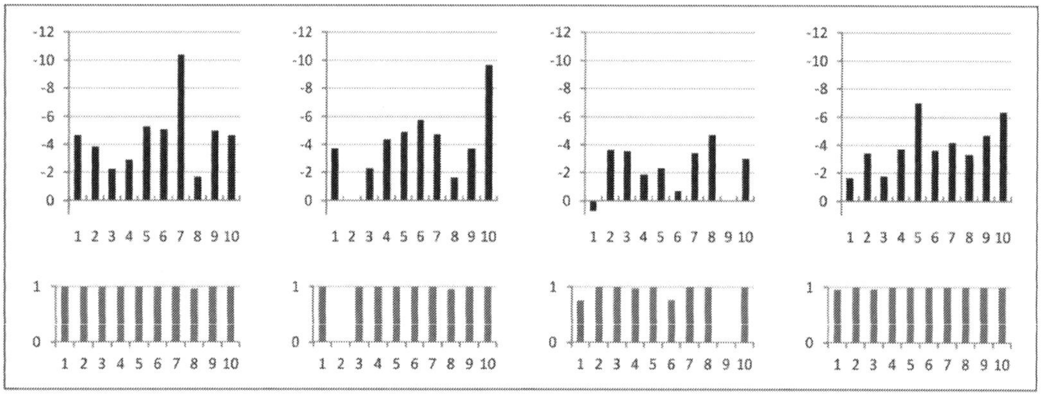

Figure 6. Summary of theZ-scores between the range of the data EDs and the range of the surrogate EDs of subjects 1 (a), 2 (b), 3 (c), and 4 (d). The associated confidence intervals (red bars) are shown directly below the Z-scores: subject 1 (e), 2 (f), 3 (g), and 4 (h). The labels along the$x$-axis are similar to that in figure 4.

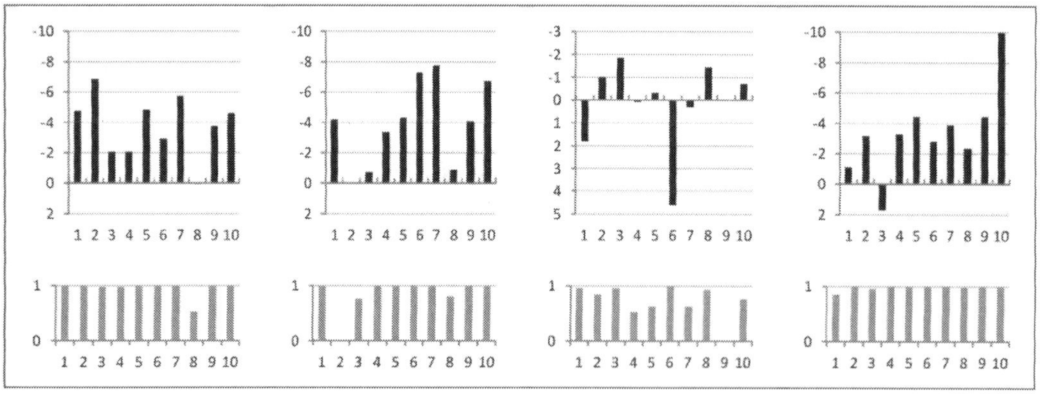

Figure 7. Summary of theZ-scores between the standard deviation of the data and the surrogate EDs of subjects 1 (a), 2 (b), 3 (c), and 4 (d). The associated confidence intervals (blue bars) are shown directly below the Z-scores: subject 1 (e), 2 (f), 3 (g), and 4 (h).The labels along the$x$-axis are similar to that in figure 4.

Another interesting result of the ED analysis concerns the Z-score calculations comparing the range and the standard deviation of the data and surrogate EDs. Figure 6 and 7 provide summaries of the Z-scores between the ranges and the standard deviations of the data EDs and the surrogate EDs, as determined by the following equations:

$$Z_{range} = \frac{range\ of\ ED_{data} - \langle range\ of\ ED_{surr} \rangle}{stdev\ of\ the\ range\ of\ ED_{surr}}. \tag{2}$$

$$Z_{stdev} = \frac{stdev\ of\ ED_{data} - \langle stdev\ of\ ED_{surr} \rangle}{stdev\ of\ the\ stdev\ of\ ED_{surr}}. \tag{3}$$

In a more descriptive manner, physicists refer to the Z-score as the *number of sigmas* that the original data differs from the surrogates [22].That is, with the assumption that the ranges and the standard deviations of ED_{surr} are normally distributed, a Z-score of 2 or greater signifies a probability of better than 97% that the original data and the surrogates do not belong to the same class. Inspection of figures 6and 7 reveals that majority of both the Z_{range} and the Z_{stdev} are indeed greater than 2, thereby suggesting that successive measurements of ERPs contain more significant dynamical connections than that expected of random data.

3.2. Kolmogorov-Smirnov analysis

Results of the Kolmogorov-Smirnov analysis support the findings of the Euclidean distance analysis discussed previously, i.e., that the individual-subject ERP waveforms are longitudinally stable across all five experiment sessions. More specifically, only 2.75% of the combined total run pairs (of all four subjects, five electrode channels, and two tasks) have KS values that significantly deviate from 1. In other words, 97.25% of all run pairs are statistically similar to each other, indicating stability of auditory ERPs over time. Figure 8 displays the KST results of all the 11 test-retest pairs (out of the total 400 run pairs) that exhibit significant deviation from 1.On the contrary, only the KS results for subject 1, task 1, electrode channel Fz are shown in figure 9 as representative of all the other 389 test-retest pairs with KS values of 1 or approximately equal to 1.

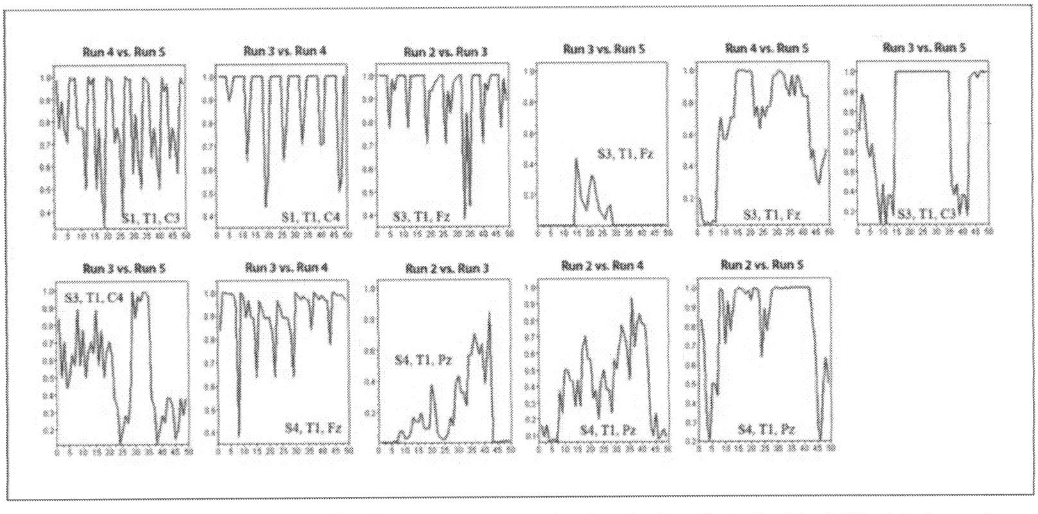

Figure 8. Plots of the KST results for the 11 test-retest pairs that deviate from the ideal. The labels on the y-axis are the KST values while the x-axis labels correspond to the indexed trial pairs per test-retest pair, as provided in table 3.

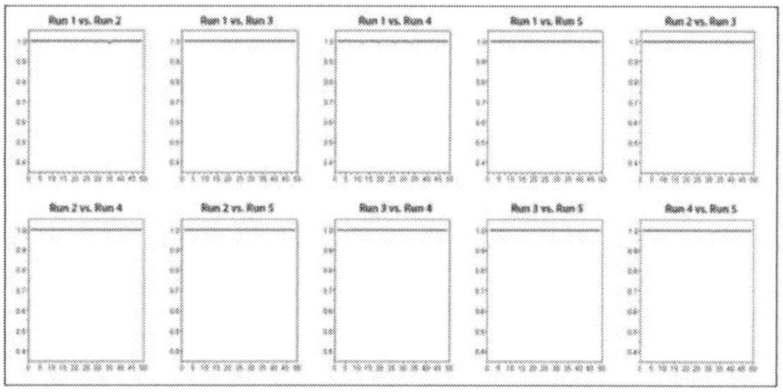

Figure 9. Plots of the KST results for subject 1, task 1, electrode channel Fz. The labels on the$x$- and$y$-axis are identical to that in figure 8.

4. Conclusion

Nonlinear dynamical analysis of multi-day, multichannel auditory event-related potentials show that in the absence of a change in medical condition, the averageERPs of healthy human adults are longitudinally stable—as evaluated by both the Euclidean distance and the Kolmogorov-Smirnov test. Given their excellent within-session reliability and test-retest stability, ERPs can therefore be legitimately used in clinical trials, diagnosis or screening and essentially, as biomarkers or as indices of stable neurophysiological traits.

References

[1] Gros C 2013 *Complex and Adaptive Dynamical Systems: A Primer* (Berlin: Springer-Verlag)
[2] Miller M ed 2008 *Brain Facts: A Primer on the Brain and Nervous System, 6th Edition* (Washington: Society for Neuroscience) pp 4–8, 36
[3] Kolb B and Whishaw I Q 2011 *An Introduction to Brain and Behavior3rd Edition* (New York: Worth Publishers)
[4] Fox N A, Hane A A and Edgar K P 2006 *Developmental Psychopathology2nd Edition* vol 2: *Developmental Neuroscience*, ed Cicchetti D and Cohen D J (New Jersey: John Wiley & Sons, Inc.) pp 381–91
[5] Sanei S and Chambers J A 2007 *EEG Signal Processing* (West Sussex, England: John Wiley & Sons, Ltd.) pp 127-31
[6] Cassidy S M, Robertson I H and O'Connell R G 2012 *Psychophysiology* **49** 659
[7] Luck S J, Mathalon D H, O'Donnell B F, Hämäläinen M S, Spencer K M and Javitt D C 2011 *Biol. Psychiatry* **70** 28
[8] Hammerer D, Li S C, Völkle M, Müller M and Lindenberger U 2013 *Psychophysiology* **50** 111
[9] Segalowitz S J and Barnes K 1993 *Psychophysiology* **30** 451
[10] Kinoshita S, Inoue M, Maeda H, Nakamura J, and Morita K 1996*Physiol. Behav.* **60** 1087
[11] Sandman C A and Patterson J V 2000 *Clin. Neurophysiol.* **111** 1427
[12] Walhovd K B and Fjell A M 2001 *NeuroReport* **12** 3149
[13] Fabiani M, Gratton G and Coles M G H 2000 in *Handbook of Psychophysiology 2nd Edition*, ed Cacioppo J T, Tassinary L G and Berntson G G (Cambridge: Cambridge University Press) pp 53–75
[14] Roth W T *et al.* 1995 in *Psychopharmacology: The Fourth Generation of Progress* (New York: Raven Press)
[15] Davidson R J, Jackson D C and Larson C L 2000 in *Handbook of Psychophysiology 2nd Edition*, ed Cacioppo J T, Tassinary L G and Berntson G G (Cambridge: Cambridge University Press) p 27
[16] Hornero R, Abásolo D, Escudero J and Gómez C 2008 *Phil. Trans. R. Soc. A* **367** 317
[17] Hwang H and Ahn H K 2011 *Advances in Neural Information Processing Systems* **24**
[18] Ciaccio A D, Coli M and Ibañez J M A ed 2012*Advanced Statistical Methods for the Analysis of Large Data-Sets* (Berlin: Springer-Verlag) pp 7, 207

[19] Celucci C J, Albano A M, and Rapp P E 2005 *Phys. Rev.*E **7 1** 066208

[20] Lopes R H C, Reid I and Hobson P R 2007*Proc. of Science: XI International Workshop on Advanced Computing and Analysis Techniques in Physics Research*

[21] Whitnall C, Oswald E and Mather L 2011 *Smart Card Research and Advance Applications* **7079** 234

[22] Albano A M, Celucci C J and Rapp P E (unpublished) *Dynamical Measures of Complex Behavior* pp 80–8, 102–5

[23] Albano A M, Passamante A and Rapp P E 1995 *Phys. Rev.* E **52** 196

[24] Henry B, Lovell N and Camacho F 2000 in *Nonlinear Biomedical Signal Processing: Dynamic Analysis and Modeling* vol 2, ed Akay M (New Jersey: John Wiley & Sons, Inc.) pp 1–3, 23–4

[25] Strozzi F, Tenrreiro E G, Noè C, Rossi T, Serati M and Zaldívar J M 2007 *Liuc Papers n.* **200**

[26] Schreiber T and Schmitz A 1996 *Phys. Rev. Lett.*77 635

IOP Conf. Series: Materials Science and Engineering **79** (2015) 012008 doi:10.1088/1757-899X/79/1/012008

Growth of zinc oxide nanostructures on glass substrates for ethanol gas sensor application

K Emphasis[1], R Vequizo, M K Odarve, J Gambe and A Alguno

Materials Science Laboratory, Physics Department
Mindanao State University – Iligan Institute of Technology, Iligan City 9200, Philippines

E-mail: katherine.emphasis@gmail.com

Abstract. ZnO nanostructures grown on glass (SiO2) substrates have been successfully fabricated for ethanol gas sensor. These ZnO nanostructures were prepared by chemical bath deposition with various ammonium hydroxide concentrations. The response of ZnO – based ethanol gas sensor exhibited higher response at lower NH4OH concentration; this might be attributed to the uniform morphology of the ZnO nanostructures which facilitates easier interaction with the ethanol gas. On the other hand, at higher NH4OH concentration showed drastic decrease in sensor response. This might be due to the growth of broken nanorods. Scanning electron microscope (SEM) images confirmed the formation of uniform and vertically grown nanostructures with lower concentration of NH4OH. On the other hand, at higher NH4OH concentration, broken nanorods and agglomerations were present as revealed in the SEM micrographs. Energy dispersive x-ray spectroscopy (EDS) measurements suggested that the grown nanostructures were most likely composed of ZnO. The calculated energy band gap from the UV-Vis spectra confirmed that the grown ZnO nanostructures.

1. Introduction

Ethanol (CH_3CH_2OH) is one of the most significant alcohols with various applications in food industry, brewing process control as well as in the medical and clinical applications. This ethanol is also used as a solvent of various alcohol soluble active ingredients in making perfumes, paints, lacquer and explosives. Furthermore, it is also used as an alternative fuel for automobiles [1]. However, behind these vast applications of ethanol, it also imposes disadvantages. At lower concentrations of ethanol, it can cause behavioural changes or impairment of vision to human [2]. On the other hand, at higher concentration, ethanol is volatile and toxic which causes skin and eye irritation. It may also cause nausea and vomiting when ingested, even death usually occurs if the concentration of ethanol in the bloodstream exceeds about 5%. To prevent such untoward incident, it is really a must for the early detection of ethanol in the environment. One of the best solutions for early detection of ethanol is the use of chemical sensors. Apparently, there are many ethanol sensors that are available in the industry, such as SnO_2, WO_3, InO_3, TiO_2 and ZnO. These materials have been extremely used as ethanol sensors due to their excellent sensing properties.

Zinc oxide is a good material for ethanol sensing compared to others because of its feasibility to be used as an ultrahigh sensitive sensors [3], high chemical stability, suitability to doping, non-toxicity,

[1] To whom any correspondence should be addressed.

abundance in nature and low cost [4]. These are various techniques that have been applied to synthesize ZnO nanostructures such as metal organic chemical vapour deposition (MOCVD) [5], pulse laser deposition (PLD) [6], molecular beam epitaxy (MBE) [7], radio frequency sputteringand chemical bath deposition. Among these techniques, chemical bath deposition is an attractive method for obtaining the nanostructures and has been extensively employed for the formation of metal oxide thin films and nanostructures which offers large area batch processing deposition and requires only solution containers and substrate mounting devices. Several studies reported that the ZnO nanostructures are very dependent on the NH_4OH concentration being utilized [8,9,14]. There were initial reports that these ZnO nanostructures grown on glass substrates can be fabricated as an ethanol sensor [10-12]. They partially found out that ZnO can be used as a good sensor device. They failed to establish the connection on the effect of utilizing different concentrations of the precursor that could eventually affect the morphology of the grown nanostructures. These morphological changes will have significant effect on the response of the fabricated ZnO-based ethanol gas sensor.

In this work, ZnO nanostructures are synthesized on glass (SiO_2) substrate with different ammonium hydroxide precursors and elucidate its morphological structures. The response of these fabricated ethanol gas sensor will be evaluated.

2. Experimental procedure

The ZnO nanostructures were synthesized on glass (SiO_2) substrates using chemical bath deposition technique. These substrates are thoroughly cleaned with acetone, ethanol and diluted hydrochloric acid to eradicate the impurities present on the substrates. The reagents used were aqueous zinc sulphate ($ZnSO_4$) and aqueous ammonium hydroxide (NH_4OH) solution. The concentration used for $ZnSO_4$ was maintained constant at 0.03 M while for NH_4OH concentration, it was varied to 1.0 M and 3.0 M concentrations. In these concentrations, the resulting ZnO nanostructures adhered well on amorphous glass substrates. The surface morphology and elemental composition of the grown nanostructures were characterized using scanning electron microscopy – energy dispersive x-ray spectroscopy (SEM-EDS), respectively. Optical reflectance was carried out in the wavelength 200-1000 nm by using ultraviolet-visible (UV-Vis) spectroscopy. Two-point probe method was also utilized to obtain the electrical characteristics of the synthesized ZnO nanostructures. Figure 1 depicts the sensor configuration while Figure 2 shows the schematic diagram for the two-probe method.

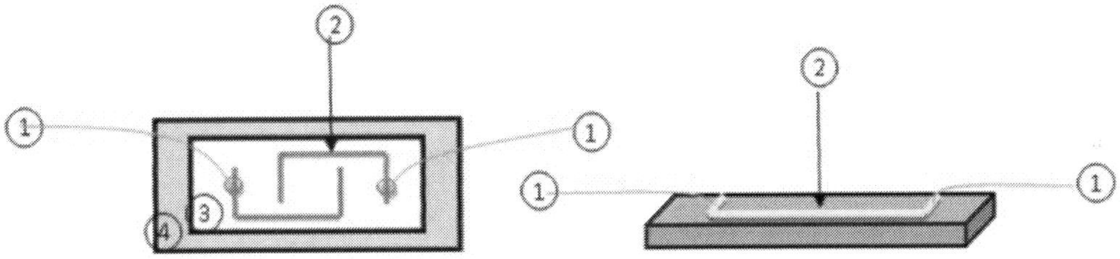

Figure 1. The sensor configuration where in (1) copper wires as the electrodes, (2) silver paste, (3) SiO2 substrate and (4) microscopic glass slide.

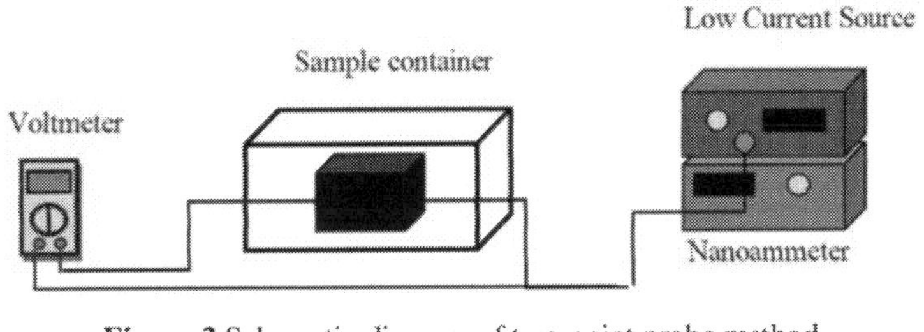

Figure 2. Schematic diagram of two-point probe method.

3. Results and discussion

3.1. Response of ZnO-based sensor device to ethanol gas

Figure 3 shows the response on ethanol gas of ZnO - based sensor using glass substrate prepared from 0.03M $ZnSO_4$ and different NH_4OH (1.0 M and 3.0 M) concentrations. It can be observed that the ZnO nanostructures deposited on glass substrate with 1.0 M NH_4OH responded to the ethanol gas andit has the higher sensitivity of about 33.75%. This high response to ethanol gas might be attributed to the uniformity of the nanorods. The uniformity of the grown nanostructures exhibit higher response due to higher surface area being exposed to ethanol gas. On the other hand, at 3.0 M NH_4OH revealed a lower sensitivity of about 4.21%. This might due to the morphology of the nanostructures as revealed in Fig. 5 (b). It can be further observed that the 1.0 M NH_4OH concentration was not able to recover to its original state after exposure to ethanol gas.

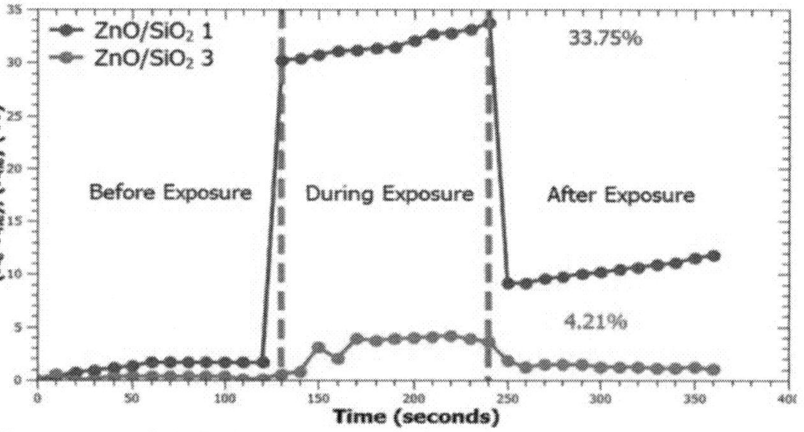

Figure 3. The response of the ZnO-based sensor under ethanol environment with 1.0 M and 3.0 M NH_4OH concentrations.

3.2. I-V curve of the ZnO-based sensor

The current-voltage (I-V) characteristics of the ZnO – based sensors were measured to determine the recovery of the fabricated sensor after exposure to ethanol gas. It is believed that when a sensor is fully recovered to its original state, the I-V curve has no changes before and after exposure to ethanol gas. It can be seen in Figure 4 (a) that the I-V characteristics of ZnO/SiO_2 with 1.0 M NH_4OH concentration exhibit a drastic decreased on the resistances of about 17.45%, before and after exposure to ethanol gas. This indicates that the ZnO-based sensor was not able to return to its original state. The same observation occurs at 3.0 M NH_4OH concentration as shown in Figure 4 (b).

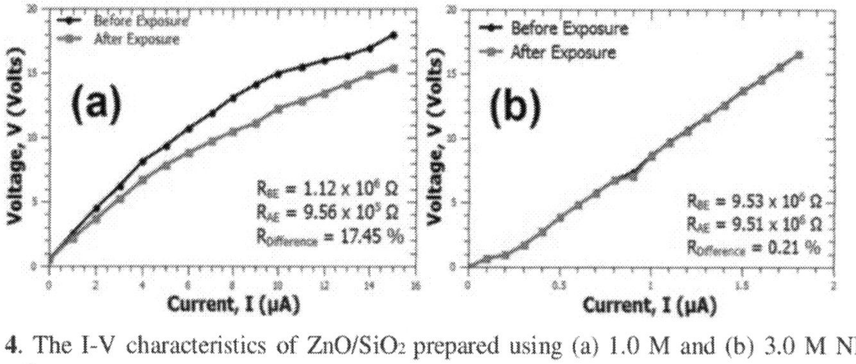

Figure 4. The I-V characteristics of ZnO/SiO_2 prepared using (a) 1.0 M and (b) 3.0 M NH_4OH concentrations.

3.3. Surface morphology of the grown ZnO nanostructures

In Figure 5 showed the surface morphology of ZnO nanostructures prepared at 1.0 M NH_4OH and 3.0 M NH_4OH concentrations. The 1.0 M NH_4OH concentration exhibits uniform nanorods. However, at higher NH_4OH concentration of about 3.0 M, agglomerations and broken nanorods can be observed. This phenomenon might be due to the higher concentration of OH^- ions that may trigger dissolution and etching [9,16].

Figure 5. SEM images for ZnO deposited on glass substrates prepared with (a) 1.0 M and (b) 3.0 M NH_4OH concentrations.

3.4. Energy band gap

The optical band gaps of ZnO nanostructures were obtained through the use of the UV-Vis reflectance spectra [15]. Figure 6 depicts the extrapolated optical band gap of the samples prepared with 1.0 M and 3.0 M NH_4OH concentrations. Through linear fitting the extrapolated band gap for 1.0 M NH_4OH is 3.34 eV while the energy band gap of 3.0 M NH_4OH concentration gives 3.28 eV. These values were in good agreement to the reported band gap energy of ZnO [13].

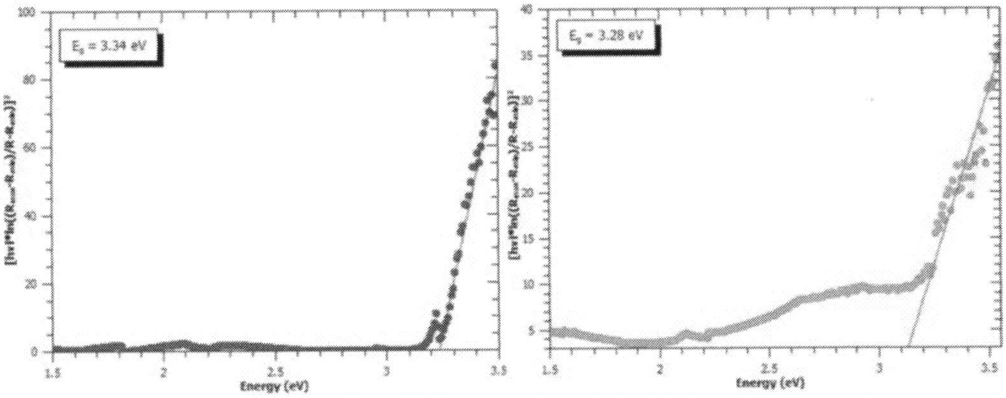

Figure 6. Variation of $h\nu(\ln(R_{max}-R_{min})/(R-R_{min}))$ vs ($h\nu$) of the fabricated ZnO samples grown at (a) 1.0 M and (b) 3.0 M NH_4OH concentrations.

4. Summary

The response of ZnO nanostructures to ethanol gas was successfully investigated. It shows that 1.0 M NH_4OH concentration exhibits higher response as compared to 3.0 M NH_4OH. It has been observed that this response might be due to the uniform morphology of the samples as revealed it the SEM micrographs. It was shown that the sensitivity of ZnO nanostructures grown on glass substrates prepared from 1.0 M NH_4OH concentration exhibits the highest sensitivity due to the uniformity of

ZnO nanostructures grown on the substrate surface. At higher concentration of NH_4OH, there was a small response of fabricated ZnO nanostructures sensor as exposed to ethanol gas. SEM micrographs revealed a non-uniform growth of the nanorods, instead agglomerations and broken nanorods were found. The fabricated ZnO – based sensor with lower concentrations of NH_4OH showed the highestresponse to ethanol gas. In contrast, the fabricated ZnO – based sensor with higher concentration of NH_4OH showed a low response to ethanol gas. The calculated energy band gap of ZnO nanostructure is in good agreement to the reported value.

5. References

[1] Vaishanv V, Patel P and Patel N 2005 *Thin Solid Films* **490** 94–100
[2] Strang K 2009*Ethanol: the Dose, Effects, and Side Effects of the World's Second Most Popular Drug* 1-2
[3] Choopun S 2010 *Sensors and Actuators B* **144** 67–72
[4] Zhou X 2010 *Physica* E **42** 2021–25
[5] Biswas P, Kundu S, Banerji P and Bhunia S 2013 *Sensors and Actuators* B*: Chemical* **178** 331–8
[6] Wongrat E, Pimpang P and Choopun S 2009 *Applied Surface Science* **256** 968–71
[7] Iwata K, Fonsa P, Niki S, A Yamada, K Matsubara, K Nakahara, T Tanabe and H Takasu 2000 *Journal of Crystal Growth* **214/215** 50–4
[8] Manulat S and Alguno A 2011 *Growth mechanisms of zinc oxide nanostructures and thin films deposited in Si(100) and Si(111) substrates* Master Thesis, Mindanao State University-Iligan Institute of Technology, Tibanga, Iligan City, Philippines
[9] Rico C A, Alguno A, Gambe J, Candidato R, Odarve M K, Sambo B R, Bagsican F R and Vequizo R 2013 *International Journal of Materials, Mechanics and Manufacturing* **1**(3)
[10] Chen I C 2007 *Sensors and Actuators* B **126** 473–77
[11] Mamutin V, Veskin V, Davydov V, Ratnikov V, Shubina T, Inanov S, Kopev P, Karlsteen M, Soderwall U and Willander M 1999 *Phys. Status Solidi* **176** 247
[12] Jiangqiang X *et al.*2008 *Sensors and Actuators* **2008** 334–9
[13] Marotti R E 2004 *Solar Energy Materials & Solar Cells* **82** 85–103
[14] N. Tabudlong 2008 *Fabrication and Characterization of Chemically Deposited ZnO Thin Films on Amorphous silica and p-Si (100)* Master Thesis, Mindanao State University-Iligan Institute of Technology, Iligan City, Philippines
[15] Kumar V, Sharma SK, Sharma TP and Singh V 1999 *J. Opt. Mater.* **12** 115
[16] Emphasis K and Alguno A 2012 *Responseof ethanol gas sensors using inc oxide nanostructures grown on silicon-on-insulator and glass substrates* Undergraduate Thesis, Mindanao State University-Iligan Institute of Technology, Tibanga, Iligan City, Philippines

Acknowledgment
K Emphasis would like to acknowledge the Commission on Higher Education (CHED) for the scholarship grant. We would like to thank Philippine Council for Industry, Energy and Emerging Technology Research and Development – Department of Science and Technology (PCIEERD-DOST) for the laboratory equipment.

IOP Conf. Series: Materials Science and Engineering **79** (2015) 012009 doi:10.1088/1757-899X/79/1/012009

Automated Car Park Management System

JP Fabros [1], D Tabañag[2],A Espra[3],OJ Gerasta[4]

Mindanao-State University-Iligan Institute of Technology
Iligan City, Philippines

johnpatrick.fabros@g.msuiit.edu.ph, arceile.espra@g.msuiit.edu.ph,
darlyn.tabanag@g.msuiit.edu.ph, olgajoy.labajo@g.msuiit.edu.ph,

Abstract. This study aims to develop a prototype for an Automated Car Park Management System that will increase the quality of service of parking lots through the integration of a smart system that assists motorist in finding vacant parking lot. The research was based on implementing an operating system and a monitoring system for parking system without the use of manpower. This will include Parking Guidance and Information System concept which will efficiently assist motorists and ensures the safety of the vehicles and the valuables inside the vehicle. For monitoring, Optical Character Recognition was employed to monitor and put into list all the cars entering the parking area. All parking events in this system are visible via MATLAB GUI which contain time-in, time-out, time consumed information and also the lot number where the car parks. To put into reality, this system has a payment method, and it comes via a coin slot operation to control the exit gate. The Automated Car Park Management System was successfully built by utilizing microcontrollers specifically one PIC18f4550 and two PIC16F84s and one PIC16F628A.

1. Introduction

Fully automated systems are being adopted in industries across the world at a rapid rate. Control systems are replacing manual operators and fully automated machines are replacing human labor. Less personnel and smarter machines means less operating and labor costs while increasing the quality of the products or services offered.

This study will increase the quality of service of parking lots by integrating a smart system which assists motorists in finding vacant parking slots [1]. It has a Parking Guidance and Information system concept which will efficiently assist motorists and ensure the safety of the vehicles and the valuables inside the vehicle.

In this parking system, a database lists the plate numbers of users and records the duration of their use of the parking lot. An Optical Character Recognition (OCR) system is also implemented to recognize the plate numbers of the users as they enter the parking lot [2,3]. Though this limits the use of the parking lot, it increases the security of the lot which has been deemed as a reasonable trade-off.

JP Fabros, D Tabañag, A Espra, OJ Gerasta
are with EECE Dept. of Mindanao-State University-Iligan Institute of Technology
Iligan City, Philippines

2. Parking System

Figure 1 shows the whole system of the project which is composed of PIC micro, sensors, PC, stepper motor, LED, and coin slot. Figure 2 is the behavioral simulations of the main circuit board using Proteus.

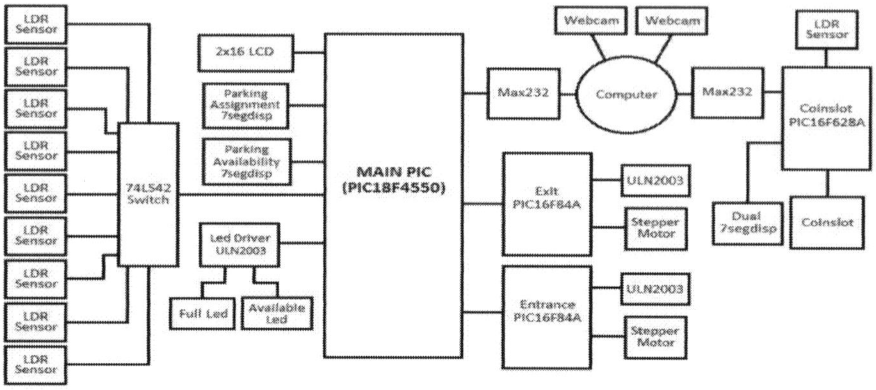

Figure 1 Automated Car Park Management System

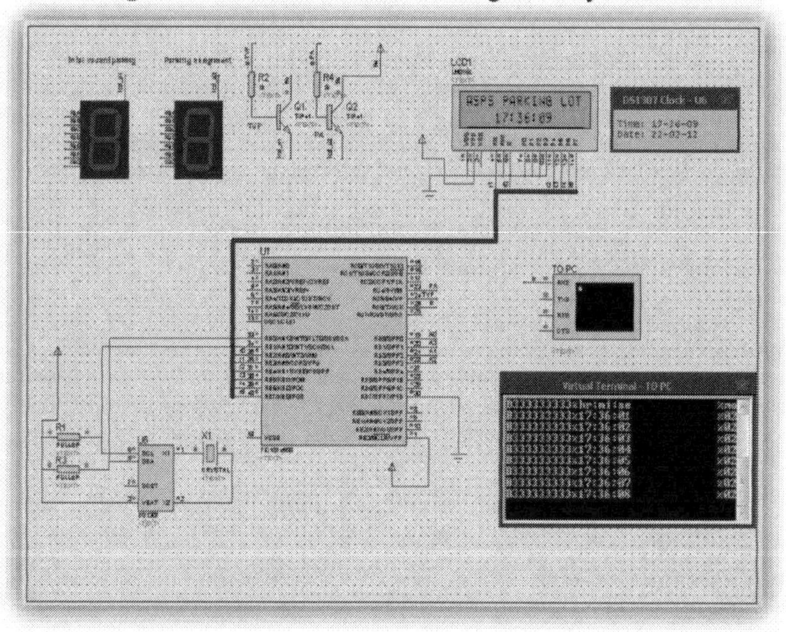

Figure 2 Main Board Simulation

Two seven-segment displays are situated at the front of the car park prototype to inform motorists of parking lot availability as shown in Figure 3. One display shows the total number of unoccupied parking slots in the car park prototype. The other shows the parking slot that is assigned to the motorist immediately at the front gate of the car park. This is to help motorists avoid wasting time in locating an available parking slot. To complement these displays, there will be two sets of LEDs designated for parking slot availability in the car park. When there is at least one parking lot that is available, the first set of LEDs will be lit. When all the parking lots have been occupied, LEDs designated for the car park being full will be lit. These LEDs are viewable outside of the car park prototype to inform motorists of parking lot availability.

Sensors underneath each lot as shown in Figure 4 determine if the lot is available or not. If the sensor for that particular parking lot is activated, it implies parking is done in that lot. This lot is then taken off the list of recommended parking lots which is displayed at the seven-segment display above the entrance.

Two stepper motors serve to control the entrance and exit gates. The operation of each stepper motor is controlled by separate PIC16F84A microcontrollers. The entrance microcontroller is in turn be activated by the main microcontroller and the exit microcontroller will be controlled by the payment system microcontroller. Once the LDR sensors at the gates are activated, the main microcontroller or the payment system microcontroller will send a signal to the gate microcontrollers to open or close the appropriate gate. An LCD will also be placed at the front of the car park prototype that displays the time and name of the car park prototype.

Figure 3 Parking Lot Prototype

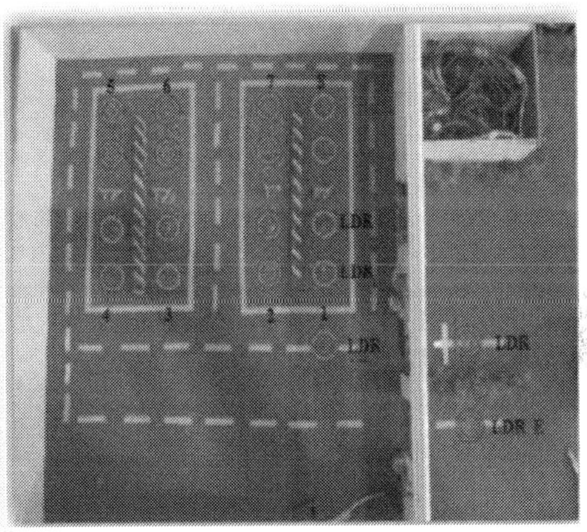

Figure 4. LDR Position

3. Coin Slot Payment System

The coin slot payment system is created as an independent block before being integrated into the prototype as shown in Figure 5.

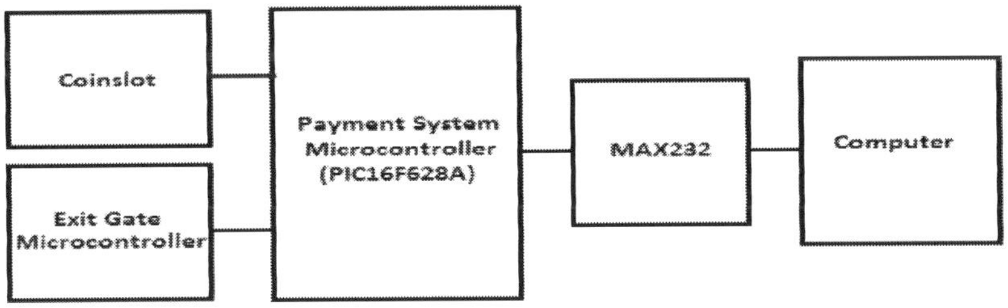

Figure 5 Coin slot Payment System Architecture

The payment system microcontroller manages the automation of payment and exit. The MATLAB GUI is responsible for observing the vehicles' parking durations. When the program detects a vehicle leaving from its designated parking slot, the changes will be reflected in the GUI. The parking duration of that particular vehicle will be sent to the payment system microcontroller through the MAX232 interfacing. The microcontroller is deactivated until data is sent to it.

The parking charges are incremented by 5 at a fixed rate. The total charge incurred during the customer for his parking duration will be displayed on a dual seven-segment display. When the customer drops coins into the coin slot, the corresponding value of the coin will be deducted from the charge displayed in the dual seven-segment display. The microcontroller then sends a signal to the exit gate microcontroller signaling it to open. The stepper motor will then control the gate to open. An LDR sensor at the front of the exit gate will determine when the gate should be closed. When the LDR sensor is triggered by the exiting vehicle, the gate is then brought down. At the same time when the customer is able to pay his charges in full, the microcontroller is brought back to its deactivated state until another vehicle exits the car park.

4. Design of Management GUI in MATLAB

The MATLAB GUI will serve as the monitoring tool for the Automated Car Park System. It is used for convenient management and data logging of the events of the system. Also, this is used as plate number recognition function.

4.1 GUI as monitoring tool

The availability of the parking lot is shown along with the basic information of the motorist's usage of that parking lot. The eight parking lots will be shown with their respective graphical axes as shown in Figure 6. These will show the screenshots of the vehicles front area which includes the plate number. Along with these screenshots will be the time-in, time consumed, and tentative parking fee of each vehicle.

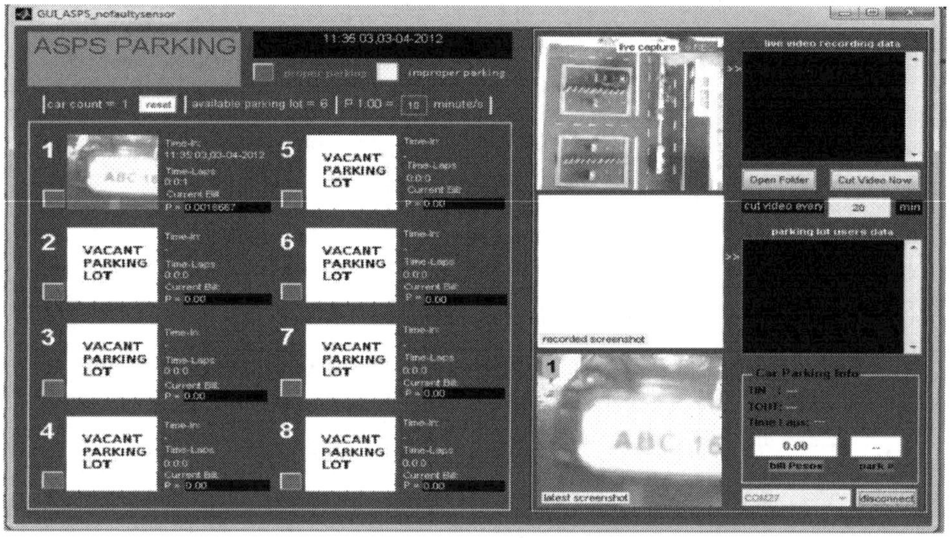

4.2 GUI as Plate Recognition

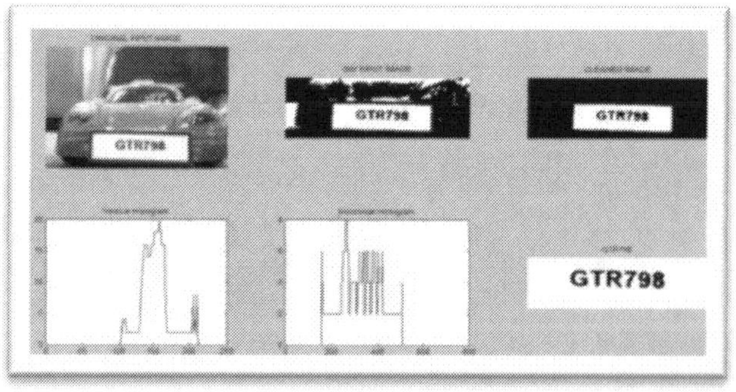

Figure 7 Plate Number Recognition Function

IOP Conf. Series: Materials Science and Engineering **79** (2015) 012009 doi:10.1088/1757-899X/79/1/012009

Figure 7 screenshot shows the operation of the plate number recognition function. The first image is the input image. The second image is the image darkened and converted to black and white. The small area connected white components are removed from the image and what is left is most probably the plate number area. The third image shows the image after the small white parts are removed. The coordinates of this large white area is then extracted and saved using the values of the vertical and horizontal histograms. Lastly, only the plate number region is left. This is shown on the last image. This is then processed through segmentation and template matching.

The label above the image shows the result of the image processing. The extracted string is shown to be "GTR798" which coincides with the plate number image. Therefore the plate number recognition has some degree of accuracy.

4. 3 Design of OCR Function in MATLAB

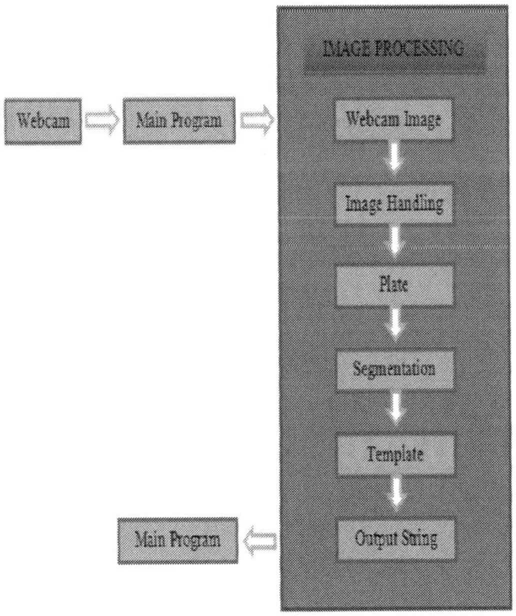

Figure 8 Image Processing Block Diagram

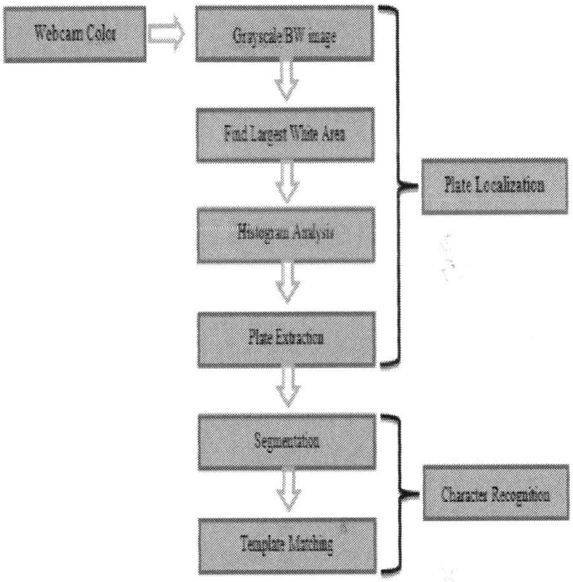

Figure 9 MATLAB Algorithm

The system is developed as a complement to the automatic smart parking system prototype as shown in Figure 8. When the ASPS system detects an incoming car, a front view snapshot of the car will be taken. This image is sent to the main program to be processed by the image processing system.

For ease of use, the system has been created in MATLAB as a function named processplate.m. The Main Program will call for the function to process the webcam image. After processing the image, the output which is the plate number will then be available in string format. The string will be used for identification of individual cars and will be easily usable for the GUI of the Main Program.

4.4 MATLAB Algorithm

4.4.1 Grayscale/BW image

The input image is taken from the webcam, and this will first be in RGB color format. An RGB color image is understood by MATLAB as an m-by-n-by-3 array consisting of the individual values for the red, green, and blue components.

4.4.2 Find Largest White Area

After the image is darkened and converted into black and white, the image becomes an image of white blobs on a dark background. The most probable plate number region is the area that has the highest number of interconnected white pixels. The small white regions are removed from the image using the bwareaopen function: testimage = bwareaopen(testimage, 15000);

We are left with a patch of white that has the highest area. The row and column coordinates of this region is determined using a scanning algorithm. Each pixel is scanned by row until the logical value 1 is found which represents the white region.

4.4.3 Histogram Analysis

Histograms representing the sum of the differences of neighboring pixels column-wise and row-wise will be used to indicate the number of transitions from black to white and vice versa along a row or column as shown in Figure 10. The distinguishing characteristic of a plate number is that it is a dark row of letters across a white background, that is, it represents quick transitions from light to dark to light. For an image with no other structure with this property, an area with a high sum of pixel differences is a very likely candidate for the number plate area.

First, the vertical histogram is created by running the algorithm through each row from end to end. The code starts at the leftmost value of a row and subtracts this with the value exactly next to it. The absolute value is taken and added to the sum of pixel differences for that row. This process repeats with the 2nd, 3rd, 4th value and so on until it reaches the second rightmost value.

This process is also repeated with the succeeding rows and the final histogram will consist of the pixel differences of each row. To get the horizontal histogram, the same process used to get the vertical histogram need only be changed to consider the columns instead of rows. Both horizontal and vertical histogram length will have equal number of rows and columns respectively.

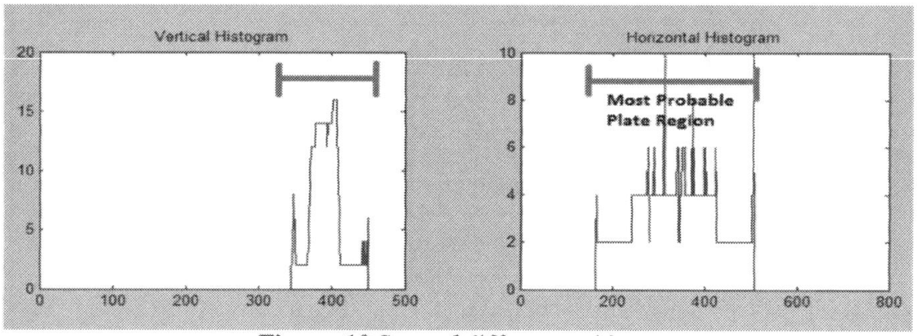

Figure 10 Sum of differences histogram

4.4.4 Plate Extraction

After getting the histograms, the region indicated by the histograms to have the highest number of transitions from white to black to white is most probably the number plate region. The coordinates of this region is saved. From a copy of the original and unedited black and white image, the image found on the coordinates indicated by the histogram will be extracted.

4.4.5 Segmentation

After the plate region has been extracted, we are left with an image of consecutive letters. An essential step in character recognition is separating these letters from one another as shown in Figure 11b. The algorithm used to separate the letters from one another assumes that the plate number is horizontal. With this, the algorithm traverses through each column of the image and looks for black pixels which are the letters themselves. When a black pixel is recognized in one column, it signals the starting edge of a letter. The algorithm continues until it reaches a column in which there are no more black pixels. This signals the rightmost edge of that letter and this letter is saved separately. The process continues until the last column is reached and all the letters saved separately. This process is then repeated for the separate letters, but this time the code is run vertically.

(a)  (b) 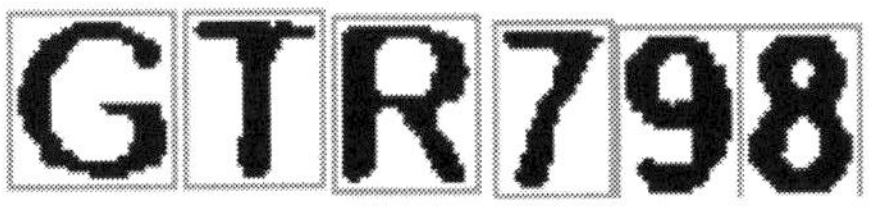

Figure 11 (a) Horizontal Scan, (b) Segmented Letters

4.4.6 Template Matching

The final step in character recognition involves recognizing the individual segmented letters. A template matching technique is employed to recognize these characters. Since template matching involves comparing a test image with a template, a database of templates has been created. Every letter of the alphabet and the numbers 0 to 9 each have their own template. Each segmented letter then of the plate number is scaled to equate the templates and compared one by one to the templates. The comparison with the highest match is considered to be the equivalent character of the image. To determine how good a match between the test image and the template is, a correlation function is used. The algorithm uses the normalized cross correlation of the matrices of the test image and the template to determine the accuracy of the match. The process is done for each of the segmented letters and the results are combined to acquire the string format of the plate number.

(a) 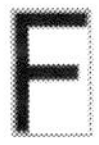 (b) 

Figure 12 (a) Character Template, (b) Segmented Letter

5.0 Summary

The practicality of the design is evident in that it lessens the stress of the motorists in finding a parking lot. It acts the way it is supposed to and shows motorists even from afar if the car park has vacant parking lots or not. It then recommends to the entering vehicle which parking lot is vacant. Since the whole system is automated, the need for manual operators is lessened, ergo; the labor cost is also lessened.

The automated car park prototype was successfully built by utilizing microcontrollers specifically the one PIC18F4550, two PIC16F84s and one PIC16F628A. The main microcontroller which is the PIC18F4550 successfully controls functions of the whole prototype. The prototype is put through a series of tests to identify the efficiency of the system. Despite its limitations, the prototype does the jobs it was meant to do excellently

References

[1] M. M. Rashid, A. Musa, M. Ataur Rahman, and N. Farahana, A. Farhana, M. M. Rashid, A. Musa, M. Ataur Rahman, and N. Farahana, A. Farhana, "Automatic Parking Management System and Parking Fee Collection Based on Number Plate Recognition", International Journal of Machine Learning and Computing, Vol. 2, No. 2, April 2012

[2] Sulaiman, N., Mustafa, M. , Hawari, K., "Development of Automatic Vehicle Plate Detection System",in proc. Conf. System Engineering and Technology (ICSET), 2013 IEEE 3rd International Conference.

[3] Ayer Keroh ; Idris, A. ," Study of Car Park Control System Using Optical Character Recognition", Computer. Vision & Robot. (CoVisBot) Lab., Tech. Univ. of Malaysia Melaka.

Physico-chemical effects of supercritical carbon dioxide post polymerization treatment on HCl-doped polyaniline prepared via oxidative chemical polymerization

J G Fernando[1,2,4], R M Vequizo[2], M K G Odarve[2], B R B Sambo[2], A C Alguno[2], R M Malaluan[2], R T Candidato Jr.[2], J E Gambe[2], M Jabian[2], G J Paylaga[2], F R G Bagsican[2] and H Miyata[3]

[1] Physics Department, College of Science and Mathematics, Western Mindanao State University, Zamboanga City 7000, Philippines
[2] Department of Physics, College of Science and Mathematics, Mindanao State University-Iligan Institute of Technology, Iligan City 9200, Philippines
[3] Department of Physics, Faculty of Science, Niigata University, 2-Nocho, Ikarashi, Niigata City, Japan

E-mail: jgfernando82@gmail.com

Abstract. Polyaniline films doped with varying HCl concentrations (0.2 M, 0.6 M and 1.0 M) were synthesized on glass substrates via oxidative polymerization of aniline. The films were treated with supercritical carbon dioxide (SC-CO_2) at 30 MPa and 40 °C for 30 minutes. Their structural, optical and morphological properties were studied and compared to conventionally prepared polyaniline films using FTIR analysis, UVVis spectroscopy and scanning electron microscopy. It was observed that supercritical carbon dioxide (SC-CO_2) could interact with PANI films that consequently altered the bandgaps and changed the film thickness. SC-CO_2treatment also increased the oxidation level of polyaniline films and modified the morphology of polyaniline film doped with 1M HCl.

1. Introduction

Supercritical fluid (SF) is a peculiar material because it exhibits properties that are separately inherent to liquids and gases. For instance, it can diffuse through solids like gas and at the same time dissolve a number of substances like a liquid solvent [1]. SF's properties (e.g., solubility, viscosity and density) can also be tuned by controlling its pressure or temperature [2], and have been exploited in synthesis [3], extraction [4] and material processing [5]. Among SF components, carbon dioxide is most widely used because it is non-toxic, less expensive and has convenient critical temperature and pressure (7.38 MPa and 31.1 °C) [6].

Supercritical carbon dioxide (SC-CO_2) has been extensively used in processing polymers resulting to improved properties or new products. The plasticization and swelling of polymers induced by SC-CO_2makes it possible to prepare polymer blends [7], infuse additives such as dyes and metal complexes (impregnation) [8], extract contaminants [9] and modify morphology [10]. While there have been a number of studies on SC-CO_2 processing of polymeric materials, none has been reported

[4] To whom any correspondence should be addressed.

on polyaniline. More so, the influence of SC-CO_2 on the physico-chemical properties of polyaniline has not yet been elucidated which restricts better predictability during processing.

In the present paper, we synthesized HCl-doped polyaniline film on glass substrates via chemical oxidative polymerization. The prepared films were treated with supercritical carbon dioxide and the influence of SC-CO_2 on the structure, morphology and optical property of PANI films was investigated.

2. Methodology

2.1. Synthesis

Polyaniline films on glass substrates were prepared by oxidative chemical polymerization of aniline with ammonium peroxydisulfate as oxidant in an acidic medium. Aniline and ammonium peroxydisulfate were dissolved separately in hydrochloric solutions (0.2 M, 0.6 M and 1.0 M HCl). One side of the substrate was sealed to prevent film formation on both sides of the substrate. The substrates were then placed on the reaction vessel where aniline and ammonium peroxydisulfate were mixed to start the polymerization. The samples were then removed after 30 minutes; rinsed with 0.2 m HCl, ethanol and water; dried; and removed from the attachment for SC-CO_2 treatment and characterization. Only half of the films (one film per synthesis) were treated with SC-CO_2

2.2. Supercritical Carbon Dioxide Processing

The films were placed in the extracting chamber and treated with supercritical carbon dioxide for 30 minutes using supercritical carbon dioxide extraction machine (Akico). The pressure and temperature were maintained at 40°C and 30 MPa, respectively.

2.3. Characterization

The UV-Visible absorption spectra of the films were measured in the range 400 to 900 nm by Lambda 35 UV-Vis spectrometer (Perkin Elmer, UK). Fourier transform infrared spectra (FTIR) with a 1 cm^{-1} resolution were obtained in the range 650-4000 cm^{-1} by using Spectrum 100 FTIR spectrometer (Perkin Elmer, UK). The morphologies of the samples were examined using scanning electron microscope (JEOL).

3. Results and Discussion

3.1. FTIR spectra

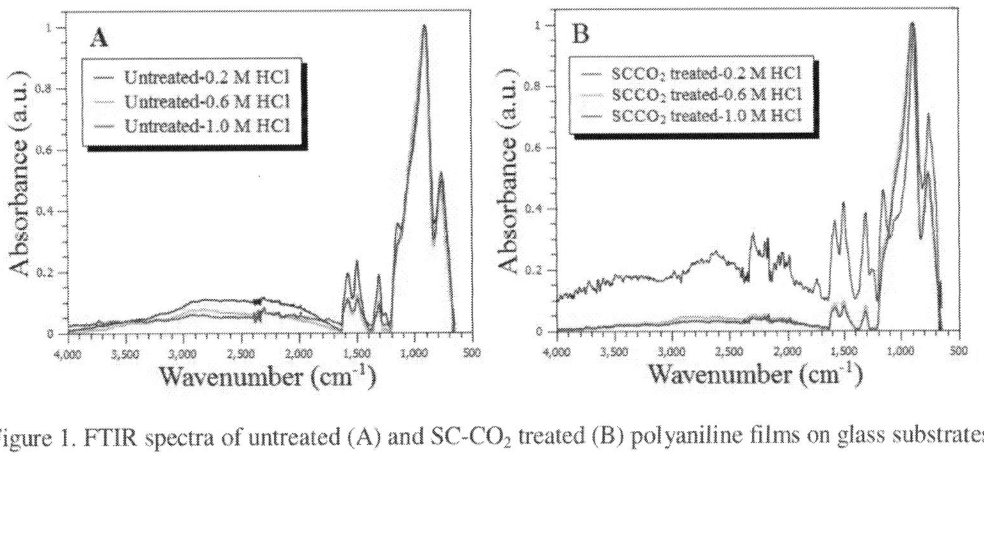

Figure 1. FTIR spectra of untreated (A) and SC-CO_2 treated (B) polyaniline films on glass substrates.

The FTIR spectra of untreated polyaniline films are presented in figure 1. It shows extended absorption band above 2000 cm^{-1} which is characteristic of conducting polyaniline films. In the range 1400-1700 cm^{-1}, two main bands were observed with maxima located at 1571 and 1492 cm^{-1},which are assigned to C=C stretching of the quinoid (Q) and benzenoid (B) rings, respectively [11]. A shoulder appears near the quinoid ring (1608 cm^{-1}) that corresponds to –C=C- ring stretching. This peak becomes more evident as HCl concentration increases because protonation induces conformational changes in the polymer which makes the vibration more infrared-active [12]. Below 1400 cm^{-1}, the band assigned to C-N stretching of secondary aromatic amine is situated at 1303 cm^{-1}. Two other bands intrinsic to conducting polyaniline were found at 1243 and 1148 cm^{-1} that correspond to C-N^{+o} stretching and vibrational mode of –NH^{+}=. A shoulder is also detected at 1333 cm^{-1}. A less intense peak typical to a PANI base is noted at 1374 cm^{-1}. The intense peaks in the range 650-1200 cm^{-1} are linked to the glass substrates.

For SC-CO_2 treated films, the main characteristic absorption bands of polyaniline were also observed at 1571 (C=C stretching of quinoid ring), 1492 (C=C stretching of benzenoid ring), 1303 (C-N stretching of secondary aromatic amine) and 1148 cm^{-1} (vibration mode of –NH^{+}=). The maxima of these bands are not shifted with respect to untreated films except for the film protonated with 0.2 M HCl in which the –NH^{+}= band shifted by 12 cm^{-1} to 1136 cm^{-1}. The presence of C-N^{+o} stretching vibration band at 1248 cm^{-1} indicates that SC-CO_2 treated films are also conducting like the untreated films. The band at 1248 cm^{-1} splits into a doublet for 0.2 M HCl; broadens for 0.6 M HCl and maintains its shape for 1.0 M HCl. The shoulder of C-N stretching band is unshifted for all HCl concentrations. The band at 1374 cm^{-1} corresponding to base form of polyaniline becomes less prominent. New bands were also observed near 2300 cm^{-1} and 1730 cm^{-1} in the spectra of film protonated with 0.2 M HCl. The former is attributed to asymmetric stretch of carbon dioxide [13] and the latter is associated to carbonic acid.

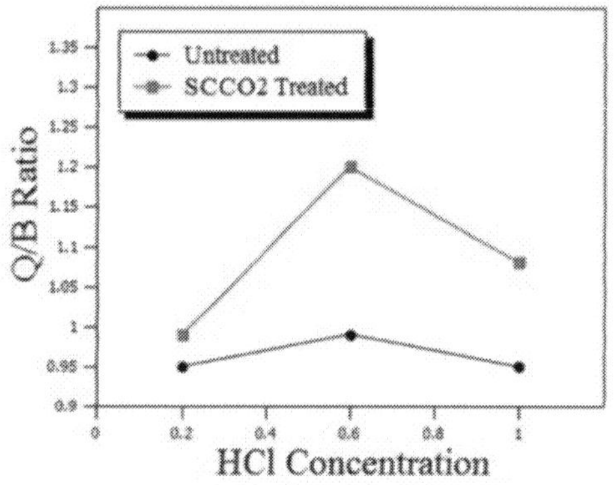

Figure 2. Q/B ratio of untreated and SC-CO_2 treated polyaniline films on glass substrates.

The bands in the region 1400-1650 cm^{-1} were normalized and deconvoluted so that the integrated intensity ratios (Q/B) of quinoid and benzenoid peaks can be compared. The Q/B ratio is a measure of the extent of oxidation of the films [14] i.e. 1 being the ideal Q/B ratio for emeraldine state. The Q/B ratios for untreated films protonated with 0.2 M, 0.6 M and 1.0 M HCL are 0.95, 0.99 and 0.95, respectively (figure 2). These values are close to unity which implies that the films are emeraldine. For SC-CO_2 films, the Q/B ratios are 0.99, 1.20 and 1.02 for 0.2 M, 0.6 M and 1.0 M HCL protonation, respectively. The Q/B ratios for treated films were found higher than the untreated films suggesting that SC-CO_2 at elevated pressure can remove electrons from polyaniline. Oxidation of polyaniline is known to convert benzenoid ring to quinoid ring and this accounts for higher amount of quinoid rings.

3.2. UV-VIS Spectra

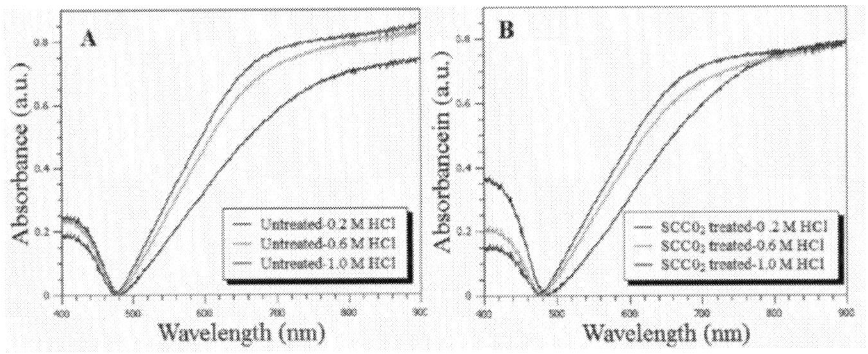

Figure 3. UV-VIS absorbance spectra of untreated (A) and treated (B) PANI films doped with different HCl concentrations.

From the UV-VIS absorption spectra (figure 3), the thicknesses of the films were determined using the absorbance at 400 nm [15]. Untreated films synthesized with 0.2 and 0.6 M HCl are thicker than the film prepared with 1.0 M HCl (figure 4). This variation in the film thickness is linked to the relative amount of aniline and anilinium cations during the initiation period of polymerization. At low HCl concentrations, aniline molecules dominate the anilium cations and form more initiation centers on the surface of the substrates [16]. The high density of initiation centers more likely orient the PANI macromolecules to grow in the direction perpendicular to the substrates resulting to thicker films [17].

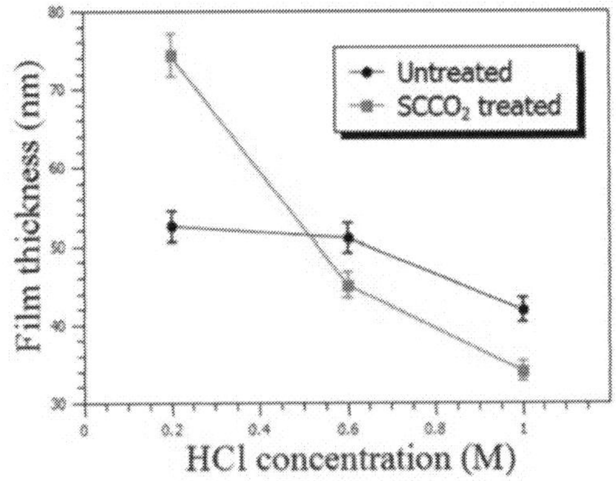

Figure 4. Thickness of untreated and treated PANI films.

In SC-CO_2 treated films, the thickness of the films protonated with 0.6 and 1.0 M HCl decreasedas compared to untreated films. This may be due to extraction of low-molecular-weight organic intermediates and oligomers or even degradation of the polymer matrix. On the contrary, the thickness of the film with lowest HCl concentration increased after it was treated with SC-CO_2. We propose that this effect is due to dissolution of SC-CO_2in the polymer matrix that caused it to swell[2].

Despite having the least thickness among untreated films, the film protonated with 1.0 M HCl has the highest absorbance in the range 800 to 900 nm. This suggests that the film is highly protonated since the absorption at longer wavelength is linked to the presence of polarons [18]. As expected, the film protonated with 0.2 M HCl shows the least absorbance in 800-900 nm range indicating the least level of protonation among the samples. The presence of a peak near 420 nm (polaron-π^* transition), one of the characteristic peaks of polyaniline emeraldine salt, verifies the protonation of the films.

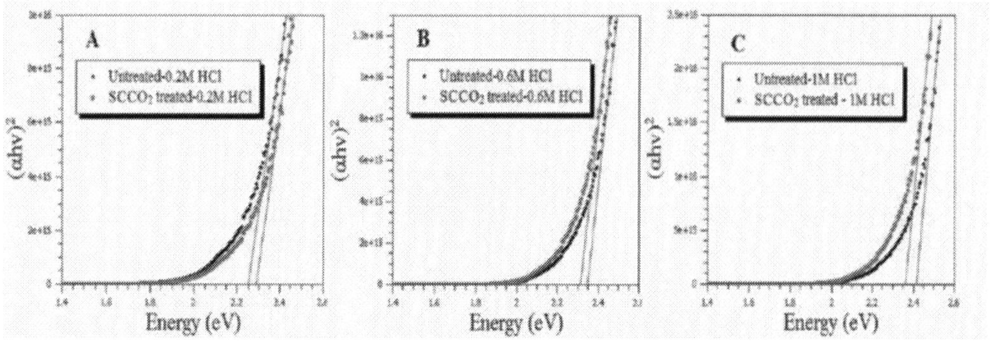

Figure 5. Plot of $(\alpha h\nu)^2$ versus photon energy for PANI films doped with 0.2 M (A), 0.6 M (B), and 1.0 M (C) HCl. Dashed lines are the extrapolated lines from which the bandgaps were determined.

Similarly, SC-CO_2treated films exhibit high absorbance in the range 800-900 nm implying that the films remained protonated after the treatment. The rise/drop of absorbance with respect to untreated films in this range is thickness related since it coincides directly with the change in absorbance at 400 nm.

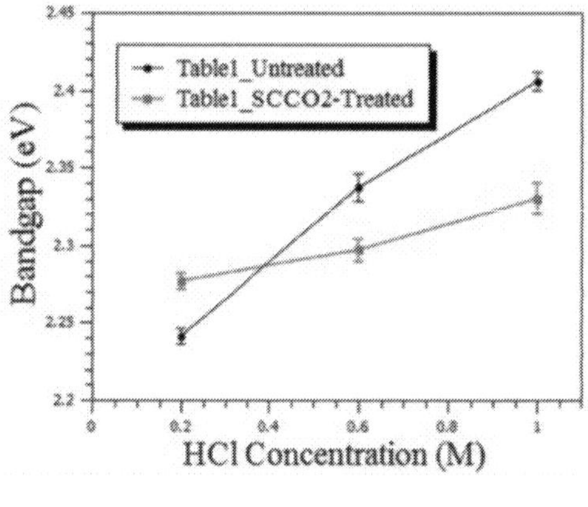

Figure 6. Bandgaps of untreated and SC-CO_2 treated films doped with HCl at different molarities.

The energy bandgaps were determined from the absorption spectra using the Tauc relation

$$(\alpha h\nu)^2 = A(h\nu - E_g), \tag{1}$$

where A is a constant, α is the absorption coefficient, $h\nu$ is the photon energy and E_g is the bandgap [19]. The plot of $(\alpha h\nu)^2$against the photon energy and the extrapolated linear part of the graph are shown in figure 5. The bandgaps of untreated films (2.241, 2.337 and 2.406 eV for 0.2, 0.6 and 1.0M HCl, respectively) are close to the reported experimental value (2.21 eV)for doped polyaniline [20]and seen to increase with increasing HCl concentration (figure 6). This transition energy is recognized as the electronic excitation to the polaron band [21].

The bandgaps of SC-CO_2 treated films protonated with higher HCl molarities (2.297 and 2.330) tend to be lower than that of untreated films (figure 6) due to their enhanced quinoid character. The quinoid form has weaker resonance effect that allows electronic transitions more readily than the aromatic form [22].However, the bandgap of the film doped with 0.2 M HCl increased to 2.277 eV that may be due to stronger interaction of carbon dioxide with the polymer chains. Carbon dioxide may have interacted with the polymer via the N-H structurethat leads to modification of the polymer structure [11].

3.3. Morphology

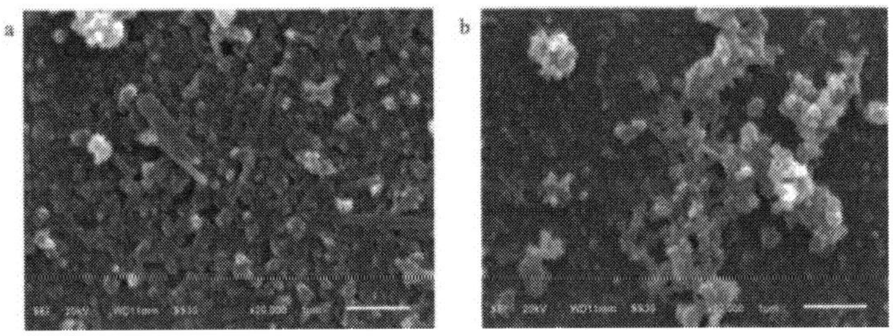

Figure 7. SEM images of untreated (a) and SC-CO_2 treated (b) PANI films doped with 0.2 M HCl.

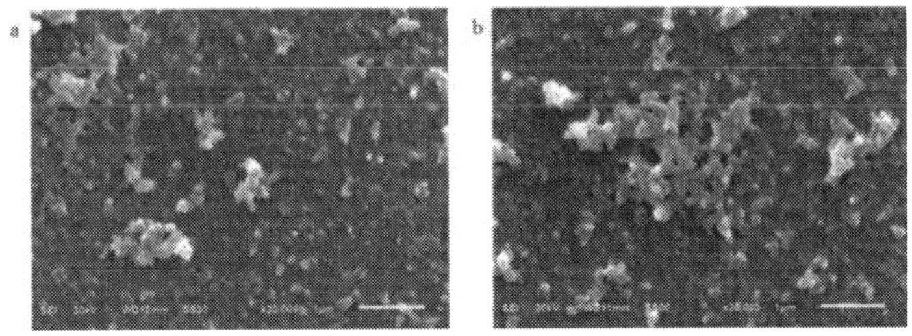

Figure 8. SEM images of untreated (a) and SC-CO_2 treated (b) PANI films doped with 0.6 M HCl.

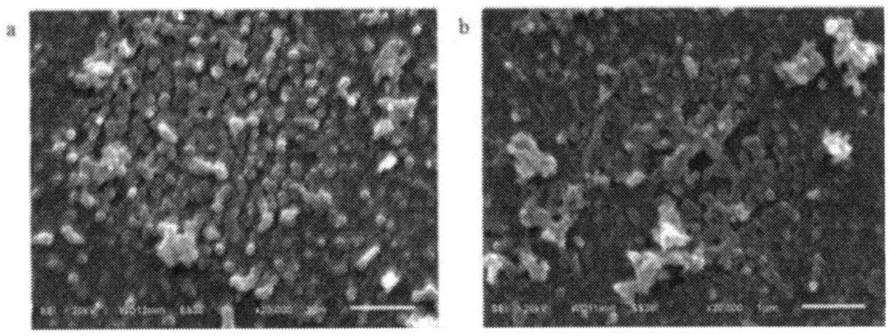

Figure 9. SEM images of untreated (a) and SC-CO_2 treated (b) PANI films doped with 1.0 M HCl.

SEM micrographs of untreated films reveal that their surfaces are not entirely smoothdue to the presence of precipitates adhering on the grown film. All untreated films completely covered the glass substrates except for 0.2 M HCl, i.e., a small portion of the film displays nanostructures that are not well knitted. A mixture of dendritic nanorods and granular nanostructures were observed on the surface of the film protonated with 0.6 M HCl and circular patches composing of granular nanostructures in 1.0 M HCl.

The precipitates still appear on the surface of the films after they were treated with SC-CO_2whichsuggests the inability of the treatment to remove these agglomerates. The loosely knitted nanostructures in 0.2 M HCl film were replaced with continuous film that completely covered the substrate. This may support our claim that SC-CO_2 can reorganize the polymer chains. Voids began to appear in films protonated with 0.6 M HCl and its density($2.4x10^6/cm^2$) drastically increased in 1.0 M HCl protonated film. These observations seem to suggest that the surface quality of the film protonated with high HCl concentration deteriorates when treated with SC-CO_2.

4. Conclusion

Post-polymerization SC-CO_2 treatment was shown to affect the physical and chemical properties of HCl-doped polyaniline films on glass substrates. The treatment tends to decrease the thickness of the films synthesized using high HCl molarities and increase the thickness for low molarity. The bandgaps were also altered and the quinoid characters of the films were enhanced after the treatment. Voids were formed on the surface of the treated film doped with 1.0 M HCl.

5. References

[1] Sapkale G N, Patil S M, Surwase U S and Bhatbhage P K 2010 *Int. J. Chem. Sci.* **8(2)** 729-43
[2] Kazarian S G *Polymer Science* 2000 **42** 78-101
[3] Du J, Zhang Z, Han B, Liu Z and Wan M 2005 *Synth. Met.* **155** 523-526
[4] Reverchon E and De Marco I 2006 *J. Supercrit. Fluids* **38** 146-166
[5] Woods H, Silva M, Nouvel C, Shakesheff K and Howdle S 2004 *J. Mater. Chem.* **14** 1663-78
[6] Nalawade S, Picchioni F and Janssen L 2006 *Prog. Polym. Sci. 31* 19-43
[7] Tang M, Wen T, Du T and Chen Y 2003 *Eur. Polym. J* **39** 143-149
[8] Kikic I and Vecchione F 2003 *Curr. Opin. Solid State Mater. Sci.* **7** 399-405
[9] Vandenburg H, Clifford A, Bartle K, Carroll J, Newton I, Garden L, Dean J and Costley C 1997 *Analyst* **122** 101R-115R
[10] Elvassore N, Baggio M, Pallado P and Bertucco A 2001 *Biotechnol. Bioeng.* **73** 449-57
[11] Sajeev U S, Mathai C J, Saravanan S, Ashokan R R, Venkatachalam S and Anantharaman M R 2006 *Bull. Mater. Sci.* **29** 159–163
[12] Sedenkova I, Miroslava T and Stejskal J 2008 *Polym. Degrad. Stab.* **93** 2147-57
[13] Yuan Y and Teja A 2011 *J. Supercrit. Fluids* **56** 208-12
[14] Arenas M C, Sanchez G, Nicho M E, Elizalde-Torres F and and Cataño VM 2011 *Appl. Phys. A*
[15] Stejskal J, Sapurina I, Prokes J and Zemek J 1999 *Synth. Met.* **105** 195-202
[16] Stejskal J, Sapurina I and Trchova M 2010 *Prog. Polym. Sci.* **35** 1420-81
[17] Sapurina I, Yu A, Volchek B Z, Trchova M, Riede A and Stejskal 2002 *Synth. Met.* **129** 29-37
[18] Blinova N, Sapurina I, Klimovic J and Stejskal J 2005 *Polym. Degrad. Stab.* **88** 428-34
[19] Abdulla H and Abbo A 2012 Int. *J. Electrochem. Sci.* **7** 10666 - 78
[20] Yakuphanoglu F, Basaran E, Senkal B F and Sezer E 2006 *J. Phys. Chem. B.* **110** 16908
[21] Stafstrom S, Bredas J L, Epstein A J, Woo H S, Tanner D B, Huang W S and MacDiarmid AG 1987 *Phys. Rev. Lett.* **59** 1464-1467
[22] Bredas JL 1985 *J. Chem. Phys.* **82** 3808.

Acknowledgment

Authors would like to thank Mr. Lloyd Arvin M. Malaluan for the SC-CO_2 treatment. We would like to thank the Commission on Higher Education for the scholarship grant of JGF and for the funding of the project.

Enhanced RF to DC converter with LC resonant circuit

L J Gabrillo, M G Galesand J A Hora[1]

Microelectronics – Power IC Laboratory, College of Engineering, MSU-Iligan Institute of Technology, Iligan City, Philippines

Email: jefferson.hora@g.msuiit.edu.ph

Abstract. Presented in this paper is an experimental comparison of the conventional and proposed design circuit of a radio frequency (RF) energy harvesting. RF to DC energy harvester simply consists of antenna and rectifier block for receiving electromagnetic radiation signal and to produce a DC voltage, respectively. In addition to this conventional circuit, the proposed design includes LC tank circuit as receiving block of a well-designed antenna radio frequency receiver. Proper choice of an antenna type, realizing of point contact Germanium diodes as rectifier and correct design values for the LC passive components, greatly improved the measurement of the maximum output power, giving approximately a 100% increase compared to the conventional method. Experimental results of the enhanced RF to DC converter measured a maximum output power of 1.80 mW at a distance of 77.84 meters from a TV signal tower operating at 165 MHz. Thus, the harvested signal was enough to supply a low power wireless device applications without battery maintenance.

1. Introduction

Developing efficient methods for extracting DC power from RF signals has become an important necessity for a number of applications involving self-powered devices and sensor nodes [1]. In recent years, development of technology is rapidly growing allowing the decrease of certain characteristics of a system or device like size and power consumption. However, these devices are powered by bulky batteries that need to be recharge or replace every so often.

Extracting energy from the environment particularly RF signal is not only an eco-friendly way of generating free energy but can be a solution to minimize the consumption and usage of battery in most wireless devices. Wireless power transfer will be possible for low-power applications only since the expected received power will be low at about 1mW.

However, converting RF energy from RF signals at different frequencies is relatively difficult issue particularly when RF signals have relatively low power levels [1,2]. The operating range of such self-powered devices has been severely limited by the failure of existing power extraction techniques to successfully extract power from radio frequency signals having relatively low power levels. This is mainly because of the problems of extracting DC power from electromagnetic radiation which consist of two main parts: collecting the incident radiated power, and converting the collected power to DC signals enough to power self-powered devices[3].

[1] To whom any correspondence should be addressed.

Thus, proper choice and design of antenna and a rectifying circuit is needed to solve this problem. Having an efficient antenna for matching the desired frequency and a high performance rectifying circuit will greatly affect and improve the output power generated [3].

With the challenging issue of improving the generated output power generated in a conventional RF to DC converter, this paper explored the addition of properly designed tank or tuner circuit to improve input matching of the desired radio frequency received before rectification process. Also, different antenna type will be tested to determine the maximum output power generated of the overall system.

2. Design architecture

This section shows the comparison of the blocks present in the conventional and the proposed radio frequency energy scavenging method.

2.1.Conventional design block

The block diagram of the conventional RF to DC converter is shown in figure 1. In consists of an antenna, input matching circuit, rectifier, output matching circuit and a load [4].A method for performing a power extraction includes receiving an electromagnetic radiation signal using antenna, rectifying the signal produce DC voltage and providing the DC voltage to a circuit.

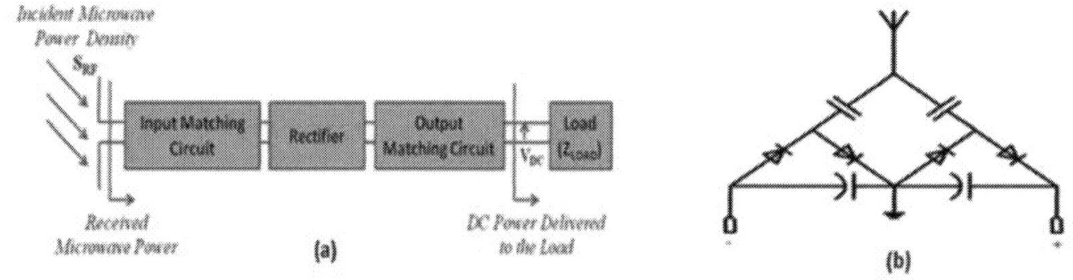

Figure 1. Conventional RF to DC Converter Block: (a) block diagram, (b) circuit.

2.2. Proposedcircuit design

The electromagnetic radiation captured by the antenna will be optimized if a proposed tuner circuit in addition to the traditional method is employed. This tuner or tank circuit converts the selected carrier frequency and its associated bandwidth into a fixed frequency that is suitable for further processing and in consideration to the required impedance matching network for optimizing the performance of the overall RF energy extraction circuit. Shown in figure 2 the block diagram and circuit of the proposed design.

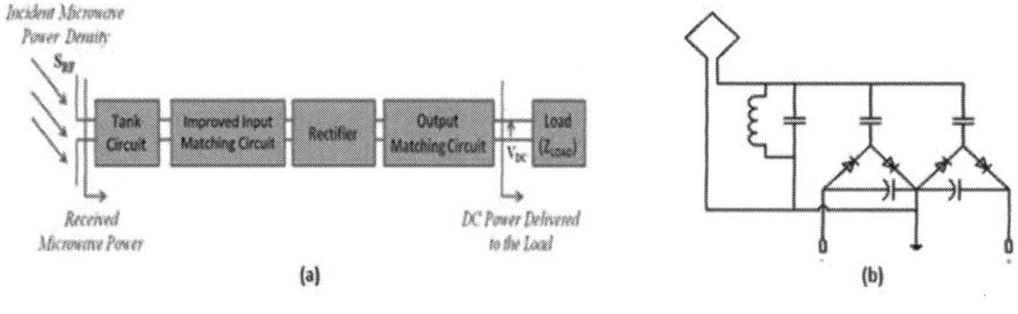

Figure 2. ProposedRF to DC Converter Block: (a) block diagram, (b) circuit.

3. Block components design

The proper choice of antenna and diode type has been greatly considered in this paper. Two types of antenna were considered as experimental comparison to the chosen loop antenna used in the proposed design of RF energy harvesting. Also, germanium diode was primarily chosen over silicon type diode due to the threshold voltage consideration. The following section will discuss the details of the components employed in this paper.

3.1. RF antenna

The proposed RF to DC energy harvester circuit used a loop antenna tuned at 165MHz television station. Loop antennas take many different forms such as a rectangle, square, triangle, ellipse, circle, and many other configurations. Because of the simplicity in analysis and construction, the circular loop is the most popular and has received the widest attention [5]and which was also employed in this paper.

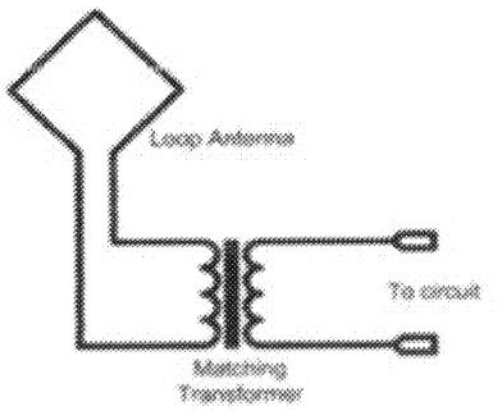

Figure 3. RF loop antenna component.

The geometrical construction of the loop antenna to tune at the required 165 MHz frequency of the transmitter has been designedwith the graph of the axial directivity, $ka = c/\lambda$ and the radiated electric fields formula:

$$E_r = E_\theta = 0 \quad (1)$$

$$E_\phi = \eta \frac{(ka)^2 I_0 \sin\theta}{4r}[1 + \frac{1}{jkr}]e^{-jkr} \quad (2)$$

3.2. Tank or tuner circuit

A tuner is a subsystem that receives radio frequency (RF) transmissions like radio broadcasts and converts the selected carrier frequency into a fixed frequency, and that is, 165 MHz in this paper. The proposed circuit employed a simple tank circuit consisting of a capacitor and inductor connected in parallel. This creates a resonant circuit which corresponds to an alternating current at one frequency. The calculated values of the capacitor and inductor in resonant to the frequency drawn by the known transmitter TV station was based on the equivalent frequency and the cylindrical air-core type inductor formula, shown in equation (3) and (4), respectively.

$$f_0 = \frac{\omega_0}{2\pi} = \frac{1}{2\pi\sqrt{LC}} \quad (3) \qquad L = \frac{N^2 \mu A}{l} \quad (4)$$

3.3. Germanium and silicon diode for rectifier block

Even when the forward voltage is just below 0.7V for Silicon or 0.3V for Germanium, diodes will start to conduct minute currents. For Germanium diodes, it will conduct at a forward voltage of only about 0.15V, however, a silicon diode will not start to conduct until a forward voltage of 0.6V is reached [6].

Thus, primarily due to low threshold voltage of the Germanium diodes and sometimes known as signal diodes, this type of diode is well suited for RF signal harvesting. As shown if figure 4, an example of demodulator circuit that recovers the audio signal from the amplitude modulation (AM) radio waves can utilized either Silicon and Germanium diode as a rectifier to remove unwanted half of the AM signal, having the difference of the frequency of operation and voltage used.

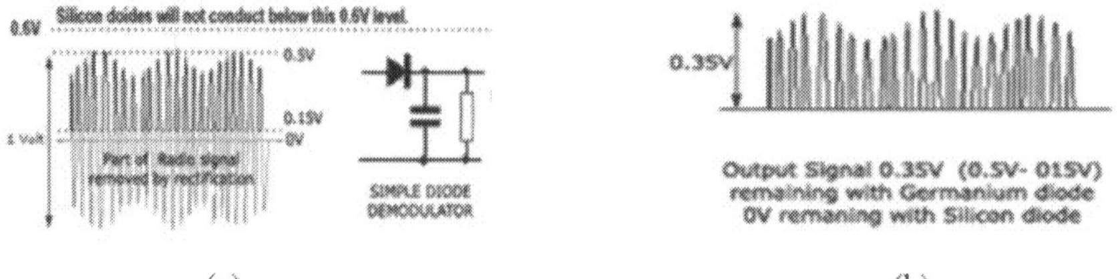

(a) (b)

Figure 4. (a) Test response of Si, (b) Test response of Ge.

3.4. Capacitor filter and voltage multiplier

A design of value of capacitor is very important after the rectification of the signal and which is essential on both input and output matching blocks. This capacitor filter networks will smoothen the DC rectified signal into ripple voltages of the proposed RF to DC energy harvester circuit, not to mentioned it main importance in the output of power supplies, in the resonant circuits that tune radios to particular frequencies, in electric power transmission systems for stabilizing voltage and power flow, and for many other purposes [7].

On the other hand, a voltage multiplier was also employed in the proposed design to boost up the voltage generated from a few volts to acceptable several volts as it was generally used in a high-energy physics experiments and lightning safety testing. This paper utilized a half-wave series multiplier consisting network of capacitors and diodes.

4. Experimental results

A fixed design of the rectifier circuit using germanium diode and well calculated capacitor values for the matching output were fixed in the test set-up. Correspondingly, three types of antenna, namely: straight, yagi-uda and loop antenna were designed and tuned accordingly to receive radio frequency signal from a TV station operating at 165 MHz. Also, the voltage and current measurement of the conventional RF to DC converter were considered with the three test locations, with an approximate distance of 77, 246 and 411 meters away from the transmitter TV station. The used of google earth application approximated this distance from one point to another as shown in figure 5.

Figure 5. Example of approximated transmitter & receiver distance measurement.

From the test measurement of the different types of antenna employed and at a different distance location from the transmitter TV station, it was proven that loop type antenna in an adjacent position or placement showed a superior measurement of voltage and current which can able to power one single LED load. Accordingly, antenna position or placement for the three types was considered. Test and measurement were made for the vertical or horizontal position Straight antenna; facing or parallel position Yagi-uda antenna; and facing, parallel and adjacent placement Loop antenna. The loop type antenna measured 4.93V at no load condition at approximately 77.84 meters away from the transmitter TV station. The maximum voltage was measured with respect to the different test location as shown in figure 6.

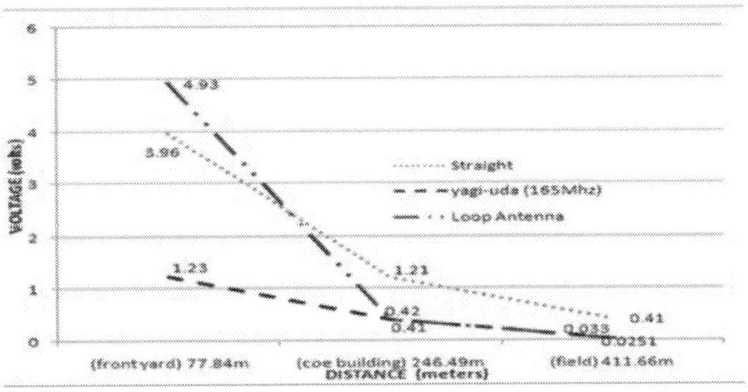

Figure 6. Maximum voltage measured of the three types of antenna.

Correspondingly, at no load condition, the maximum output power was calculated using the measured voltage & current of the three types of antenna that were designed for the conventional RF to DC converter circuit. The Loop, Yagi-uda, and Straight type antenna had measured a maximum output power of 0.887 mW, 685 mW, and 87.3 mW, respectively. The said measurement was tested at the approximate distance of 77 meters away from the transmitter signal as shown in figure 7.

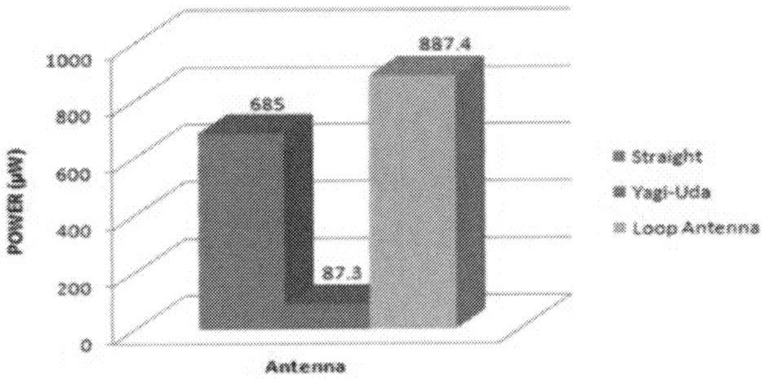

Figure 7. Maximum output power measured for the three types of antenna at 77m distance.

Considering the utilization of loop antenna, rectifier germanium diode and approximated distance of 77 meters, the conventional circuit set-up has been modified with the addition of a properly designed values LC tuner circuit. Typical and maximum measurement of the voltage and current had been recorded as shown in figure 8. Consequently, the proposed design showed a maximum output power of 1.8 mW while the conventional circuit only had 0.887 mW, that is, approximately a 100% improvement as summarized in figure 9. Also, in figure 10 shows the hardware set-up of the proposed RF to DC energy harvester.

IOP Conf. Series: Materials Science and Engineering **79** (2015) 012011 doi:10.1088/1757-899X/79/1/012011

CONVENTIONAL	Resolution	
Voltage (V)	4.86	4.93
Current (μA)	173	180
Power (μW)	840.8	887.4

PROPOSED	Resolution	
Voltage (V)	8.50	8.64
Current (μA)	202	210
Power (μW)	1717	1814

Figure 8. Measured voltage and current of the proposed and conventional RF to DC converter energy harvester circuit.

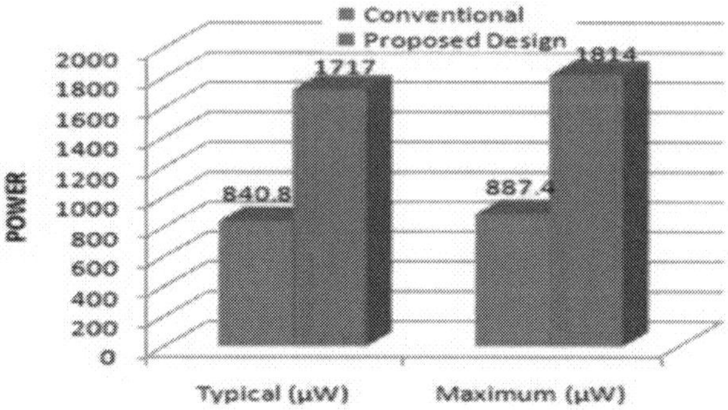

Figure 9. Output power comparison of the proposed and conventional design.

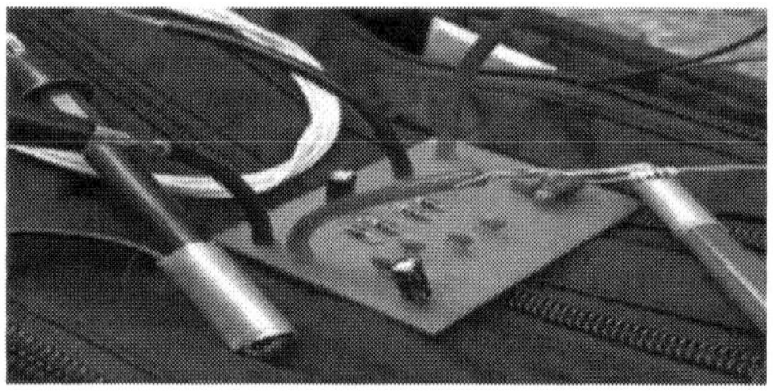

Figure 10. Actual hardware set-up of the proposed RF-DC energy harvest.

This proposed RF to DC converter circuit design with the use of loop antenna, cascaded half-wave doubler as rectifier circuit and the addition of tuner circuit was compared to other researches on radio frequency energy harvesting. Table 1 shows this comparison and the proposed design had an extreme higher power consumption at no load condition.

Table 1. Performance Comparison.

Parameters	Reference [23]	Reference [24]	This design
Voltage (V)	2.4	not mentioned	8.64
Current (μA)	51	not mentioned	210
Power (μW)	122.4	1	1814
distance from source (m)	50	40	77

5. Conclusion

The addition of a properly design values of LC tuner circuit with rectifier circuit consisting of two cascaded half-wave doubler demonstrated the highest power harvested from RF energy. Also, the use of a loop antenna contributes more in capturing ambient RF energy. The system has proved to be an RF recycler that could be very useful in indoor applications like wireless sensors.

6. References

[1] Liu G, Mrad N, Xiao G, Li Z and Ban D 2011 *International Workshop Smart Materials, Structures & NDT In Aerospace Conference*
[2] Mitcheson P D et al. 2008 *Proceedings of the IEEE* **96**
[3] Vullers R J M, van Schaijk R, Doms I, Van Hoof C and Mertens R 2009 *Solid-State Electronics* **53** 684–693
[4] Monti G and Congedo F 2012 *Progress in Electromagnetics Research* **26** 181
[5] MillerE K and LandtJ A 1980 *Proc. IEEE* **68**1396–1423
[6] Miller E K and BurkeG J1992 *Proc. IEEE* **80**24–43
[7] Balanis C A 1992*Proc. IEEE* **80**7–23

IOP Conf. Series: Materials Science and Engineering **79** (2015) 012012 doi:10.1088/1757-899X/79/1/012012

On polymer loop in a gel under external fields: analytical approach using white noise analysis

B Gemao[1,3], **J Bornales**[1] **and M Loquero**[2]

[1] Mindanao State University – Iligan Institute of Technology, Iligan City, Lanao del Norte, Philippines

[2] Central Mindanao University, Musuan, Maramag, Bukidnon, Philippines

E-mail: beverly.gemao@gmail.com

Abstract. This paper presents derivation of the probability distribution for the area enclosed by a polymer loop in a gel and under different external fields using white noise analysis. In this context, the polymer loop is represented by Brownian paths and its immersion in a gel constraints it to occupy a constant area[1]. The external fields considered are electric field, and crossed electric-magnetic fields.

1. Introduction

The Hida-Streit's infinite dimensional analysis, otherwise known as white noise analysis has been used successfully as a framework for stochastic and infinite-dimensional systems. This has been applied to several disciplines, most notably in quantum physics [2,3] and statistical mechanics[4,5]. Its applicationin solving polymer conformation problems was first introduced by C. Bernido and M. V. Bernido where they studied statistical mechanical properties of polymers with length dependent potentials[4,5], polymer chirality [5]and winding probability of entangled polymers[6].

As an extension to the works of C. Bernido and M. V. Bernido, we present in this paper the use white noise analysis in getting the probability distribution for the area enclosed by a polymer loop in a gel under different external fields. In this context, the polymer conformation is viewed as Brownian paths and the steric and topological effects of others polymers in a gel constraints the loop to enclose a fixed area[1,7]. An example of a polymer in a gel under an external field is DNA electrophoresis where the DNA is subjected by an external electric field. In this paper, we generalize the probability distribution to include other possible external fields.

This paper is organized as follows: we give a brief review of the path integral of the probability distribution for the area enclosed by a polymer loop in a gel as presented in [7]. Then we generalize the formulation to include external fields where we consider time-independent electric fieldand crossed electric-magnetic fields. We then translate the probability distribution in the language of white noise analysis and evaluate the distribution using the *T-* and *S-transforms* [5]. We note that the same polymer system with and without electric field was also studied by Khandekar and Wiegel [7,8] using straight forward evaluation of the path integral describing the system.

[3] To whom any correspondence should be addressed.

2. Path integral for the distribution of the area enclosed by a polymer loop under external fields

An unconstrained polymer may be viewed as a polymer in a dilute solution where the interaction between the solvent and the polymer is negligible. The Wiener integral representation of the probability density of this polymer with ends fixed at $\boldsymbol{r_0}$ and $\boldsymbol{r_1}$ is given by [7]

$$\boldsymbol{P(r_1, N|r_0, 0)} = \int_{r_0}^{r_1} \exp\left[-\frac{1}{l^2}\int_0^N \left(\frac{dr}{dv}\right) dv\right] d[r(v)], \tag{1}$$

where for non-interacting polymer, l is the distance between two monomers, and v is the number of monomers with increasing value from 1 to N.

For a polymer in a gel or a polymer solution, the steric and topological effects of others polymers constraints the polymer to enclose a constant area A [1,7] making the probability distribution for the area enclosed as

$$P(A, N) = C\int_{r_0}^{r_1} \delta\left[A - \int_0^N \frac{1}{2}\left(x\frac{dy}{dv} - y\frac{dx}{dv}\right) dv\right] \exp\left[-\frac{1}{l^2}\int_0^N \left(\frac{dr}{dv}\right)^2 dv\right] d[r(v)], \tag{2}$$

where C is the normalization constant and

$$A = \int_0^N \frac{1}{2}\left(x\frac{dy}{dv} - y\frac{dx}{dv}\right) dv \tag{3}$$

is the algebraic area which is enclosed by the polymer loop configuration $\boldsymbol{r}(v)$. The sign convention is that A is counted positive if the enclosed area is located to the left of the curve when traced by increasing number of monomers, v, otherwise it is negative [7].

When this polymeric system is under an external field with potential V, we can write the probability distribution in general form as

$$P(A, N) = C\int_{r_0}^{r_1} \delta\left[A - \int_0^N \frac{1}{2}\left(x\frac{dy}{dv} - y\frac{dx}{dv}\right) dv\right] \exp\left[-\frac{1}{l^2}\int_0^N \left(\frac{dr}{dv}\right)^2 dv - \int_0^N V dv\right] d[r(v)]. \tag{4}$$

The effect of the potential is added on the *'kinetic'* part of the path integral.

Rewriting the Dirac delta as Fourier integral, we have the probability density in equation (4) as

$$P(A, N) = \frac{C}{2\pi}\int_{-\infty}^{+\infty} \tilde{P}(g, N) \exp(igA)\, dg \tag{5}$$

where

$$\tilde{P}(g, N) = \int_{r_0}^{r_1} \exp\left(-\int_0^N \tilde{L} dv\right) d[r(v)] \tag{6}$$

and

$$\tilde{L} = \frac{1}{l^2}\left[\left(\frac{dx}{dv}\right)^2 + \left(\frac{dy}{dv}\right)^2\right] + \frac{1}{2}ig\left(x\frac{dy}{dv} - y\frac{dx}{dv}\right) + V. \tag{7}$$

We then evaluate the probability distribution, equation (5), using white noise analysis. In this method, the paths are parametrized in terms of Brownian motion. Essential steps of the evaluation are presented in the next section.

3. Probability distribution in the framework of White Noise Analysis

In writing the probability distribution in equation (5) to the language of white noise analysis, we first parametrize the paths as

$$x(L) = x_0 + lB_x(l) \text{ and} \tag{8}$$

$$y(L) = y_0 + lB_y(l). \tag{9}$$

where $B_x = \int_0^N \omega_x dv$ and $B_y = \int_0^N \omega_y dv$ are Brownian motions in the x and y coordinates respectively parametrized by $v$that runs from 0 to N, $\omega(v)$ is the white noise variable, l is the distance between monomers and L is the length of the polymer loop.

In white noise analysis, the integral over the paths $d[r(v)]$ becomes an integral over the Gaussian white noise $d\mu(\omega)$ with relation given by [5]

$$d[r(v)] \rightarrow N_{xy}\exp\left[\frac{1}{2}\int_0^N (\omega_x{}^2 + \omega_y{}^2)dv\right]d\mu(\omega_x)d\mu(\omega_y), \tag{10}$$

where $N_{xy}$is an appropriate normalization constant. We further note that in the process of path parametrization done in equations (8) and (9), only the initial point is fixed. Hence we use the Donsker delta functions

$$\delta(x(L) - x_1) = \delta(x_0 + lB_x(l) - x_1)\text{and} \tag{11}$$

$$\delta(y(L) - y) = \delta\big(y_0 + lB_y(l) - y\big) \tag{12}$$

to fix the endpoints. For a polymer loop, the final and initial points are the same simplifying equations (11) and (12) as$\delta(B_x(l))$ and $\delta(B_x(l))$ respectively. These Donsker delta functions are then written into their Fourier integral representations.

These are the initial preliminary steps in writing a Wiener integral representation of the polymer loop in the language of white noise analysis. The bulk of the evaluation of the probability distribution is done via the *T-transform*, the mathematical tool used to evaluate the integration over the Gaussian white noise measure $d\mu(\omega)$.

In the following subsections, we consider different systems in applying white noise analysis.

3.1. Polymer in a gel in the absence of an external field

When there is no external field, the distribution for the area enclosed by the polymer loop is given by

$$P_0(A, N) = \frac{C}{2\pi}\int_{-\infty}^{+\infty} \widetilde{P}_0(g, N)\exp(igA)\, dg \tag{13}$$

where

$$\widetilde{P}_0(g, N) = \int_{r_0}^{r_1} \exp\left(-\int_0^N \tilde{L}dv\right) d[r(v)] \tag{14}$$

and

$$\tilde{L} = \frac{1}{l^2}\left[\left(\frac{dx}{dv}\right)^2 + \left(\frac{dy}{dv}\right)^2\right] + \frac{1}{2}ig\left(x\frac{dy}{dv} - y\frac{dx}{dv}\right). \tag{15}$$

Writing this distribution in the language of white noise analysis yields

$$P(A, N) = \frac{C_1}{8\pi^3}\int_{-\infty,g}^{+\infty}\int_{-\infty,p_x,p_y}^{+\infty}\int_{r_0}^{r_1} \exp\left[-\frac{1}{2}\int_0^N (\omega_x{}^2 + \omega_y{}^2)dv - \frac{igl^2}{2}\int_0^N (B_x dB_y - B_y dB_x)dv\right]$$

$$\text{x}\exp\left\{\int_0^N \left[\frac{-igl}{2}(x_0\omega_y - y_0\omega_x) + \big(ip_x l\omega_x + ip_y l\omega_y\big)\right]dv\right\}\exp(igA)\, d\mu(\omega)dp_x dp_x dg, \tag{16}$$

where C_1 is the collective normalization constant, the variablesp_x and p_y came from the Fourier integral representations of the Donsker delta functions in equations (11) and (12), and$\int_0^N \frac{1}{2}(B_x dB_y - B_y dB_x)\, dv$ corresponds to the Levy's stochastic areawhich is equal to A[9].

Notice that the probability distribution in equation (16) is in two dimensional Brownian motion. This can be realized in the probability space of one-dimensional white noise as presented in [9]resulting to

$$P_0(A,N) = \frac{C_1}{8\pi^3} \int_{-\infty,g}^{+\infty} \int_{-\infty,p_x,p_y}^{+\infty} \int_{r_0}^{r_1} \exp\left[-\frac{1}{2} < \omega,\omega > -igl^2 < \omega, F_s\omega >\right]$$

$$\text{x} \exp\left[\left(i < \omega, \left(p_y l + \frac{1}{2} g l x_0\right) X_{[-N,0]} + \left(p_x l + \frac{1}{2} g l y_0\right) X_{[0,N]} \cdot >\right)\right] d\mu(\omega) dp_x dp_x \exp(igA) dg. \quad (17)$$

The term$< \omega, F_s\omega >$which can be expanded as$\int_{R^2} \omega(v) F_s(v,v')\omega(v)' dvdv'$ corresponds to the Levy's stochastic area [9].

We note that equation (17) can be written as

$$P(A,N) = \frac{C_1}{8\pi^3} \int_{-\infty,g}^{+\infty} \int_{-\infty,p_y}^{+\infty} \int_{-\infty,p_x}^{+\infty} T(\Phi) dp_x dp_x \exp(igA) dg\,, \quad (18)$$

where$T(\Phi)$is the T transform of the functional $\Phi(\omega)$:

$$T(\Phi) = \int \exp\left(i < \omega, \xi >\right) \Phi(\omega)\, d\mu(\omega) \quad (19)$$

with

$$\Phi(\omega) = \exp\left[-\frac{1}{2} < \omega,\omega > -igl^2 < \omega, F_s\omega >\right] \quad (20)$$

and

$$\xi = \left(p_y l + \frac{1}{2} g l x_0\right) X_{[-N,0]} + \left(p_x l + \frac{1}{2} g l y_0\right) X_{[0,N]}. \quad (21)$$

Calculations of the T-transform of Φ and evaluation of the integrals in equation (18) yields

$$P_0(A,N) = \left[2Nl^2 \cosh^2\left(\frac{2\pi A}{Nl^2}\right)\right]^{-1}. \quad (22)$$

This agrees with the result obtained by Wiegel and Khandekar [7] where they used straightforward integration over $d[r(v)]$.

3.2. Polymer in a gel subjected to a uniform electric field

In this system, we consider the uniform electric field to be along x and y axes. The contribution of the electric field on the probability distribution of the area enclosed by the polymer loop is added on the *'kinetic'* part making the expression for the probability distribution as

$$P_E(A,N) = C \int_{r_0}^{r_1} \delta\left[A - \int_0^N \frac{1}{2}\left(x\frac{dy}{dv} - y\frac{dx}{dv}\right) dv\right] \exp\left[-\frac{1}{l^2}\int_0^N \left(\frac{dr}{dv}\right)^2 dv + q\int_0^N \mathbf{E}\cdot \mathbf{r} dv\right] d[r(v)], \quad (21)$$

where $\mathbf{E} = E_1\hat{\imath} + E_2\hat{\jmath}$, $\mathbf{r} = x\hat{\imath} + y\hat{\jmath}$ and $q$is$\frac{q'}{k_B T}$withq' as the charge of each monomer and T as the absolute temperature.

Writing the Dirac delta function in its Fourier integral form, the probability distribution becomes

$$P_E(A,N) = \frac{C}{2\pi} \int_{-\infty}^{+\infty} \tilde{P}_E(g,N) \exp(igA)\, dg, \quad (22)$$

where

$$\tilde{P}_E(g,N) = \Gamma_E \int_{r_0}^{r_1} \exp\left\{-\int_0^N \frac{1}{l^2}\left[\left(\frac{dx'}{dv}\right)^2 + \left(\frac{dy'}{dv}\right)^2\right] dv + \int_0^N \frac{1}{2}\lambda\left(x'\frac{dy'}{dv} - y'\frac{dx'}{dv}\right) dv\right\} d[r(v)] \quad (23)$$

and

$$\Gamma_E = \exp\left[-\frac{2}{l^2}\alpha(x'_N - x'_0) - \frac{2}{l^2}\beta(y'_N - y'_0) - \frac{N}{l^2}(\alpha^2+\beta^2) - \frac{qN^2}{2}(E_1\alpha + E_2\beta)\right]. \quad (24)$$

Using the preliminary steps, presented in equations (8-12), we have the probability distribution in white noise analysis as

$$\boldsymbol{P_E(A,N)} = \frac{\boldsymbol{C_1}}{\boldsymbol{8\pi^3}} \int\limits_{-\infty,g}^{+\infty} \int\limits_{-\infty,p_x,p_y}^{+\infty} \int\limits_{r_0}^{r_1} \boldsymbol{\exp\left[ip'_y(y'_0 - y'_1) + ip'_x(x'_0 - x'_1) + igA\right]}$$

$$\text{x}\exp\left[-\frac{2}{l^2}\alpha(x'_N - x'_0) - \frac{2}{l^2}\beta(y'_N - y'_0) - \frac{N}{l^2}(\alpha^2 + \beta^2) - \frac{qN^2}{2}(E_1\alpha + E_2\beta)\right]$$

$$\text{x}\exp\left\{\int_0^N \left[-\frac{1}{2}\left(\omega_{x'}{}^2 + \omega_{y'}{}^2\right) + \frac{\lambda l}{2}\left(x'_0\omega_{y'} - y'_0\omega_{x'}\right) + ip_{x'}l\omega_{x'} + ip_{y'}l\omega_{y'}\right]dv\right\}$$

$$\textbf{x}\ \boldsymbol{\exp\left\{\int_0^N \left[\frac{\lambda l^2}{2}(B_{x'}dB_{y'} - B_{y'}dB_{x'})\right]dv\right\}d\mu(\omega_{x'})d\mu(\omega_{y'})dp_{x'}dp_{y'}dg}, \quad (25)$$

where C_1 is the collective normalization constant, the variables p_x and p_y came from the Fourier integral representations of the Donsker delta functions in equations (11) and (12), and $\int_0^N \frac{1}{2}(B_{x'}dB_{y'} - B_{y'}dB_{x'})\,dv$ corresponds to the Levy's stochastic area[9].

We again note that the probability distribution in equation (25) is in two dimensional Brownian motion. This can be realized in the probability space of one-dimensional white noise as inequation (17) resulting to

$$\boldsymbol{P_E(A,N)} = \frac{\boldsymbol{C_1}}{\boldsymbol{8\pi^3}} \int\limits_{-\infty,g}^{+\infty} \int\limits_{-\infty,p_x,p_y}^{+\infty} \int\limits_{r_0}^{r_1} \boldsymbol{\exp\left[-\frac{1}{2} < \omega,\omega > +\lambda l^2 < \omega, F_s\omega >\right]}$$

$$\text{x}\exp\left[-\frac{2}{l^2}\alpha(x'_N - x'_0) - \frac{2}{l^2}\beta(y'_N - y'_0) - \frac{N}{l^2}(\alpha^2 + \beta^2) - \frac{qN^2}{2}(\text{E}_1\alpha + \text{E}_2\beta)\right]$$

$$\text{x}\exp\left\{\left(i < \omega, \left(p_{y'}l + \frac{1}{2}i\lambda l x'_0\right)X_{[-N,0]} + \left(p_{x'}l + \frac{1}{2}i\lambda l y'_0\right)X_{[0,N]} >\right)\right\}$$

$$\textbf{x}\ \boldsymbol{\exp\{+ip'_x(x'_0 - x'_1) + ig\text{A}\}\,d\mu(\omega)dp_{x'}dp_{y'}dg}. \quad (26)$$

Furthermore, equation (26) can be rewritten, as in equation (17), as

$$\boldsymbol{P_E(A,N)} = \frac{\boldsymbol{C_1}}{\boldsymbol{8\pi^3}} \int_{-\infty,g}^{+\infty} \int_{-\infty,p_x,p_y}^{+\infty} \boldsymbol{T(\Phi)dp_{x'}dp_{y'}\exp(igA)dg} \quad (27)$$

with$T(\Phi)$ given in equation (19) but with functional

$$\boldsymbol{\Phi(\omega) = N_0\exp\left[-\frac{1}{2} < \omega,\omega > \ +\lambda l^2 < \omega, F_s\omega >\right]} \quad (28)$$

and

$$\boldsymbol{\xi = \left(p_{y'}l + \frac{1}{2}i\lambda l x'_0\right)X_{[-N,0]} + \left(p_{x'}l + \frac{1}{2}i\lambda l y'_0\right)X_{[0,N]}}. \quad (29)$$

Performing the *T-transform* of Φ and integrating over$dp_{x'}$ anddp_yin equation (27) yields

$$\boldsymbol{P(A,N)} = \frac{\boldsymbol{C_1\lambda}}{\boldsymbol{32\pi^4 sin\left(\frac{\lambda l^2 N}{4}\right)}} \int\limits_{-\infty,g}^{+\infty} \boldsymbol{\exp\left[-\frac{\lambda}{2}\alpha\left(x_0 y_1 - y_0 x_1\right) + -\frac{N}{l^2}(\alpha^2 + \beta^2) - \frac{qN^2}{2}(E_1\alpha + E_2\beta) + igA\right]}$$

$$\textbf{x}\,\boldsymbol{\exp}\left\{-\frac{\lambda}{4\,tan\left(\frac{\lambda l^2 N}{4}\right)}\left[(x'_0 - x'_1)^2 + (y'_0 - y'_1)^2\right] - \frac{2\alpha}{l^2}(x'_N - x'_0) - \frac{2\beta}{l^2}(y'_N - y'_0)\right\}\boldsymbol{dg}. \quad (30)$$

Rewriting this back to *x* and *y* variables using the relations [8]

$$x = x' + \alpha v \text{ and} \quad (31)$$

$$y = y' + \beta v \quad (32)$$

with$\alpha = \frac{qE_2}{\lambda}$ and $\beta = -\frac{qE_1}{\lambda}$results to

$$P(A,N) = \frac{C_1\lambda}{32\pi^4 sin\left(\frac{\lambda l^2 N}{4}\right)} \int_{-\infty,g}^{+\infty} \exp\left[\frac{qN}{2}\left(E_1x_0 + E_2y_0\right) + \frac{N}{l^2}(\alpha^2+\beta^2)\right]$$

$$\text{x} \exp\left\{-\frac{\lambda N^2}{4\,tan\left(\frac{\lambda l^2 N}{4}\right)}(\alpha^2+\beta^2) + iga\right\} dg, \quad (33)$$

where $\lambda = -ig$.

From equation (33), we can consider two limiting cases. For weak electric field, the probability density is given by the expression

$$P_{weakE}(A,N) \cong P_0(A,N)\left[1 + \frac{4\pi a E^2}{N^2 l^4} - 3NlP_0(A,N)\right]. \quad (34)$$

wherea $= \frac{\pi(Nl^2)^4 Q^2}{23040 k_B^2 T^2}$ and $P_0(A,N)$ represents the distribution for the area enclosed by the polymer loop in the absence of electric field given in equation (22).

For strong electric field, the probability density is

$$P_{strongE}(A,N) \cong \frac{24k_BT\left(\frac{5}{\pi}\right)^{\frac{1}{2}}}{(Nl^2)^{\frac{3}{2}}QE} \exp\left[-\frac{bA^2}{E^2}\right], \quad (35)$$

where $b = \frac{2880 k_B^2 T^2}{Q^2 N^2 l^4}$.

Equations (34) and (35)agree with the works of Khandekar and Wiegel in reference [8].

3.3. Polymer loop in a gel subjected to crossed electric-magnetic fields

In this system, the polymer loop which encloses a constant area is subjected to crossed electric-magnetic fields. For this case, the probability distribution can be written as

$$P_{EB}(A,N) = C\int_{r_0}^{r_1} \delta\left[A - \int_0^N \frac{1}{2}\left(x\frac{dy}{dv} - y\frac{dx}{dv}\right)dv\right] \exp\left[-\frac{1}{l^2}\int_0^N \left(\frac{dr}{dv}\right)^2 dv\right]$$

$$\text{x} \exp\left[+q\int_0^N (\mathbf{E}\cdot\mathbf{r} + \mathbf{B}\cdot\dot{\mathbf{r}})dv\right] d[r(v)], \quad (36)$$

where$q$is$\frac{q'}{k_BT}$withq' as the charge of each monomer, $\mathbf{E} = E_1\hat{\imath} + E_2\hat{\jmath}$ and $\mathbf{B} = \frac{q}{2}(H\,\text{x}\dot{\mathbf{r}})$ with H as the uniform magnetic field along the z axis.

Following the same procedure prescribed in the previous subsections, equation (36) can be written as

$$P_{EB}(A,N) = \frac{C}{2\pi}\int_{-\infty}^{+\infty} \tilde{P}_{EB}(g,N)\exp(igA)\,dg \quad (37)$$

where

$$\tilde{P}_{EB}(g,N) = \int_{r_0}^{r_1} \exp\left(-\int_0^N \tilde{L}dv\right) d[r(v)] \quad (38)$$

and

$$\tilde{L}_{EB} = \frac{1}{l^2}\left[\left(\frac{dx}{dv}\right)^2 + \left(\frac{dy}{dv}\right)^2\right] + \frac{1}{2}\lambda\left(x\frac{dy}{dv} - y\frac{dx}{dv}\right)$$

$$-q(E_1 x - E_2 y) + \frac{qH}{2}\lambda\left(x\frac{dy}{dv} - y\frac{dx}{dv}\right); \lambda = -ig. \tag{39}$$

This expression can be simplified by using the relations

$$x = x' + \alpha v \text{ and} \tag{40}$$

$$y = y' + \beta v, \tag{41}$$

where $\alpha = \frac{qE_2}{\lambda}$ and $\beta = -\frac{qE_1}{\lambda}$.

The probability distribution in terms of these new variables is then

$$P_{EB}(A,N) = \frac{C}{2\pi}\int_{-\infty}^{+\infty} \Gamma_{EB} \int_{r_0}^{r_1} \exp\left\{\int_0^N \left[-\frac{1}{l^2}\left(\frac{dx'}{dv}\right)^2 - \frac{1}{l^2}\left(\frac{dy'}{dv}\right)^2\right] dv\right\}$$

$$\exp\left\{\int_0^N \left[\frac{\lambda}{2}\left(x'\frac{dy'}{dv} - y'\frac{dx'}{dv}\right)\right] dv\right\} d[r(v)]dg, \tag{42}$$

where

$$\Gamma_{EB} = \exp\left[-\frac{2}{l^2}\alpha(x'_N - x'_0) - \frac{2}{l^2}\beta(y'_N - y'_0) + igA\right]$$

$$\text{x} \exp\left[-\frac{N}{l^2}(\alpha^2 + \beta^2) - \frac{qN^2}{2}(E_1\alpha + E_2\beta) - \frac{igN}{2}(\alpha y'_N - \beta x'_N) + \frac{qHN}{2}(\alpha y'_N - \beta x'_N)\right]. \tag{43}$$

Equation (42) can be expressed in terms of white noise variables, that is, following the procedure prescribed in the previous sections. So we have

$$P_{EB}(A,N) = \frac{C_1}{8\pi^3}\int_{-\infty,g}^{+\infty}\int_{-\infty,p_x p_y}^{+\infty}\int_{r_0}^{r_1} \Gamma_{EB} \exp\left[\int_0^N [(\omega_{x'}{}^2 + \omega_{y'}{}^2) + ip_{x'}l\omega_{x'} + ip_{y'}l\omega_{y'}]dv\right]$$

$$\text{x} \exp\left\{\int_0^N \left[\frac{(-ig+qH)l}{2}(x'_0\omega_{y'} - y'_0\omega_{x'}) + ip_{x'}l\omega_{x'} + ip_{y'}l\omega_{y'}\right] dv\right\} dp_{x'}dp_{y'}$$

$$\text{x} \exp\left\{(-ig+qH)l^2 \int_0^N \left[\frac{1}{2}(B_{x'}dB_{y'} - B_{y'}dB_{x'})\right] dv\right\} d\mu(\omega_{x'})d\mu(\omega_{y'})dg, \tag{44}$$

where C_1 is the collective normalization constant.

In the probability space of one-dimensional white noise as presented in reference [9], the probability distribution may be written as

$$P_E(A,N) = \frac{C_1}{8\pi^3}\int_{-\infty,g}^{+\infty}\int_{-\infty,p_x p_y}^{+\infty}\int_{r_0}^{r_1} \Gamma_{EB} \exp\left[-\frac{1}{2} < \omega,\omega > +(\lambda + qH)l^2 < \omega, F_s\omega >\right]$$

$$\text{x}\, exp\left\{i\langle\omega, \left(p_{y'}l + \frac{1}{2}i\lambda l x'_0\right)X_{[-N,0]} + \left(p_{x'}l + \frac{1}{2}i\lambda l y'_0\right)X_{[0,N]}\rangle\right\} d\mu(\omega)dp_{x'}dp_{y'}dg. \tag{45}$$

Integration over $d\mu(\omega)$ can then be evaluated using the *T-transform* of Φ and we can write equation (45) as

$$P_E(A,N) = \frac{C_1}{8\pi^3}\int_{-\infty,g}^{+\infty}\int_{-\infty,p_x p_y}^{+\infty} T(\Phi)dp_{x'}dp_{y'}\exp(igA)dg\,, \tag{46}$$

where $T(\Phi)$ is given by equation (19). For this case, the white noise functional Φ has the form

$$\Phi(\omega) = N_0 \exp\left[-\frac{1}{2} < \omega, \omega > +(\lambda + qH)l^2 < \omega, F_s\omega >\right] \tag{47}$$

with

$$\xi = \left(p_{y'}l + \frac{1}{2}i\lambda l x_0'\right) X_{[-N,0]} + \left(p_{x'}l + \frac{1}{2}i\lambda l y_0'\right) X_{[0,N]}. \tag{48}$$

Calculations of the *T-transform* of Φresults to

$$T(\Phi) = \left[\cos\frac{(\lambda + qH)l^2 N}{4}\right]^{-1} \exp\left\{\frac{-1}{(\lambda + qH)l^2} \tan\left(\frac{(\lambda + qH)l^2 N}{4}\right)\left[p_{y'}^2 l^2 - i(\lambda + qH)l^2 {x_0'}^2 p_{y'}\right]\right\}$$

$$\text{x}\exp\left\{\frac{-1}{(\lambda+qH)l^2} \tan\left(\frac{(\lambda+qH)l^2 N}{4}\right)\left[p_{x'}^2 l^2 - i(\lambda + qH)l^2 {y_0'}^2 p_{x'}\right] - \frac{(\lambda+qH)^2 l^2({x_0'}^2 + {y_0'}^2)}{4}\right\}. \tag{49}$$

The integrations over $dp_{x'}$ and $dp_{y'}$in equation (46) are done using one-dimensional Gaussian integral. Rewriting the probability distribution, equation (52), back to x and y variables and identifying the normalization constant [8] results to

$$P_{EB}(A, N) = K(E, H) \int_{-\infty,g}^{+\infty} \frac{u}{\sinh u} \exp\left[\frac{4iuA}{Nl^2} - \frac{N^3 l^2 q^2 E^2}{72} u^2\right] du, \tag{50}$$

where $u = Nl^2(g + qHi)$ and

$$K(E, H) = \frac{8\sin\left(\frac{Nl^2 qH}{4}\right)}{N^2 l^4 qH\pi} \exp\left[\frac{N^3 l^2 q^2 E^2}{48} + qHA + \frac{E^2 N}{l^2}\left(-\frac{1}{H^2} + \frac{Nl^2 q}{4H\tan\left(\frac{Nl^2 qH}{4}\right)}\right)\right]. \tag{51}$$

Similar in section 3.2, the structure of equation (51) is complicated and so limiting cases are considered [8]. We further note that the integrand in equation (51) only affects electric field which means limiting cases are for electric field only.

In the presence of constant magnetic field and weak electric field, the probability density is given by the expression

$$P_{E_w B}(A, N) \cong K(E, H)\{p_0(\xi) - \beta[-\pi^2 p_0(\xi)] + 3p_0^2(\xi)\}, \tag{52}$$

where$\beta = \frac{Nl^2 Q^2 E^2}{720 k_B^2 T^2}$, $\xi = \frac{4A}{Nl^2}$ and $p_0(\xi) = \frac{\pi^2}{2\cosh\left(\frac{\pi\xi}{2}\right)}$. And the probability distribution in the presenceof magnetic field and strong electric field is given by

$$P_{E_s B}(A, N) \cong K(E, H)\left(\frac{k_B T}{lQE}\sqrt{\frac{720\pi}{N}}\right) \exp\left[-\frac{4(720)k_B^2 T^2}{N^3 l^6 Q^2}\left(\frac{A^2}{E^2}\right)\right]. \tag{53}$$

For both limiting cases, results agree with reference [8] when magnetic field is switched off.

4. Conclusion

In this paper, we have obtained the expression for the probability distribution of the area enclosed by a polymer loop under different external fields: constant electric field and crossed electric-magnetic fields. The method used in the evaluation of the probability distribution is white noise analysis where the polymer loop is represented by various paths of a Brownian motion.

5. References

[1] Breretron M G and Butler *J. Phys.* A **20** 3955
[2] Cunha M, Drumond C, Leukert P, Silva J L and Westerlamp W 1995*Arin. Phys.***4** 53–67
[3] Gorthaus M and Streit L 1997 *J. Phys. Math***32**, 3278
[4] Bernido C C and Carpio-Bernido M V 2002 *Proceedings of the 3rdJagna International*

Workshop **18**

[5] Bernido C C and Carpio-Bernido M V 2003 *J. Phys.* A:*Math. Gen.* **36** 1–11

[6] Bernido C C and Carpio-Bernido M V 2005 *Phys. Lett.* A **339** 232

[7] Khandekar DC and Wiegel FW 1988 *Stat. Phys.* **53** 1073

[8] Wiegel F W and Khandekar DC 1992 *Int. J. Mod. Phys.* B **6** 3147

[9] de Falco D and Khandekar D C 1988 *Stoch. Proc. Appl.* **29** 257–66

IOP Conf. Series: Materials Science and Engineering **79** (2015) 012013 doi:10.1088/1757-899X/79/1/012013

Dynamics of passive tracers in a bath of self-propelling granular particles

E Hamoy[1] and M N Confesor[2,3,4]

[1] Department of Physics, Mindanao University of Science and Technology, CM Recto Avenue, Lapasan, Cagayan de Oro City 9000, Philippines

[2] Department of Physics, Mindanao State University – Iligan Institute of Technology, Andres Bonifacio Avenue, Tibanga, Iligan City 9200, Philippines

[3] Complex Systems Initiative, Mindanao State University – Iligan Institute of Technology, Andres Bonifacio Avenue, Tibanga, Iligan City 9200, Philippines

E-mail: marknolan2006@gmail.com

Abstract. We report on our experimental investigation of the dynamics of a passive tracer in a bath of active self-propelling granular particles. We found a caging like dynamics of the passive tracer such that for low active particle concentrations the passive tracer exhibits longer periods of inactivity. For increasing active particle concentration the occurrence of short period inactivity increases.

1. Introduction

Active matter consists of systems wherein constituent particles undergo self-propulsion due to energy supplied to each of the particle. Due to the constant supply of an energy source and that this energy is also dissipated, active matter is a good example of a system far from equilibrium. For instance it was found out that the mobility of bacteria is affected by the amount of food present in the surrounding medium [1]. Self-propelling colloidal particles also constitutes active matter [2]. Pattern formation is an interesting consequence of a bath containing active particles. Biological examples range from flocking type motion of birds and fishes to spiral wave dynamics of the motion of a microorganism, Dictyostelium discoideum [3-4]. Although a general framework wherein one can understand different active matter systems is still lacking, there has been consistent effort to find a general theory of active matter. Recently physicists have found interest on active matter system due to the interesting prospect of applying results in statistical mechanics to some properties of active matter [5-6].

A particular problem of interest is how to characterize active bath. Two approaches have been gaining steam on this. First is to do single particle tracking of one of the active particles. For the trajectories one can obtain correlations that can be used to describe the system [7]. The second approach is putting a passive tracer in the active medium. The mean square displacement of the passive tracer is then used to describe the dynamics of the active medium [8].

A granular bath containing self-propelling granular particles constitutes an active bath [7]. Already in our previous experiment we found that the active granular bath can be used to extract energy to perform work [9]. In this paper we are interested on a general way to characterize active granular bath. Our preferred approach is to use a passive tracer. There are many types of active

[4] To whom any correspondence should be addressed.

granular particles and characterization of these active baths by the same passive tracers could reveal a general picture of the dynamics of active matter.

2. Experiment methodology

The experimental set-up consists of a bath of active particles on an acrylic platform driven out of equilibrium by a shaker (see figure 1). The active particle has a head and a tail part. The head of the active particle is a rectangular prism shaped acrylic (6.7 x 6.7 x 10mm) and this is coupled to 7 loosely connected metal beads (5mm dia.) forming the tail. We varied the number of active particles used in this experiment from 3, 10, 20, 40, 50, and 68 particles. This corresponds to area fractions Φ= 0.0063, 0.0211, 0.0421, 0.0842, 0.105 and 0.1433 where the area fraction is defined as the ratio of the total body area of the active particles to the area accessible to the active particles A= 212.42 cm^2. To characterize the random motion of the bath we use a Styrofoam disk (7mm thickness and 26mm dia.) with a reference dot (5mm dia.) for tracking.

a

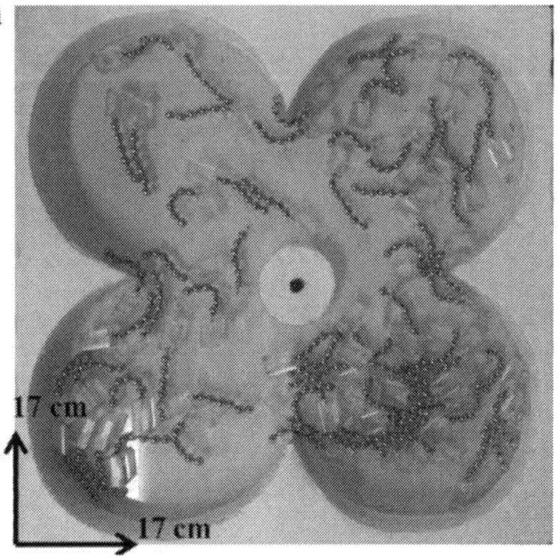

b

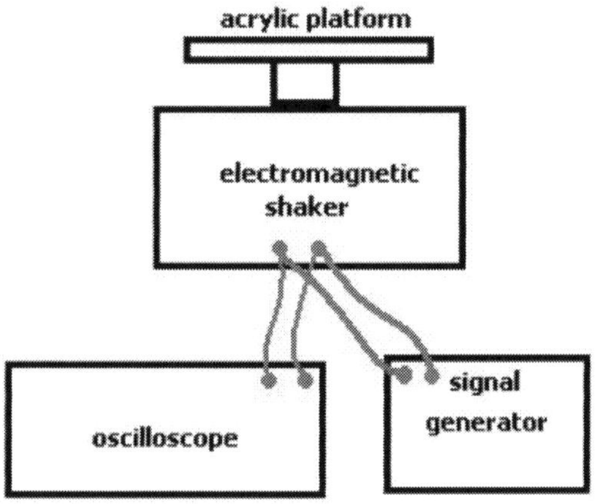

Figure 1.(a) Active particles and passive particle in the vibrated acrylic substrate. (b) Schematic diagram of experimental set-up.

Active particles require a constant supply of energy to move around the acrylic platform. This is achieved by vibrating the shaker with a constant frequency set at 28 Hz. At this frequency the active particles move head first then turn randomly which allows it to explore the bounds of the platform. The amplitude of oscillation is kept constant throughout the whole experiment.

The boundary of the acrylic platform (170 x 170 x 5mm) used in this experiment has curved shapes to prevent jamming of active particles that would happen on the corners of a square platform. The curved shapes ensure the smooth movement of the active particles throughout the acrylic platform.

A camera placed on top of the vibrating acrylic platform monitors the each experimental run at 30 frames per second. The video recording of each experiment is deconstructed into a series of images using Virtualdub and then the images are processed by our own particle tracking software written in IDL. Our software detects the intensity difference in the grey scaled images; using this information it can calculate for the position of the reference dot on a passive tracer via centroid algorithm. The result from this procedure gives information of the passive particle's position in pixels. Using a scaling factor we can convert the number of pixels into units of distance. We obtain tracking resolution better than 0.1 mm. We gathered 2000 images per experimental condition.

3. Results and discussion

When we turn ON the shaker we observe the active particles exploring the bounds of the acrylic platform sometimes colliding with each other, the walls of the boundary or colliding with the

Styrofoam disk that we are using as a passive tracer for the experiment. When the active particle(s) collide with the tracer, momentum is imparted onto the tracer causing it to move. The tracer's displacement varies depending on the net impulse delivered to it.

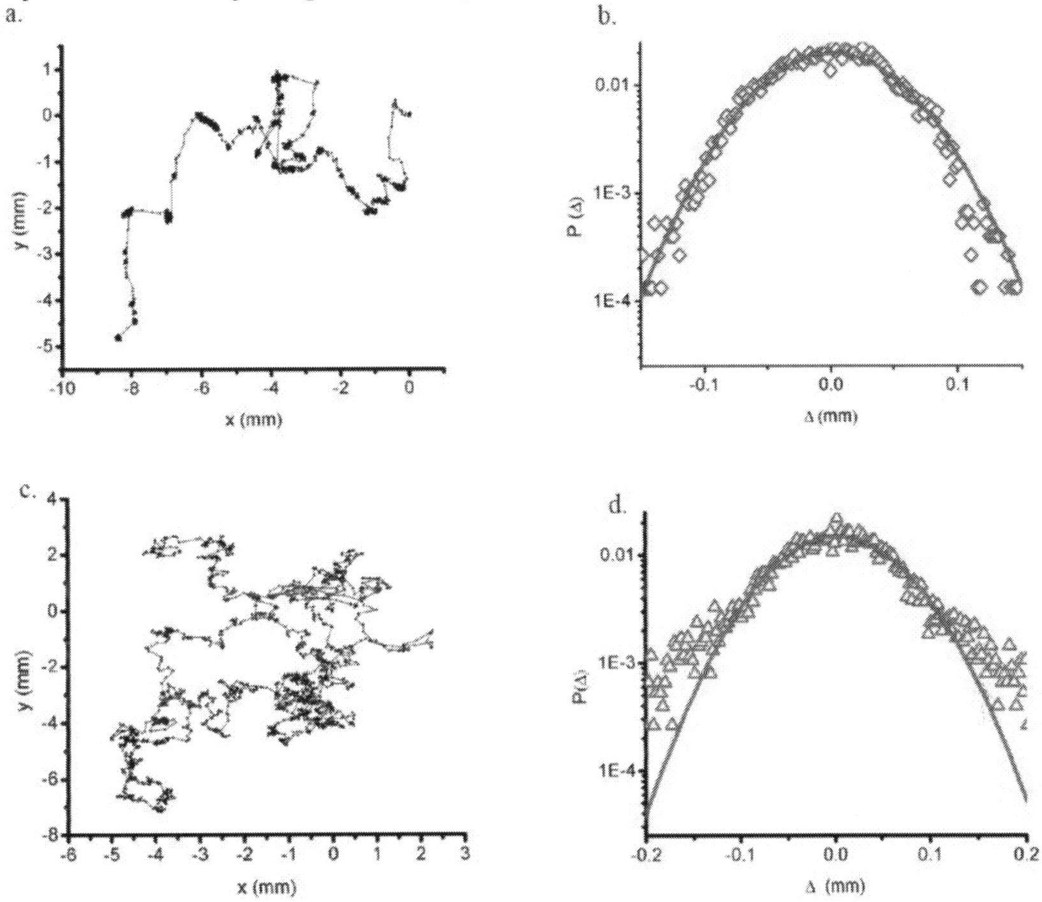

Figure 2.(a) Sample trajectory of the passive particle for Φ= 0.0211 b) Probability distribution of the displacements of the passive tracer corresponding to that in (a). c) trajectory of tracer Φ= 0.105 d) Probability distribution of the displacements of the passive tracer corresponding to that in (c). The non-Gaussian tails are evident. In the probability distribution solid curves in (b) and (c) corresponds to Gaussian fitting.

A sample trajectory of our passive tracer is shown in figure 2 a, c. We calculated the incremental change in displacements and plotted the probability distribution of the displacement in Figure 2b for the case when Φ = 0.0211. The displacements were computed using the formula $\Delta i = [(x_{i+1}-x_i)^2-(y_{i+1}-y_i)^2]^{1/2}$ where (x_i, y_i) is the particle position computed at a frame i. We found that the displacements fitted a normal curve with a mean around zero. However we found deviation to the Gaussian distribution for higher concentration of active particles where we were able to see non-Gaussian tails, figure 2c, such as the case when Φ = 0.105. Currently the precision of our data is not sufficient to see a clear trend on this property of the probability distribution. We note that the presence of Gaussian and non-Gaussian tail in a probability distribution is a signature of a system far from equilibrium.

3.1. Dwell time statistics during caging

From the trajectory we found that the passive tracers undergo a caging-like dynamics, i.e. they are immobile at some time until hit by an active particle whence they move. In the following section we examine the cage like dynamics manifested by the passive tracer at different active particle concentration. Figure 3a is the trajectory of the passive tracer for Φ= 0.0063. We can see there are

IOP Conf. Series: Materials Science and Engineering **79** (2015) 012013 doi:10.1088/1757-899X/79/1/012013

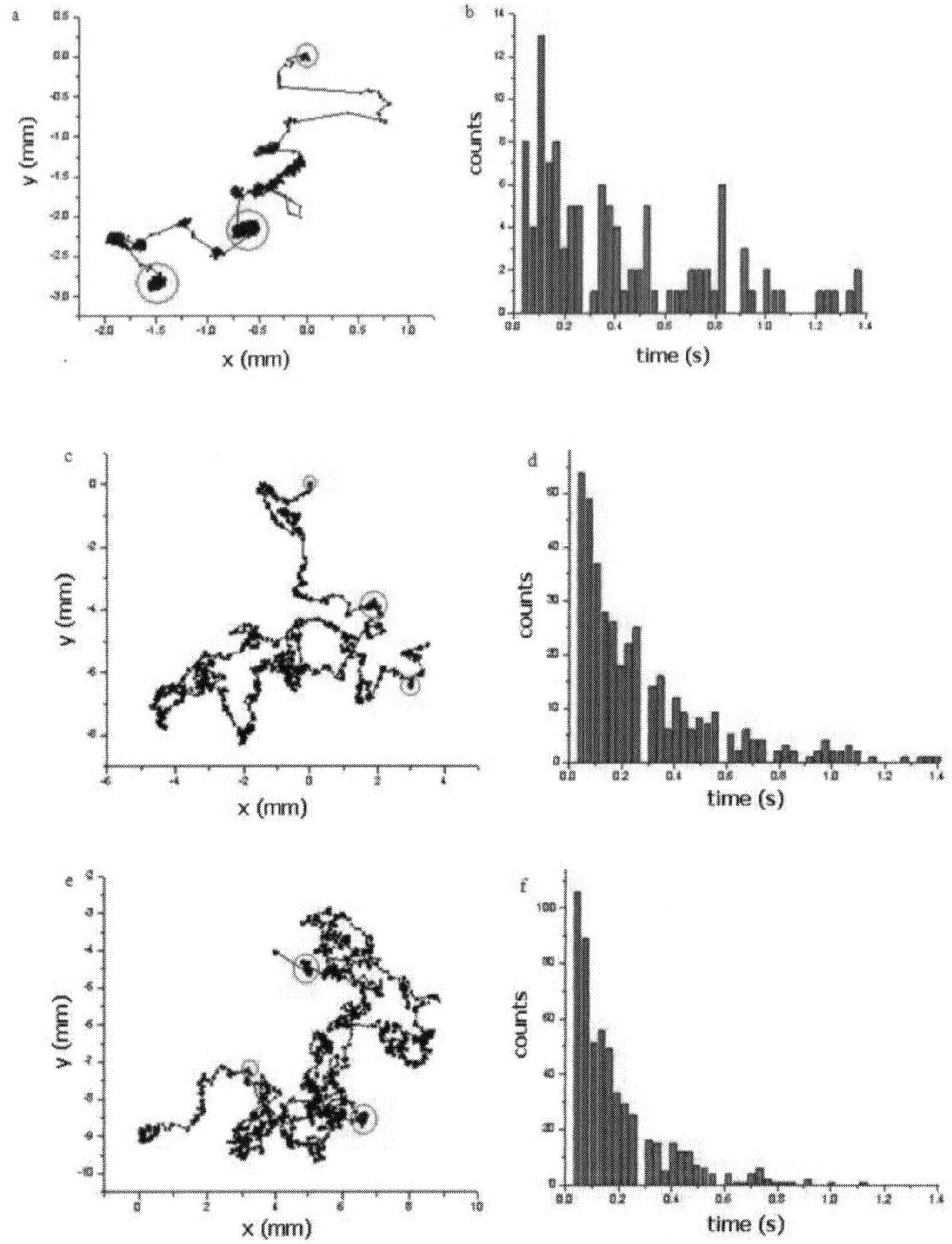

Figure 3. Trajectories (left panel) and dwell time statistics (right panel) of the passive tracer at different number of active particles. (a) Φ= 0.0063, (c) Φ= 0.0842, and (3) Φ=0.1433.

periods of movement (run) and stagnation (caging). The curve describing the path of the passive tracer becomes very dense if little or no motion occurs. Examples of the inactivity are highlighted in the passive tracer trajectories in Figure 3a. To characterize this type of dynamics we measure the dwell time, i.e. the total time the particle is inside the cage. We set a threshold of displacement of < 0.1 mm between succeeding particle positions to determine if the particle is in a cage. The dwell time is then the total time spent by the tracer when its displacement is below the threshold. Figure 3b is the histogram for the distribution of time intervals of the inactivity of the tracer. We see that there are instances of relatively long inactivity for the tracer in a bath of 3 active particles.

The long intervals of inactivity are attributed to the low probability of impact between the passive tracer and the active particles.

For larger concentrations of active particles such as in figure 3c and figure 3e the periods of inactivity are smaller as evidenced by the more convoluted trajectory of the tracer. This is supported by the histograms of the dwell times in figure 3d and figure 3f. Both histograms in figure 3d and figure 3f show a decrease in large time scale inactivity owing to the increase in probability of collisions. Another noticeable attribute of the histograms is the increase in short time scale inactivity as the number of active particles increase. To highlight this observation we plot the mean dwell time as a function of the number of particles in figure 4. We found that the mean time exponentially decays with Φ.

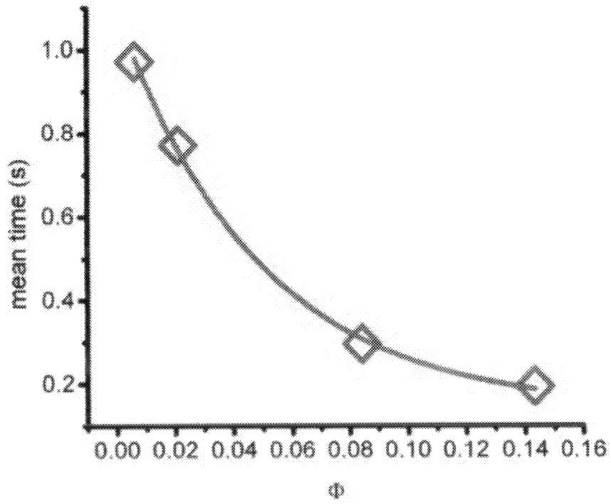

Figure 4. Exponential decay fit of mean inactivity time of the passive tracer for Φ. The solid line is an exponential fit $y = A_1*exp(-x/t_1)$with A = 1.046 and t_1= 34.42

4. Conclusion

We found that the length of inactivity of the passive tracer decreases with increasing Φ which implies higher probability of collisions of the tracer with the self-propelling particles at higher concentration of the latter. Moreover we found that for higher values of the area fraction the probability distribution of Δ yields a Gaussian part with a non-Gaussian tail. Non-Gaussian tails are not seen in the probability distribution of the displacement for $\Phi < 0.02$. This work can be extended to other types of active granular particles. If the dynamics is consistent with this work we might be able to find a general picture of active granular particles.

5. References

[1] Bray D 2002 *PNAS***99** 7–9
[2] Howse J *et al.* 2007*Phys. Rev. Lett.***99** 048102
[3] Kessler D and Levine H 1993*Phys. Rev.* E**48** 4801–4
[4] Reynolds C 1987 *ACM SIGGRAPH Computer Graphics***21** 25–34
[5] Crizol A, Stanley H E and Vicsek T 1997*J. Phys.* A: *Math. Gen.***30** 1375–85
[6] Toner J, Tu Y and Ramaswamy S 2005 *Annals of Physics***318** 170–244
[7] Kudrolli A 2010 *Phys. Rev. Lett.***104** 088001
[8] Angelani *et al.* 2011*Phys. Rev. Lett.***107** 138302
[9] Hamoy E and Confesor M N 2014 *in Preparation*

Acknowledgement

E.N.H. would like to acknowledge the financial support from DOST-SEI ASTHRDP. MNPC would like to thank Prof CK Chan and Prof Pik-Yin Lai for discussions on Non-Equilibrium systems such as the present study. The authors would like to thank the OVCAA of MSU-IIT through Dr. Edgar Ignacio for the equipment support grant.

Terahertz transitions in quasi-metallic carbon nanotubes

R R Hartmann[1,3] and M E Portnoi[2]

[1] Physics Department, De La Salle University, 2401 Taft Avenue, Manila, Philippines

[2] School of Physics, University of Exeter, Stocker Road, Exeter, EX4 4QL, UK

E-mail: Richard.Hartmann@dlsu.edu.ph, M.E.Portnoi@exeter.ac.uk

Abstract. We study interband dipole transitions across curvature-induced narrow gaps in quasi-metallic single-walled carbon nanotubes. The curvature effects not only open a gap in the nanotube energy spectrum but also allow optical transitions, which happen to be in the highly-desired terahertz frequency range. Applying a magnetic field along the nanotube axis allows one to tune the frequency peaks in the spectral density of absorption.

1. Introduction

Despite wide-range continuous efforts lasting for several decades [1–4], creating reliable, portable, tuneable sources and sensitive detectors of terahertz (THz) radiation remains one of the most formidable tasks of modern device physics. One of the recent trends in THz technology [2] is to use carbon nanotubes [5–6] (CNTs) as building blocks of high-frequency devices. There is a growing number of proposals using CNTs for THz applications including several schemes [7–13] put forward by the authors of the present work.

In what follows, it shall be shown that the same intrinsic curvature of quasi-metallic CNTs which opens the gap in the nanotube energy spectrum also allows strong optical transitions in the THz range. We show that the frequency peaks in the spectral density of absorption can be tuned by the application of a magnetic field directed along the nanotube axis. To illustrate that intrinsic curvature allows strong THz transitions, we shall study a subclass of these tubes, known as zigzag quasi-metallic CNTs (see Refs. [5–6] for classification of CNTs). However, the approach used herein can be generalized to all quasi-metallic tubes, all of which are expected to be optically active, since it is the curvature-induced gap, which gives rise to the strong THz transitions.

2. The band structure of quasi-metallic single walled nanotubes with curvature

Rolling a graphene sheet to form a CNT decreases the bond length to the nearest neighbors and rotates the $2p_z$ orbitals, this results in the re-hybridization of the π and σ orbitals. All of the aforementioned effects result in the modification of the hopping parameters of the tight-binding Hamiltonian. In what follows we shall consider the dominant effect, that of bond length contraction. A full calculation taking into account re-hybridization shall be a subject of future study. Let us begin with the electron energy spectrum of graphene:

$$\xi(k) = s\left|\sum_{i=1}^{3} t_i \exp(i\mathbf{k}\cdot\mathbf{u}_i)\right|, \qquad (1)$$

[3] To whom any correspondence should be addressed.

where $s = +1$ for the conduction band and $s = -1$for the valence band, and $\mathbf{u}_i$are the nearest neighbors vectors, which for a graphene sheet are

$$\mathbf{u}_1 = \left(\frac{a}{\sqrt{3}}, 0\right), \ \mathbf{u}_2 = \left(-\frac{a}{2\sqrt{3}}, -\frac{a}{2}\right), \ \mathbf{u}_3 = \left(-\frac{a}{2\sqrt{3}}, \frac{a}{2}\right). \quad (2)$$

Here $a = \sqrt{3}a_{CC}$, where $a_{CC}$is the nearest neighbor distance between two carbon atoms which is given as 1.42 Å. $t_i$are the tight-binding matrix elements associated with the i^{th}nearest neighbor, which for graphene are equivalent i.e., $t_i = t \approx -3$ eV. For convenience, we shall construct our zigzag CNTs such that motion is quantized alongk_y, i.e. consider $(n, -n)$ CNTs, where n must be a multiple of three. The electron energy spectrum for a quasi-metallic zigzag CNT is obtained from Eq. (1) by applying the periodic boundary condition: $k_y = 2\pi l/|\mathbf{C_h}|$, where $|\mathbf{C_h}|$is the magnitude of the chiral vector and for a quasi-metallic zigzag CNT, in the absence of curvature is given by the expression $|\mathbf{C_h}| = an$. l is an integer, which for the lowest (highest) conduction (valence) sub band is given by $l = \pm\frac{2n}{3}$. Hence, the electron energy spectrum for the lowest conduction sub band and the highest valence sub band of a quasi-metallic zigzag CNT is given by:

$$\xi(k_x) = 2s\left|t \sin\left(\frac{\sqrt{3}}{4}k_x a\right)\right|. \quad (3)$$

In the THz regime, Eq. (3) becomes $\xi(k_x) = s\hbar v_F |k_x|$, where $v_F = \sqrt{3}|t|/(2\hbar)$ is the Fermi velocity of graphene. Let us now consider the role of curvature. In rolling a graphene sheet

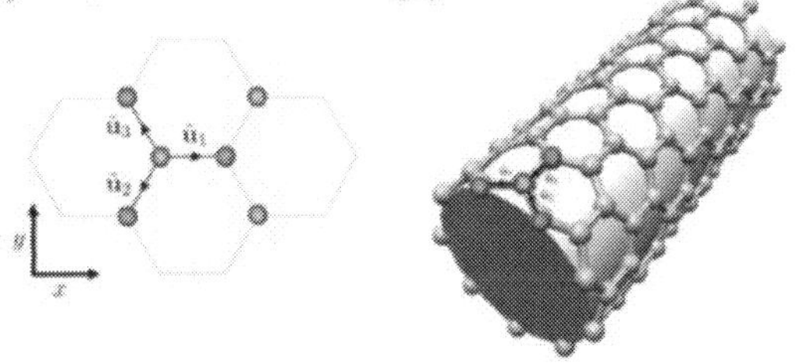

Figure 1. (a) The quasi-metallic zigzag CNT as an unrolled graphene sheet, represented by the set of effective nearest neighbor vectors$\tilde{\mathbf{u}}_i, i = 1,2,3$. (b) The quasi-metallic zigzag CNT.

to form a CNT one decreases the length of the nearest neighbor vectors. This is because the new distance is given by the chord between the two sites. One can see from Fig. 1 that upon rolling a quasi-metallic zigzag CNT, $|\mathbf{u_1}|$remains unchanged whereas the magnitude of $\mathbf{u}_2$and $\mathbf{u}_3$are reduced in comparison to that of a planar graphene sheet and their new lengths aregiven by$\sqrt{a_y^2 + \left(\frac{a}{2\sqrt{3}}\right)^2}$, where $a_y = 2R\left|\sin\left(\frac{a}{4R}\right)\right|$and R is the radius of the CNT given by$R = an/2\pi$. To understand the effects of including curvature, one can imagine the CNT as an unrolled graphene sheet, which is described by the same coordinate system as the non-curved case, but differs by a modified set of nearest neighbor vectors, $\tilde{\mathbf{u}}_i$, defined as:

$$\tilde{\mathbf{u}}_1 = \left(\frac{a}{\sqrt{3}}, 0\right), \ \tilde{\mathbf{u}}_2 = \left(-\frac{a}{2\sqrt{3}}, -a_y\right), \ \tilde{\mathbf{u}}_3 = \left(-\frac{a}{2\sqrt{3}}, a_y\right). \quad (4)$$

Therefore, the size of the chiral vector is reduced to $|\mathbf{C_h}| = 2a_y n$hence the quantization condition becomes: $k_y = \pi l/(a_y n)$.The band gap of quasi-metallic CNTs can be controlled by an applied magnetic field[14]. In the nearest-neighbor tight binding approximation the influence of a magnetic field is accounted for by adding the number $f = \Phi/\Phi_0 = eBna_y\sqrt{R^2 - a_y^2}/h$(here Φ is the magnetic flux through the polygons cross section rather than the original circular cross-section and $\Phi_0 = h/e$is the magnetic flux quantum) to the angular momentum quantum number l [6].

The effect of curvature is to break the symmetry between the nearest neighbor vectors, therefore breaking the former equivalency of t_1, $t_2$and t_3. By symmetry, the hopping parameter $t_1$does not change i.e. $t_1 = t$. The probability of hopping between sites is inversely proportional to the distance squared between hoping sites i.e. $t_i \propto \frac{1}{|\mathbf{u}_i|^2}$[15]. Hence, $t_2$and $t_3$is modified inthe same way. The modified matrix elements of hopping, $\tilde{t}_i$, are related to the original elements,t_i, by the simple relation:

$$\frac{\tilde{t}_i}{t_i} = \left(\frac{|\mathbf{u}_i|}{|\tilde{\mathbf{u}}_i|}\right)^2. \quad (5)$$

In the presence of curvature and applied field the modified electron energy spectrum for a quasi-metallic zigzag CNT is given by:

$$\xi = s|t|\sqrt{\left(\frac{\xi_g}{2t}\right)^2 + 4\lambda \sin^2\left(\frac{\sqrt{3}}{4}k_x a\right)}, \quad (6)$$

where $\lambda = -2\frac{\tilde{t}_i}{t_i}\cos\left(\frac{\pi}{n}f \pm \frac{2\pi}{3}\right)$and $\xi_g = 2|(1-\lambda)t|$. With the inclusion of curvature the spectrum is no longer linear near the crossing point (see Fig. 2) and a band gap of $\xi_g$hasappeared, whose size can be tuned by the strength of the applied field. For large radius tubesthe band gap is given by $\xi_g \approx \hbar v_F a_{CC}\left|\frac{1}{16R^2} \pm \frac{4}{\sqrt{3}a_{CC}^2}\sin\left(\frac{\pi}{n}f\right)\right|$. It should be noted that assuminga different power dependence of the transfer integral on the bond length results in the same dependence of the gap size with radius [16,17] however the magnitude of the gap varies by a geometric factor. In the absence of an applied field the energy spectrum is degenerate in l, this degeneracy is broken for any size magnetic field hence an applied magnetic field results in two separate bandgaps. For a (24,0) CNT in zero field this gap corresponds to $\approx$ 1.6 THz and in the presence of a 5 T field the two gaps are $\approx$ 2.6 THz and $\approx$ 0.5 THz which correspond to $l = \frac{2n}{3}$and $-\frac{2n}{3}$respectively.

3. Optical selection rules

In the dipole approximation, the spectral density, I_ν, is given by [10]

$$I_\nu = \frac{8\pi e^2 \nu}{3c^3}\sum_{i,f} f_e(k_i) f_h(k_f)\left|\mathbf{e}\cdot\langle\psi_f|\hat{\mathbf{v}}|\psi_i\rangle\right|^2 \delta(\xi_f - \xi_i - h\nu), \quad (7)$$

where $\psi_i$and $\psi_f$are the eigenfunctions of the electrons in the initial and final states, $\xi_i$and $\xi_f$are their associated energies, and $k_i$and $k_f$are their associated wave vectors, $\hat{\mathbf{v}}$is the velocityoperator, $\mathbf{e}$is the polarization of the excitation which we take to be propagating along the CNT axis and $\nu$is the frequency of the excitation. In the frame of the tight binding model, in thepresence of curvature and an applied field, the matrix element, $\mathbf{e}\cdot\langle\psi_f|\hat{\mathbf{v}}|\psi_i\rangle$can be written as

$$i\frac{a_{CC}}{8}\omega_{if} + i\frac{2v_F^2}{3a_{CC}\omega_{if}}\left[1 - \left(\frac{\tilde{t}_2}{t}\right)^2\right] + i\frac{8v_F^2}{3a_{CC}\omega_{if}}\left[\left(\frac{\tilde{t}_2}{t}\right)^2 \sin\left(\frac{\pi f}{n} \mp \frac{2\pi}{3}\right)\sin\left(\frac{\pi f}{n}\right)\right], \quad (8)$$

where $\hbar\omega_{if} = \xi_f - \xi_i$. Curvature not only opens the gap in the quasi-metallic zigzag CNT spectrum, but also allows dipole optical transitions at $kx = 0$ between the highest valence sub band and the lowest conduction sub band. In this regime the result should hold true across the entire class of quasi-metallic CNTs, differing only by geometrical factors. In Fig. 3 we show how the matrix element of the dipole optical transitions polarized along the CNT axis are modified in the presence of a magnetic field.

The spectral density of a zigzag quasi-metallic CNT taking into account curvature effects and applied field is obtained by substituting Eq. (8) into Eq. (7) then performing the necessary summation. In the THz regime one obtains the expression:

$$I_\nu = L f_e(k_i) f_h(k_f)\frac{\pi^3 e^2 a_{CC}^2}{3c^3\hbar^4 v_F}\frac{\left[12t^2(1-\lambda^2)+h^2\nu^2\right]^2}{\sqrt{\lambda\left(h^2\nu^2-\xi_g^2\right)}}. \quad (9)$$

In the absence of the field, the electronic (hole) energy spectrum near the bottom (top) is no longer linear due to curvature effects and the van-Hove singularity in the reduced density of states leads to a very sharp absorption maximum near the band edge and correspondingly to a very high sensitivity of the photocurrent to photon frequency, see Fig. 4. In the presence of an applied field, there are two absorption maxima corresponding to $l = \frac{2n}{3}$ and $-\frac{2n}{3}$ (see Fig. 5).

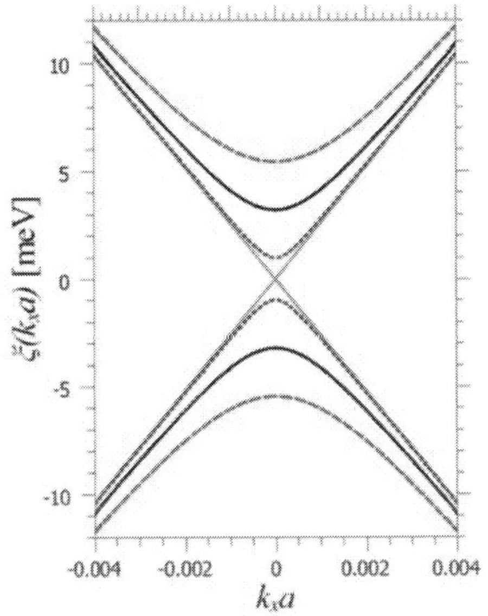

Figure 2. Detailed view of the gap for a (24, 0) CNT with (red and blue lines, corresponding to $l = \frac{2n}{3}$ and $-\frac{2n}{3}$ respectively, for the case of $B = 5$ T) and without (black line) an external magnetic field along the CNT axis. The grey line corresponds to zero field and no curvature.

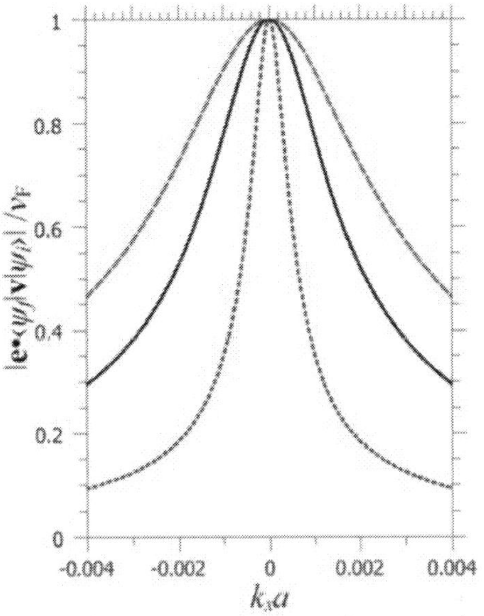

Figure 3. Dependence of the dipole matrix element for the transition between the top valence and lowest conduction sub bands on the 1D wave vector k_x, for $B = 0$ T (black line) and $B = 5$ T (red and blue lines, corresponding to $l = \frac{2n}{3}$ and $-\frac{2n}{3}$ respectively) applied along the CNT axis.

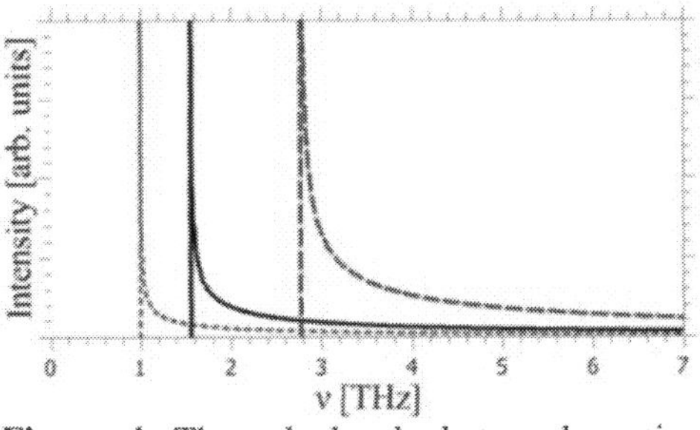

Figure 4. The calculated photon absorption spectra for a quasi-metallic zigzag CNT defined
by: (30, 0) (green short-dashed line), (24,0) (black line) and (18,0) (purple long-dashed line)

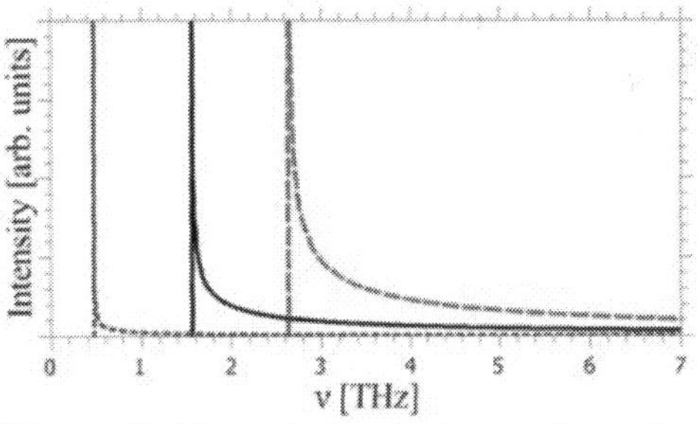

Figure 5. The calculated photon absorption spectra for a (24,0) CNT in zero field (black line) and a field of $B = 5$ T (red long-dashed and blue short-dashed lines, which correspond to $l = \frac{2n}{3}$ and $-\frac{2n}{3}$ respectively).

4. Conclusion

In the absence of curvature, at $k_x = 0$, optical transitions between the top valence sub band and the bottom conduction sub band in quasi-metallic CNTs are strictly forbidden by symmetry within the simple zone-folding model of the π-electron graphene spectrum. However, for zigzag quasi-metallic CNTs, dipole optical transitions are indeed allowed due to the gap opened in their energy spectrum by intrinsic curvature, which is of the order of THz. We propose that arrays of quasi-metallic CNTs could be used as the building blocks of THz radiation detectors, which would have a high sensitivity in the photocurrent to photon frequency. Furthermore, the band gap of quasi-metallic CNTs can also be controlled by an applied magnetic field, therefore allowing such devices to be tunable. Finally, our recent analysis [18] shows that the many-body (excitonic) effects, which dominate the optical properties of semiconducting CNTs [19], are also important in narrow-gap CNTs. However, due to the quasi-relativistic character of the free-particle dispersion near the band edge of the narrow-gap CNTs, there is a spectacular decrease in the exciton binding energy. Therefore, these effects do not alter significantly the main results of this paper.

5.0 References

[1] Ferguson B and Zhang X C 2002 *Nat. Mater.* **1** 26
[2] Dragoman D and Dragoman M 2004 *Prog. Quantum Electron.* **28** 1
[3] Lee M and Wanke M C 2007 *Science* **316** 64
[4] Nagatsuma T 2011 *IEICE Electronics Express* **8** 1127
[5] Saito R, Dresselhaus G and Dresselhaus M S 1998 *Physical Properties of Carbon Nanotubes* (London: ImperialCollege Press)
[6] Reich S, Thomsen C and Maultzsch J 2004 *Carbon Nanotubes: Basic Concepts and Physical Properties*(Berlin: Wiley)
[7] Kibis O V, Parfitt D G W and Portnoi M E 2005 *Phys. Rev.* B**71** 035411
[8] Kibis O V and Portnoi M E 2005 *Tech. Phys. Lett.* **31** 671
[9] Kibis O V, Malevannyy S V, Huggett L, Parfitt D G W and Portnoi M E 2005 *Electromagnetics* **25** 425
[10] Kibis O V, Rosenau da Costa M and Portnoi M E 2007 *Nano Lett.* **7** 3414
[11] Portnoi M E, Kibis O V and Rosenau da Costa M 2008 *Superlatt. Microstr.* **43** 399
[12] Portnoi M E, Rosenau da Costa M, Kibis O V and Shelykh I A 2009 *Int. J. Mod. Phys.* B**23** 2846
[13] Batrakov K G, Kibis O V, Kuzhir P P, Rosenau da Costa M and Portnoi M E 2010 *J. Nanophoton.* **4** 041665
[14] Fedorov G, Tselev A, Jim´enez D, Latil S, Kalugin N G, Barbara P, Smirnov D and Roche S 2007 *Nano Lett.***7** 960
[15] Harrison W A 1989 *Electronic Structure and the Properties of Solids* (New York: Dover Press)
[16] Kane C L and Mele E J 1997 *Phys. Rev. Lett.* **78** 1932
[17] Kleiner A and Eggert S 2001 *Phys. Rev.* B**63** 073408
[18] Hartmann R R, Shelykh I A and Portnoi M E 2011 *Phys. Rev.* B**84** 035437
[19] Shaver J and Kono J 2007 *Laser & Photon. Rev.* **1** 260, and references thereins

Acknowledgements

This work was supported by URCO (17 N 1TAY12-1TAY13), FP7 ITN NOTEDEV (Grant No.FP7-607521) and FP7 IRSES projects SPINMET (Grant No. FP7 - 246784), QOCaN (Grant No. FP7 316432), and CANTOR (Grant No. FP7- 612285).

IOP Conf. Series: Materials Science and Engineering **79** (2015) 012015 doi:10.1088/1757-899X/79/1/012015

Development and application of compact and on-chip electron linear accelerators for dynamic tracking cancer therapy and DNA damage/repair analysis

M.Uesaka[1], K.Demachi, T.Fujiwara, K.Dobashi, H.Fujisawa, R.B.Chhatkuli, A.Tsuda, S.Tanaka, Y,Matsumura, S.Otsuki
Nuclear Professional School/Department of Nuclear Engineering and Management, 2-22 Shirakata-shirane, Tokai, Naka, Ibaraki, 319-1106, Japan

J.Kusano, M.Yamamoto, N.Nakamura, E.Tanabe
Accuthera CO., 2-7-6 Kurigi, Asaoku, Kawasaki, 215-0033, Japan

K.Koyama, M.Yoshida
High Energy Accelerator Development Organization., 1-1 Oho, Tsukuba, Ibrakaki, 305-0810, Japan

R.Fujimori
National Institute of Radiological Sciences, 4-9-1 Anagawa, Inage, Chiba, 263-8555, Japan

A.Yasui
Institute of Development, Aging and Cancer, Tohoku University, Seiryomachi 4-1, Sendai, 980-8575, Japan

E-mail: uesaka@nuclear.jp

Abstract. We are developing compact electron linear accelerators (hereafter linac) with high RF (Radio Frequency) frequency (9.3 GHz, wavelength 32.3 mm) of X-band and applying to medicine and non-destructive testing. Especially, potable 950 keV and 3.95 MeV linac X-ray sources have been developed for on-site transmission testing at several industrial plants and civil infrastructures including bridges. 6 MeV linac have been made for pinpoint X-ray dynamic tracking cancer therapy. The length of the accelerating tube is ~600 mm. The electron beam size at the X-ray target is less than 1 mm and X-ray spot size at the cancer is less than 3 mm. Several hardware and software are under construction for dynamic tracking therapy for moving lung cancer. Moreover, as an ultimate compact linac, we are designing and manufacturing a laser dielectric linac of ~1 MeV with Yr fiber laser (283 THz, wavelength 1.06 μm). Since the wavelength is 1.06 μm, the length of one accelerating strcture is tens μm and the electron beam size is in sub-micro meter. Since the sizes of cell and nuclear are about 10 and 1 μm, respectively, we plan to use this "On-chip" linac for radiation-induced DNA damage/repair analysis. We are thinking a system where DNA in a nucleus of cell is hit by ~1 μm electron or X-ray beam and observe its repair by proteins and enzymes in live cells in-situ.

[1] To whom any correspondence should be addressed.

We try to apply this device to analyses of low dose radiation biological effect, especially Fukushima residents, and radiation sensitivity. Finally, we try to combine the pinpoint radiation therapy and radiation sensitivity analysis forward "Tailor-made Radiation Therapy".

1. Introduction

As the aging society proceeds, cancer becomes most fatal disease. Radiation therapy is covering one third of the total modality of the therapy with surgery and chemical treatment. In addition to IMRT (Intensity Modulated Radiation Therapy) and particle beam therapy (proton and carbon), pinpoint X-ray dynamic tracking therapy is now under development. Small moving lung cancer can be diagnosed in situ by keV X-ray camera and tracked by a moving linac X-ray source. For the purpose, the accelerating structure of linac should be small by using higher resonant RF such as X-band (9.3 GHz) and C-band (5.712 GHz) comparing with S-band (2.856 GHz) large strcture for IMRT. We are developing and openly use advanced S-band, X-band and laser electron linacs at University of Tokyo as shown in Figure 1. Especially, potable 950 keV and 3.95 MeV linac X-ray sources have been developed for on-site transmission testing at several industrial plants and civil infrastructures including bridges. 6 MeV linac have been made for pinpoint X-ray dynamic tracking cancer therapy [1]. We have been investigating X-ray assisted Drug Delivery Systems (DDS) with Cisplatin-micelle [2] and gold-nanoparticle-PEG (polyethylene glycol) [3]. Furthermore, we are designing and manufacturing a laser dielectric accelerator (hereafter, DLA) of ~1 MeV with Yr fiber laser (283 THz, wavelength 1.06 μm). Since the wavelength is 1.06 μm, the length of one accelerating structure is tens μm and the electron beam size is less than 1 μm. Since the sizes of cell and nucleus are about 10 and 1 μm, respectively, we plan to use this "On-chip" linac for radiation-induced DNA damage/repair analysis. We are thinking a system where DNA in a nuclear is hit by ~1 μm electron or X-ray beam and observe its repair by proteins and enzymes in live cells in-situ. We try to apply this device to analyses of low dose radiation biological effect and radiation sensitivity. Finally, we try to combine the pinpoint radiation therapy and radiation sensitivity analysis forward "Tailor-made Radiation Therapy".

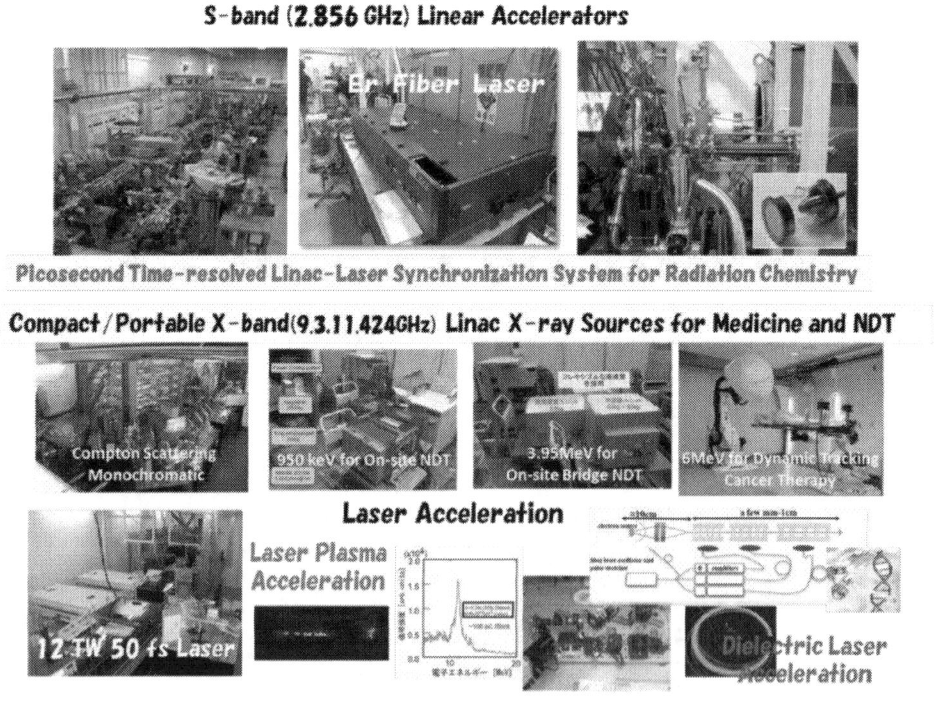

Fig.1 Staged development of advanced electron linacs at University of Tokyo

IOP Conf. Series: Materials Science and Engineering **79** (2015) 012015 doi:10.1088/1757-899X/79/1/012015

2. RF frequency and size of medical linac

Accelerating structures of S-band (2.856GHz, wavelength 105 mm), C-band (5.712GHz, 50.3 mm) and X-band (9.3GHz, 32.3 mm) linacs and cancer therapy systems using them are depicted in Figure 2 in order to explain the dependence of the size of the medical device on the accelerating RF frequency. As the RF frequency is higher, it becomes smaller so that it can be installed in the smaller moving cases. S-band linac is widely adopted for comformal RT and IMRT . X-band and C-band linacs are adopted for stereotactic RT such as Cyberknife and dynamic tracking therapy due to their compactness recently (see Figure 2).

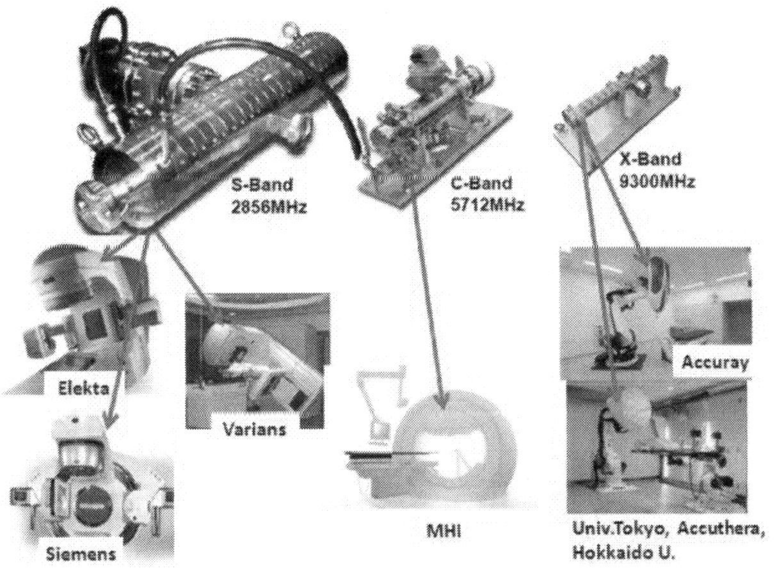

Figure 2. S-band (2.856GHz), C-band (5.712GHz), X-band (9.3GHz) accelerating structures and cancer therapy systems

3. 6 MeV medical system for dynamic tracking cancer therapy

We have developed an accelerating system capable of generating finer X-ray beams with 1mm X-ray spot size at cancers. In this system, we adopt beam scanning by controlled electric field in addition to mechanical scanning to achieve more rapid formation of radiation fields. We also aim at irradiating cancers with the X-rays from any direction in a three-dimensional space instead of on a two-dimensional plane. Accuthera CO., University of Tokyo, Hokkaido University and the National Cancer Institute are carrying out several research- and development-project under the NEDO (New Energy and Industrial Technology Development Organization) national project in Japan [4]. This project includes an X-ray micro-beam-accelerating system with the electromagnetic rapid scanning. Early cancers of 1-cm or less in diameter are the targets as well as cancers of 1-5 cm or less in diameter. This is equivalent to the practical stereotactic radiotherapy. Moreover, the dynamic tracking therapy in order to reduce irradiation dose against surrounding normal organs is available. Figure 3 shows the first 6 MeV machine for just a beam test and the fast X-Ray micro beam scanning system and a multi-axial robotic radiation and control system. Recently, we are developing an on-line MeV X-ray treatment-beam diagnostic device for safety as shown in Figure 4. We adopt the glass GEM (Glass Gas Electron Multiplier) [5,6,7] sensor which can realize both position- and dose-monitors . Position and movement of lung cancer are observed by two sets of 50 keV X-ray tubes and flat-panels.

IOP Conf. Series: Materials Science and Engineering **79** (2015) 012015 doi:10.1088/1757-899X/79/1/012015

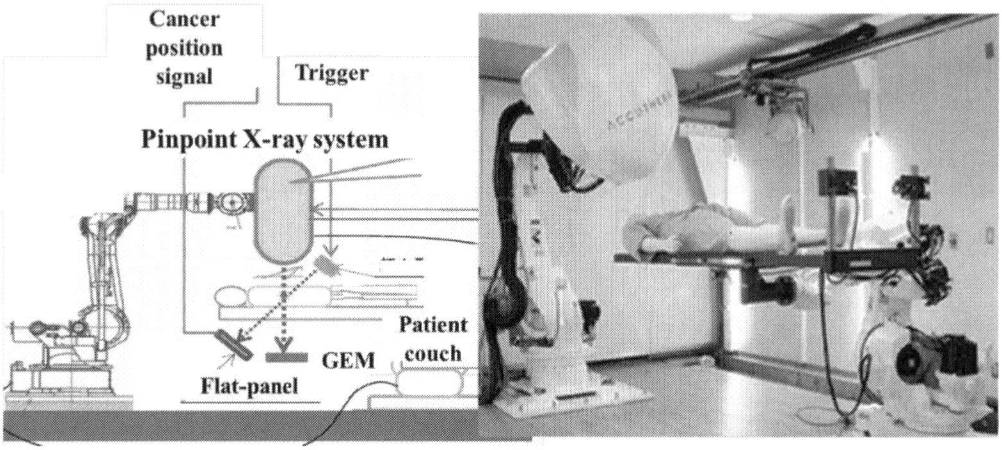

Figure 3 6 MeV pinpoint X-ray dynamic tracking therapy system

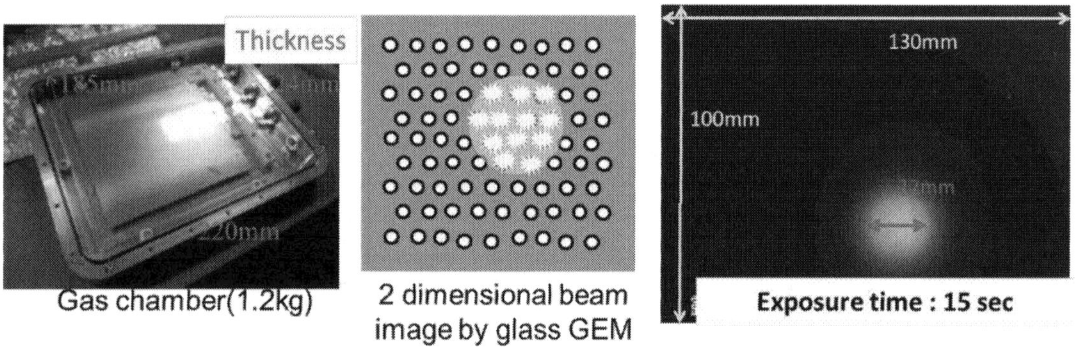

(a) Photograph (b) Schematic view (c) X-ray beam spot by 950 keV source

Figure 4 Glass GEM on-line treatment X-ray beam sensor

In order to improve the accuracy and safety of pinpoint irradiation to a moving lung cancer, we are also constructing the position/image prediction software (see Figure 5). Here we make use of MSSA (Multi-Channel Singular Spectrum Analysis) method [8] to predict future information from past ones. This method rationally picks up dominant singular aspects from the data in the past, which is understood to be more reliable and faster that Fourier Transform Analysis. This can predict the motion of the centre-of-mass of a cancer in a projected plain (see (a) of Figure 5) [9]. Even the 2-dimensional image can be predicted as show in (b) of the figure. Those image prediction would be very helpful at the monitoring screen [10]. Further, we are developing the dose calculation by the Boltzmann transport equation [11,12,13]. Pencil beam approximation where the dose distribution uses a certain modelling is widely used for its rapid calculation [14]. However, the accuracy is limited. On the other hand, Monte Carlo method can give most physically accurate distribution while it takes a lot of time. Thus, it cannot be used clinically yet. The Boltzmann scheme is rationally intermediate with acceptable accuracy and reasonable speed. We are developing our original on-line dose calculation system for the pinpoint X-ray dynamic tracking devise. We are going to apply the governmental permission in Japan and USA. .

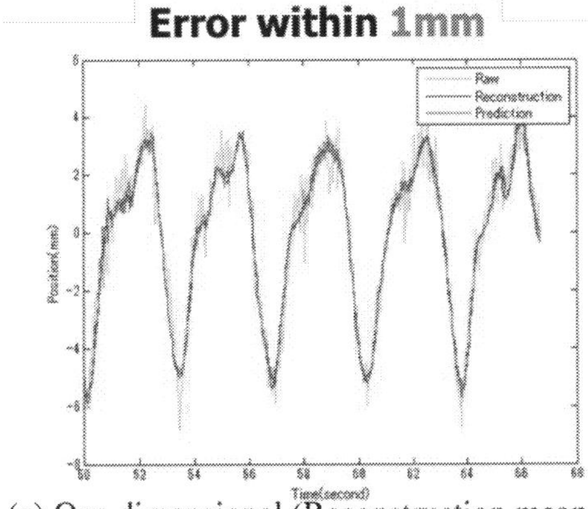

(a) One dimensional (Reconstruction means to cut electrical noises)

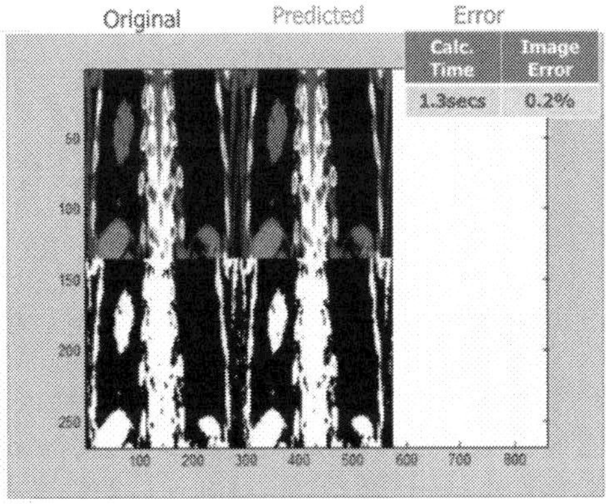

(b) Image prediction and error

Figure 5 Motion/image prediction for moving lung and cancers

4. On-chip μm electron/X-ray source for DNA damage/repair analysis

We introduce updated DNA-radiation damage/repair observation and sensitivity analysis under collaboration with NIRS (National Institute for Radiological Science) and Tohoku University. We plan to investigate the radiation sensitivity of individuals and relate it to radiation cancer therapy. This may be able to contribute to "Taylor-made" radiation therapy. We use HIMAC (Heavy Ion Medical Accelerator in Chiba) and irradiate cells by carbon beams and others. HIMAC and irradiation facility are shown in Figure 6. Updated example is given in Figure 7. We irradiate 1BR-hTERT (human fibroblast) cells by Fe ions of 500 MeV/nucleon at a dose of 1Gy. DSB (Double Strand Break) is visible by accumulation of RPA repair proteins at the DSB sites in blue-dyed nucleus. Moreover, the cell cycle is also visualized by cell cycle marker, Cyclin B antibody which indicates S and G2 phases. Now, we are investigating the difference of DNA damage/repair on radiations such as ions (C, Fe, etc.), proton and X-rays. However, the cells are not living so that the on-line observation is not available yet. Micro ion beams less than 1 μm and even single ion are used for the single shot irradiation to cells and the bystander effect is studied. However, those ions are generated by using big accelerators such as cyclotron, very narrow collimator and low intensity operation. Therefore, the positions of such beams are still random within a few μm^2 area. It is impossible to hit a targeted position in a cell.

On the other hand, on-line time-resolved DNA-UV laser damage/repair observation with live cells is available at Prof.Akira Yasui's laboratory of Tohoku University [15]. Here, 1 μm spot and 1 μm thick line of UV laser are spatially controlled and irradiation to cells and nuclei. Figure 8 and 9 show the irradiation and scanning system, and an example of the live cell imaging. Accumulation and disappearance of GFP-tagged XRCC1, single strand break marker in this study, are observed after 405nm laser irradiation with photosensitizer in U2OS cell.

In order to realize on-line time-resolved DNA damage/repair visualization system for live cells, we are designing and developing an on-chip sub-micron beam source. Figure 10 depicts that the accelerating structure becomes smaller for higher RF frequency. As an ultimate case, we adopts Yr fiber laser of 1.06 μm wavelength and 283 THz and one cell of its resonant structure becomes about 1 μm. We consider a set of comb shaped structure where two laser pulses are injected from the both sides as shown in Figure 10. The structure is made of dielectric material such as Silica glass. This kind of structure and application to beam acceleration are called photonic crystal [16] and DLA (Dielectric Laser Accelerator) [17]. The wavelength of light (λ) in a dielectric material of refraction index of n becomes λ/n. Therefore, the phase of light in the mid plane of the comb structure can be controlled by changing the depth and longitudinal length. It is known that the transverse electric field cannot accelerate a charged particle. However, we can design the optical phase in the mid plane so that an electron beam can be continuously accelerated in the longitudinal direction by applying such a comb-shaped dielectric structure. We are designing to manufacture microscopically an electron gun, several DLA structures, Yr fiber laser (283 THz, wavelength 1.06 μm) and X-ray target on one chip. Schematic configuration, the laser under development and the future image of the whole device are depicted in Figure 11. The whole device would be table-top which is equivalent to the UV laser system in Figure 8. We plan to accelerate the electrons up to ~1 MeV and the spot size and intensity of converted X-rays are ~ 1 μm and ~10^5 Bq. The intensity is similar to those of standard shielded radioisotopes so that it can be used even out of radiation controlled areas following a certain radiation safety regulation.

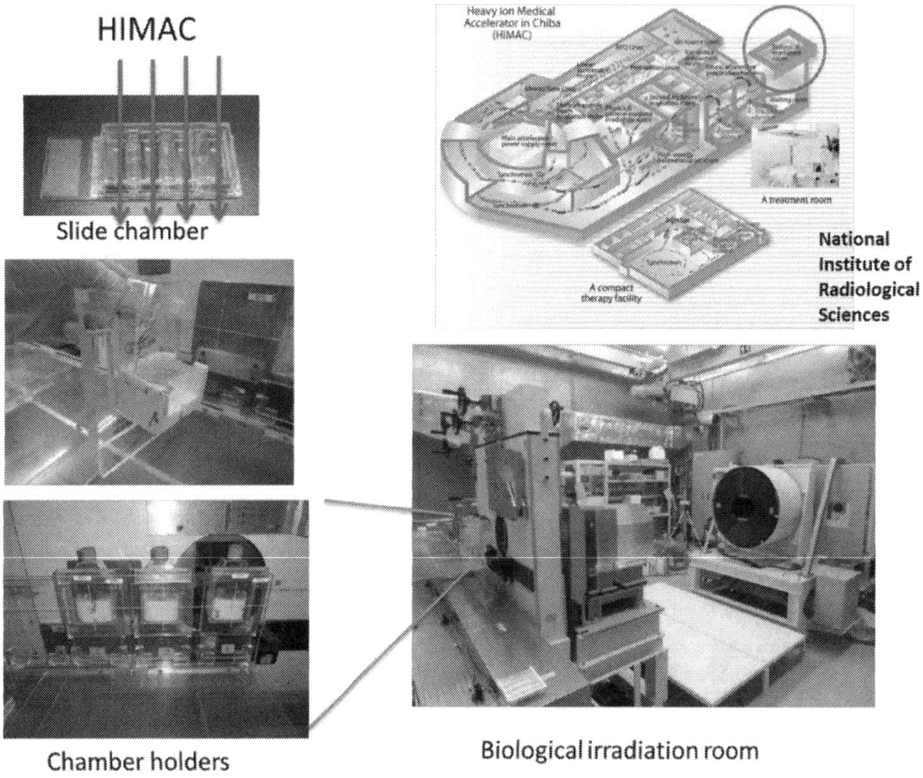

Figure 6 Cell irradiation system of HIMAC of NIRS, Japan

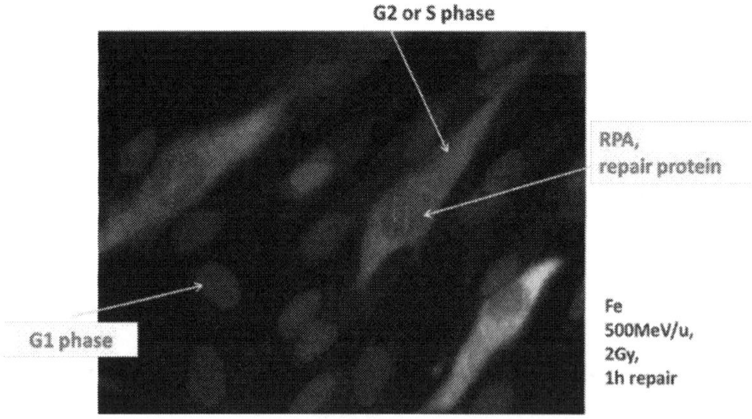

Figure 7 Visualization of DSB (Double Strand Break) and cell cycle

IOP Conf. Series: Materials Science and Engineering **79** (2015) 012015 doi:10.1088/1757-899X/79/1/012015

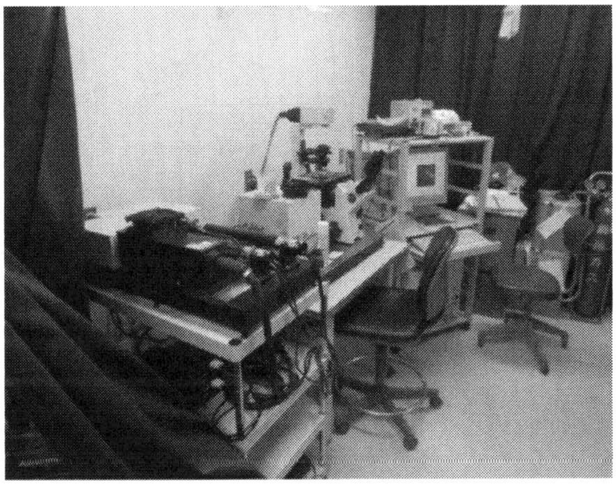

Figure 8 On-line observation system of DNA damage/repair by UV laser and optical microscopy (Prof.Akira Yasui, Tohoku University)

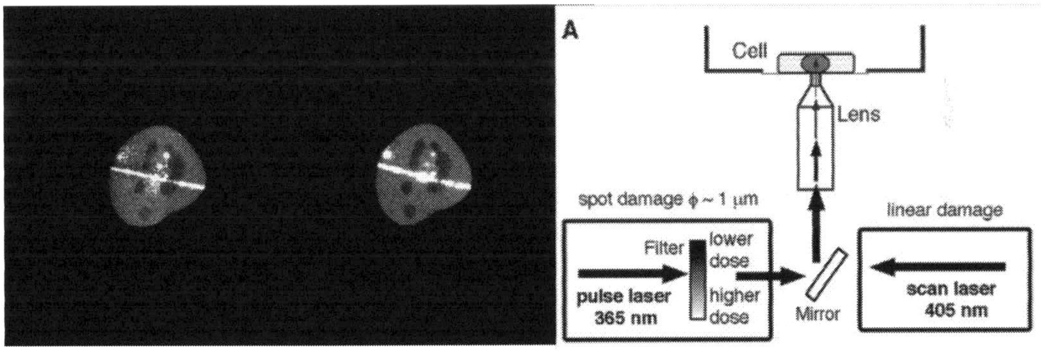

(a) 20 sec after irradiation (b) 200 sec (c) Irradiation and scanning system

Figure 9 On-line observation of DNA damage/repair by UV laser. Irradiation is done by 405nm laser 30 scans with photosensitizer, and the scanning is done by 488nm laser, by 20sec interval x10 times, and U2OS (human osteosarcoma) cells transfected by GFP-XRCC1 (Single strand break marker) are used.

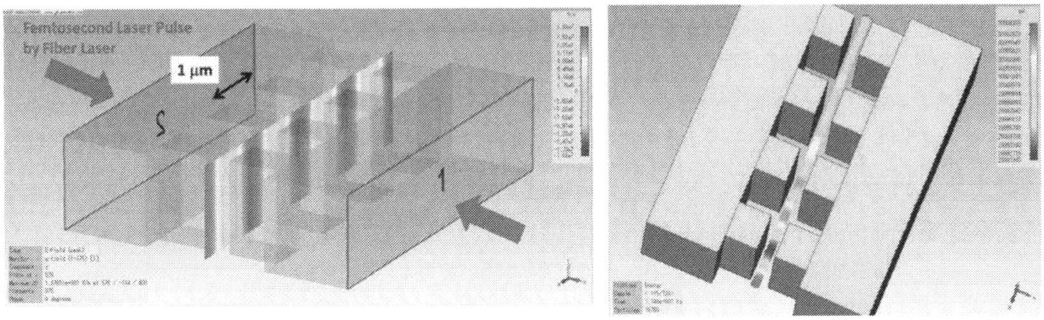

Figure 10 Schematic configuration of the regular structure of the laser dielectric accelerator

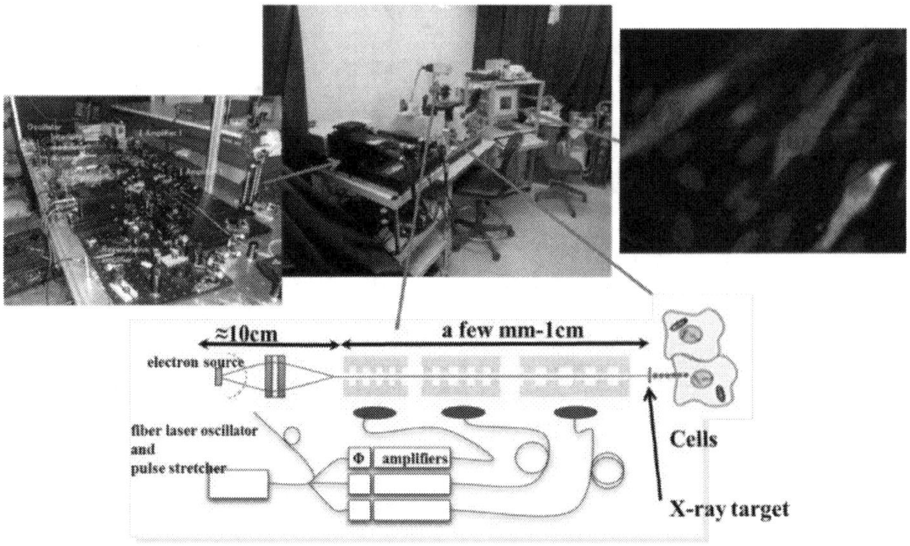

Figure 11 Image of on-chip sub-micron electron/X-ray source for on-line DNA damage/repair analysis

5. Concluding remarks

We have successfully developed the X-band electron linacs for medicine and non-destructive testing. Especially, the 6 MeV linac is used for pinpoint X-ray dynamic tracking cancer therapy. The length of the accelerating tstraucture is ~600 mm. The electron beam size at the X-ray target is less than 1 mm and X-ray spot size at the cancer is less than 3 mm. Several hardware and software are under construction for dynamic tracking therapy for moving lung cancer. Moreover, as an ultimate compact linac, we are designing and manufacturing the "On-chip" ~1 MeV DLA with table-top Yr fiber laser. We plan to use this DLA for radiation-induced DNA damage/repair analysis. We try to apply this table-top device to analyses of low dose radiation biological effect, especially for Fukushima residents, and radiation sensitivity. Finally, we try to combine the pinpoint radiation therapy and radiation sensitivity analysis forward "Tailor-made Radiation Therapy".

6. References

[1] Mitsuru Uesaka, Takuya Natsui, Lee Kiwoo , Katsuhiro Dobashi , Tomohiko Yamamoto, Takeshi Fujiwara, Haito Zhu, Kazuyuki Demachi , Eiji Tanabe, Masashi Yamamoto, Naoki Namakura, Joichi Kusano, Toshiyasu Higo, Shigeki Fukuda, Mitsuhiro Yoshida, Shuji Matsumoto, "950 keV, 3.95 MeV, 6 MeV X-band Linacs for Nondestructive Evaluation and Medicine",Nuclear Inst. and Methods in Physics Research, A, NIMA53737, S0168900211014586, 10.1016/j.nima.2011.07.026

[2] K. Mizuno、 T. Konishi、 M. Oikawa、 H. Iso、 T. Ishikawa、 H. Imaseki and M. Uesaka "Study of the Uptake of CDDP-containing Polymeric Micelles by Cells using Particle Induced X-Ray Emission" Journal of Radiation Research, Vol.52, 2, 193-198(2011)

[3] J. D. Trono、 K. Mizuno、 N. Yusa、 T. Matsukawa、 K. Yokoyama and M. Uesaka, "Size、 concentration and incubation time dependence of gold nanoparticle uptake into pancreas cancer cells and its future application to X-ray Drug Delivery System" Journal of Radiation Research,vol.52,pp103-109(2011)

[4] http://www.nedo.go.jp/english/index.html

[5] F. Sauli, "GEM: A new concept for electron amplification in gas detectors", Nuclear Instruments

and Methods in Physics Research Section A: Accelerators, Spectrometers, Detectors and Associated Equipment, Volume 386, Issues 2-3, 21 February 1997, Pages 531-534

[6] S. Fetal, C.W.E. van Eijk, F. Fraga, J. de Haas, R. Kreuger, T.L. van Vuure, J.M. Schippers, "Dose imaging in radiotherapy with an Ar-CF4 filled scintillating GEM", Nuclear Instruments and Methods in Physics, Research Section A: Accelerators, Spectrometers, Detectors and Associated Equipment, Volume 513, Issues 1-2, 1 November 2003, Pages42-46

[7] Takeshi. Fujiwara,Hiroyuki Takahashi,Mitsuru.Uesaka, "Performance of Glass GEM," Nuclear Science Symposium and Medical Imaging Conference (NSS/MIC), 2012 IEEE , vol., no., pp.1766,1767, Oct. 27 2012-Nov. 3 2012

[8] R. Vartard and M. Ghil, "Singular spectrum analysis in nonlinear dynamics,with applications to paleoclimatic time series," Physica D, 35, pp.395-424,(1989)

[9] M. Kawai, K. Demachi, H. Shiato and M. Ishikawa, "Development of Prediction System for Moving Tumor by MSSA for Chasing Radiotherapy", Proc. of JKMP-AOCMP(2011)

[10] Ritu Bhusal Chhatkuli, Kazuyuki Demachi, Masaki Kawai, Hiroshi Sakakibara, Kazuma Kamiaka and Mitsuru Uesaka, An approach for motion image prediction and reconstruction using principal component analysis, proceedings of 1st international conference on maintenance science and technology, pp103-104 (November 2012)

[11] A.Fogliata *et al.* "Dosimetric validation of the Acuros XB advanced dose calculation algorithm: fundamental characterization in water", Radiation Oncology 2011, 6:82 doi:10.1186/1748-717X-6-82

[12] E .Boman, "Radiotherapy forward and inverse problem applying Boltzmann transport equation", Kuopio university publications C. natural and environmental scineces 204, 2007

[13] L.Lorence, J .Morel and G. Valdez, "Physics guide to CEPXS: A multigroup coupled electron-photon cross section generating code", SAND89-1685, Sandia National laboratory, 1989

[14] Y.Egashira, T.Nishio, K.Hotta, R.Kohno, M.Uesaka, "Application of the pencil-beam redefinition algorithm in heterogeneous media for proton beam therapy", *Phys. Med. Biol.*, 58(4) (2013) 1169-84

[15] L.Lan, S.Nakajima, K.Komatsu, A.Nussenzweig, A.Shimamoto, J.Oshima, A.Yasui, "Accumulation of Werner protein at DNA double-strand breaks in human cells". Journal of cell science 118: 2005, pp. 4153-4162

[16] http://press.princeton.edu/titles/8696.html, http://ab-initio.mit.edu/book/photonic-crystals-book.pdf

[17] K.Koyama, et al., "Laser-driven dielectric electron accelerator for radiobiology researches", Proc. SPIE Vol. 8779, Laser Acceleration of Electrons, Protons, and Ions II, 877910 (2013).

Chemical bath deposition growth and characterization of zinc oxide nanostructures on plain and platinum-coated glass substrates for hydrogen peroxide gas sensor application

Y D J Jamasali[1] and A C Alguno

Physics Department, Mindanao State University–Iligan Institute of Technology, A. Bonifacio Avenue, Iligan City 9200, Philippines

Email: yusofdenj@gmail.com, alphaarnie@yahoo.com

Abstract. Growth of zinc oxide on plain and Pt-coated glass substrate via chemical bath deposition technique (CBD) were studied. Aqueous solutions of ammonium hydroxide (NH4OH) and zinc sulfate (ZnSO4) were used as the precursor substances in the synthesis. Ultraviolet-visible spectroscopy (UV-Vis) was performed to determine the energy band gap and X-ray diffraction (XRD) to examine crystallinity. Sensitivity measurements were carried out in order to examine its potential to be fabricated as hydrogen peroxide (H2O2) gas sensor. Experimental results in the sensitivity experiment show that in the presence of H2O2 gas, the resistance of ZnO increase which can be used as the basis for H2O¬2 detection. UV-Vis showed variation of energy band gap values but were all near the generally accepted value. XRD spectra further verify that ZnO were indeed synthesized.

1. Introduction

The large varieties of applications of hydrogen peroxide (H_2O_2) have continued to increase over the last decades. Its vapor can act as sterilant in the food industry [1,2]. Its liquid compound is used in hair bleaching and hair dyes [3] and medically important as antiseptic and disinfectant [4]. Nevertheless, H_2O_2 can be harmful in concentrations higher than 35%. It is an aggressive oxidizer and can pose several risks including explosive vapors, hazardous reactions, corrosive internal ailment, and skin disorder [5]. Over the past 15 years, three bleach plants in North America experienced catastrophic events caused by decomposition of H_2O_2 which resulted in serious injuries [6]. Also, disastrous events like the sinking of the Russian Kursk submarine in 2000 [7], and explosions in wastewater treatment tanks [8] have occurred due to H_2O_2 reactions.

To reduce risks, several researches were directed on the early detection of H_2O_2 gas. Recently, there has been a number of studies in order to fabricate H_2O_2 gas sensors employing several routes in detecting H_2O_2 such as sensors based on thermocouple [9], gold (Au) and platinum (Pt) electrodes [10], and Prussian – blue [11]. However, these materials are expensive and its processing technique is difficult to perform and required high vacuum technology. Therefore, it is imperative to find an alternative material that offers promising properties that can detect H_2O_2 gas. A good candidate as an alternative material is the semiconductor zinc oxide (ZnO) because it has strong surface sensitivity [12]. It is naturally grown n-type II-VI semiconductor. It has direct band gap of 3.37 eV at room temperature and has a large exciton binding energy of 60 meV which is necessarily adequate to overcome the thermal energy at room temperature [13].

[1] To whom any correspondence should be addressed.

There are various techniques to synthesize ZnO such as metal organic chemical vapor deposition (MOCVD) [14], molecular beam epitaxy (MBE) [15], pulse laser deposition (PLD) [16,17], and chemical vapor deposition (CVD) [18]. But these techniques are very expensive, highly toxic, high cost in environmental disposal and they need high vacuum equipment. Nonetheless, it is essential to find another technique to grow ZnO that offers low cost and does not need high vacuum equipment.

Nowadays, it is believed that chemical bath deposition (CBD) technique might be the cheapest method to deposit ZnO nanostructures. It does not require sophisticated instruments while the starting chemicals are commonly available and cheap, and the preparation parameters are easily controlled [19]. And this technique does not need high vacuum equipment.

Basically, CBD can be classified into two: unseeded and seeded processes. In unseeded, the CBD grows the nanostructures after the substrates had undergone conventional washing [20]. In seeded, before the CBD growing of nanostructures, substrates are pre–treated after conventional washing in order to create seed layers of ZnO [21].

ZnO has also been fabricated as sensor to detect various species. This includes fabricating ZnO as methane sensor [22], LPG sensor [23], ultraviolet radiation sensor [24], and calcium ion sensor [25]. But to the best of our knowledge, there has been no report that ZnO nanostructures synthesized via CBD have been utilized as H_2O_2 sensor. In this study, ZnO nanostructures are grown on plain and Platinum (Pt) coated glass substrates and then fabricated as H_2O_2 gas sensors. Sensitivity to H_2O_2 gas of the fabricated ZnO–based sensors was measured. Energy band gap determinations were carried out via UV-Vis spectroscopy.

2. Procedure

2.1. Cleaning of glass substrate

The very first step before the CBD proper is the washing of glass substrate in order to remove dirt and impurities that may have unknowingly settled on the glass substrate. This is done by washing in several substances, namely: acetone, distilled water, ethanol, distilled water again, hydrochloric acid (HCl), and lastly distilled water. These glass substrates are then dried in air at room temperature.

2.2. Platinum coating

After the glass substrates are dried, some of these were coated with platinum using a platinum auto-coater machine shown in Figure 1.

In coating a cleaned glass substrate, platinum auto-coater machine is used with pressure 4 Pa and 40 A current. The glass substrate is placed at the center-stage of the machine and the area is vacuumed for 10 seconds. The machine coated the substrate with platinum through spray sputtering for 60 s.

2.3. Seeded (two-stage) CBD

The seeded (two–stage) CBD commence with the making of Ammonium zincate bath solution which is done by combining 100 mL of 0.03 M $ZnSO_4$ and 100 mL of 1 M NH_4OH (in another trial this is 3 M) in a 250–mL beaker. This beaker is placed inside a 1000–mL beaker with water having the same level as the solution inside the 250–ml beaker. The bath is stirred at a constant rate of 360 RPM by magnetic stirrer for 30 minutes and its temperature is not varied, the same with room temperature. The solution was then maintained at 70^0C and stirred at a constant rate of 360 RPM for 5 hours. The outcome of this were white precipitates, which were most likely ZnO settled at the bottom and some on the walls of the beaker. These precipitates are carefully scooped and then placed on the cleaned plain and Pt-coated glass substrates. They are then oven-dried at temperature 200^0C. These are now the seeded substrates.

Afterwards, another Ammonium zincate bath solution is made, and stirred, as mentioned above. Then, the seeded substrates were lowered and submerged in the bath. The solution was heated and maintained at 70^0C and stirred at a constant rate of 360 RPM for 5 hours (refer to Figure 2).

Figure 1. Platinum coating. (a) The center stage, opened and the cleaned glass substrates were placed inside. (b) Pt-coating on process. The number on the display shows the countdown time (in s).

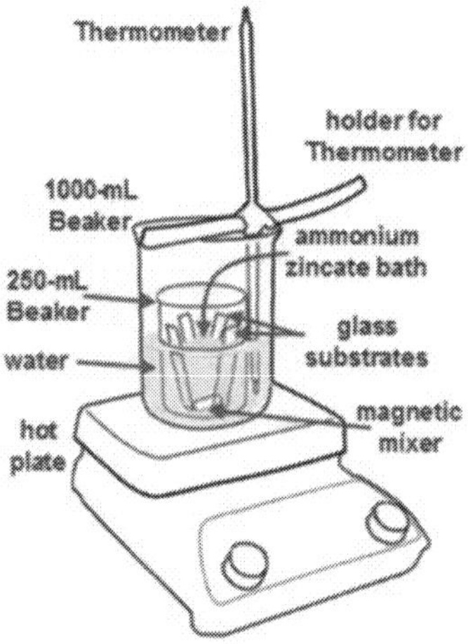

Figure 2. Essential constituents in chemical bath deposition (CBD) of zinc oxide (ZnO).

3. Chemical reactions involved

The ammonium zincate bath undergoes chemical reaction between ammonium hydroxide and an aqueous solution of zinc sulfate ($ZnSO_4$):

$$ZnSO_4 + 2NH_4OH_{(aq)} \rightarrow Zn(OH)_{2(s)} + (NH_4)_2SO_4 \quad (1) \qquad (1)$$

Addition of ammonia dissolves the precipitate following the reaction.

$$Zn(OH)_{2(s)} + 2NH_4OH_{(aq)} \rightarrow (NH_4)_2ZnO_2 + 2H_2O \quad (2) \qquad (2)$$

The overall reaction:

$$ZnSO_4 + 4NH_4OH_{(aq)} \rightarrow (NH_4)_2ZnO_{2\,(s)} + (NH_4)_2SO_4 + 2H_2O \quad (3) \qquad (3)$$

And the reaction occurring on the surface of the substrate leading to the formation of ZnO from ammonium zincate bath is [20]:

$$(NH_4)_2ZnO_{2(s)} + H_2O_{(aq)} \rightarrow ZnO_{(s)} + 2NH_4OH_{(aq)} \quad (4) \qquad (4)$$

4. Characterizations

4.1. Sensitivity measurement

Small portions with dimension of about 1 mm × 1 mm of the synthesized ZnO samples were cut and were fabricated as H_2O_2 gas sensor by connecting to it thin wires (refer to figure 3) and were put into a circuit (refer to figure 4) to monitor the current and voltage in the absence and presence of H_2O_2 gas.

In measuring the sensitivity the current pumped by the low current source is held to a fixed value. And the sensitivity is calculated by using the following formula [26]:

$$\text{Sensitivity} = \frac{|R_e - R_o|}{R_o} \times 100\ \% \qquad (5) \tag{5}$$

where R_o and R_e are the resistances of ZnO sample before and during/after exposure to H_2O_2 gas.

The IV curve measurement uses the same circuit used in sensitivity measurement. The only difference is that here, the current was continuously varied and the corresponding voltage was monitored before, during and after exposure to

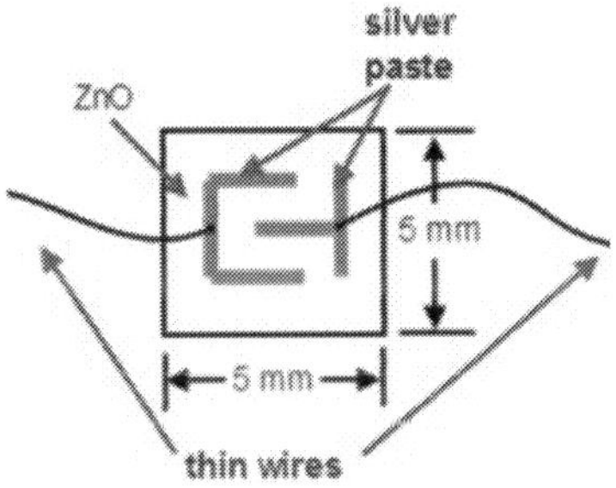

Figure 3. Schematic diagram of the ZnO sample fabricated as H_2O_2 gas sensor.

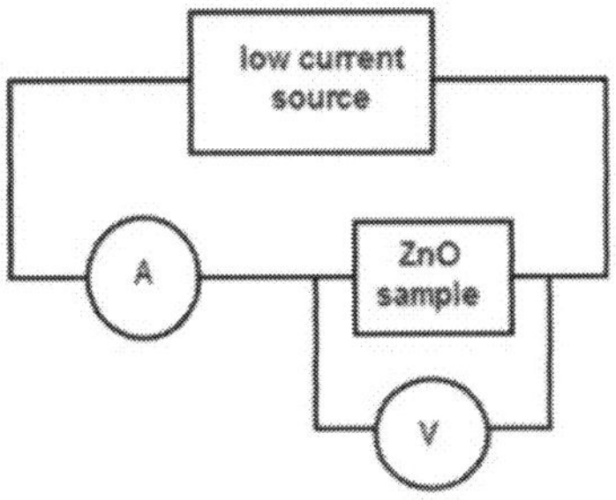

Figure 4. Circuit diagram used in sensitivity measurement.

4.2. UV-Vis spectra

The UV-Vis spectra were obtained Perkin Elmer Lambda 35 UV/Vis Spectrometer.

4.3. XRD

The structural properties were studied by X-ray Diffraction measurements (XRD) with CuK_α radiation.

The XRD spectra were obtained from National Institute of Geological Sciences (NIGS) of University of the Philippines.

5. Results and discussion

5.1. Sensitivity and IV curve

Both the uncoated and Pt-coated ZnO samples were fabricated as H_2O_2 sensors according to description shown in figure 3.

While ZnO sample was connected to the circuit described by figure 4, current and voltage were measured and the resistance was calculated during every 5 min with and without the presence of H_2O_2 gas. Figure 5 shows the sensitivity graph of the sample. Clearly, the sensitivity value does not considerably change when the sample was not yet exposed to H_2O_2 gas. However, once the H_2O_2 gas is injected to the vicinity of ZnO sample, the sensitivity value increases abruptly. And this continues to increase as time goes on. But when the source of H_2O_2 gas is shut off, the sensitivity value suddenly decreases to a lower value and approximately maintains that value.

The increase in the sensitivity value during the exposure may have been caused by the increase in the number of oxygen ions occupying the surface ZnO. These ions are the outcome of adsorption of oxygen gas molecules. This oxygen gas comes from the following chemical decomposition of H_2O_2 which happens when it is exposed to heat, light, sunlight or comes into contact with a metallic surface:

$$H_2O_2 \rightarrow H_2O + O_2 \qquad (6) \qquad (6)$$

Some electrons, instead of participating in the electron-hole recombination, are trapped by these oxygen ions(as illustrated in Fig. 6) [26]. Hence, increase in thenumber of oxygen on the surface of ZnO may trigger the increase in resistance of the ZnO sample as a whole and thus, the increase in sensitivity value according to Eq. 5.

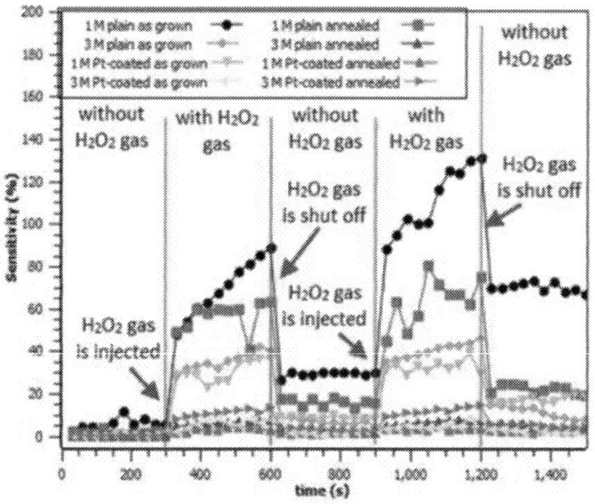

Figure 5. Sensitivity measurement of the ZnO sample.

The samples grown from the bath containing 1 M NH_4OH have greater capability of detecting H_2O_2 gas than those from 3 M NH_4OH. And ZnO grown on Pt-coated substrates have less capability of detecting H_2O_2 gas than those grown on plain substrate. The annealed samples have less responsivity than their unannealed counterparts (as grown) but they have more ability to recover.

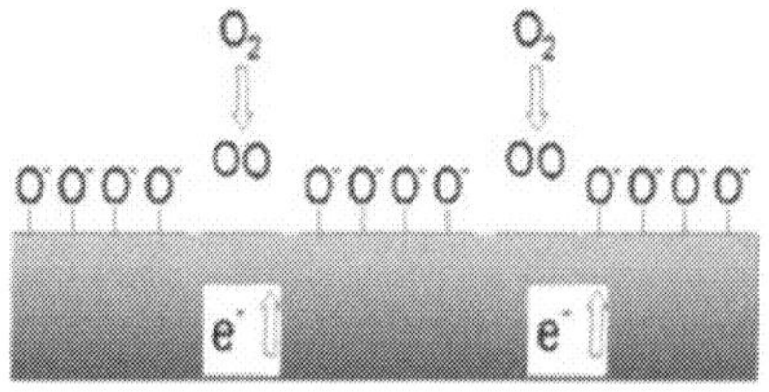

Figure 6. Schematic sketch of the ZnO film's surface with Oxygen gas becoming Oxygen anion by trapping electrons [26].

5.2. UV-Vis

The $(\alpha h\nu)^2$ versus $h\nu$ graph of all the annealed samples is shown in Figure 7. These curves are based on the UV-Vis reflectance spectra. Through linear fitting, the extrapolated energy band gap of the samples on plain glass substrates are 3.32 eV and 3.24 eV for 1 M and 3 M[2], respectively. On the other hand, those on Pt-coated glass substrates are 3.22 eV 3.21 eV for 1 M and 3 M, respectively. Clearly, the energy band gap of samples on Pt-coated are less than those on plain glass substrates. This might be due to the fact that Platinum is a good conductor. Clearly, the energy band gap of samples on Pt-coated are less than those on plain glass substrates.

This might provide an additional insight on why samples on plain glass detect more H_2O_2 during exposure than those on Pt-coated (refer to Fig. 5). This may suggest that electron-hole recombination prefers more to take place in Pt rather than in ZnO. And so, Oxygen anions may have lesser numbers on Pt-coated samples because of lesser number of electrons passing through ZnO at the surface. Thus, lesser change in resistance during the H_2O_2 exposure.

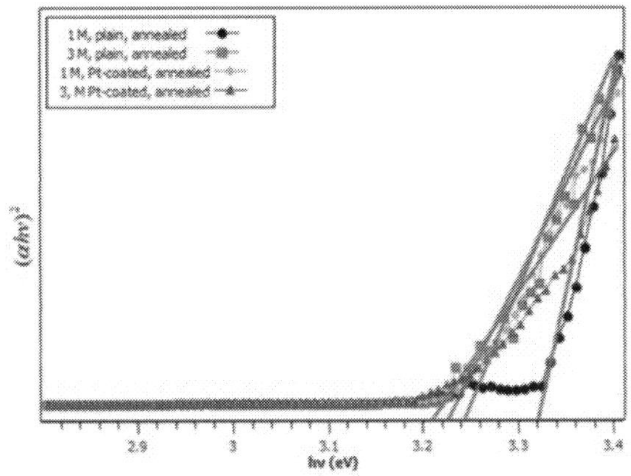

Figure 7. *UV-Vis* spectra of annealed samples.

5.3. XRD

XRD spectra of as grown ZnO samples on plain substrate are shown in Figure 8. The peaks of XRD patterns correspond to those of the theoretical ZnO patterns from JCPDS data file, with hexagonal wurtzite structure of the bulk and lattice constants: a = 3.24982 Å, c = 3.26064 Å.

Some of the characteristic peaks with their corresponding indices and interplanar distances in each sample are listed in tables 1(for 1 M) and 2 (for 3 M).

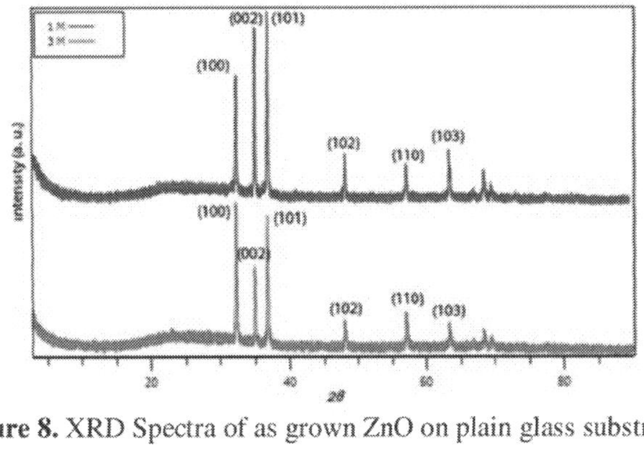

Figure 8. XRD Spectra of as grown ZnO on plain glass substrates.

On each sample, three peaks are more pronounced than the rest. These three peaks correspond to (100), (002), and (101) directions of crystallization of wurtzite ZnO nanorods. This suggests that the crystallites are highly oriented with these directions and that the growth along the directions a, b, and

[2]From this point onwards, the notation 1 M means sample from chemical bath whose NH_4OH was 1 M, and the same for notation 3 M, from bath with 3 M NH_4OH.

c-axis are favored. In the diffraction spectra, the intensity ratio in the direction of a axis and c axis (Ia/Ic) and the intensity ratio in the direction of b axis and c axis are useful. These intensities and intensity ratios are listed in table 3.

Table 1. The X-ray diffraction data results of as grown ZnO on plain glass (1 M).

Peak no.	Indices (*hkl*)	2*Θ*[deg]	*d*[Å]
1	(100)	32.433	2.7604
2	(002)	35.110	2.5559
3	(101)	36.919	2.4347
4	(102)	48.204	1.8878
5	(110)	57.210	1.6102
6	(103)	63.498	1.4650

Table 2. The X-ray diffraction data results of as grown ZnO on plain glass (3 M).

Peak no.	Indices (*hkl*)	2*Θ*[deg]	*d*[Å]
1	(100)	32.396	2.7635
2	(002)	35.075	2.5583
3	(101)	36.886	2.4368
4	(102)	48.176	1.8888
5	(110)	57.001	1.6156
6	(103)	63.478	1.4654

As the concentration of NH_4OH of the bath was increased from 1 M to 3 M, the values of Ia/Ic and Ib/Ic also increase. This suggests that there are some differences in the aspect ratios of the nanorods in these samples. Specifically, nanorods in 3 M sample are wider than those in sample 1 M.

Sample (as grown, plain substrate)	Ia (100)	Ic (002)	Ib (101)	Ia/Ic	Ib/Ic
1 M	594	812	886	0.73	1.09
3 M	690	398	632	1.73	1.59

6. Conclusion

In conclusion, ZnO were successfully grown on plain and Pt-coated glass substrates. All of the synthesized samples were able to detect H_2O_2 gas as verified by the sensitivity measurement. The resistance of the samples increases during exposure to H_2O_2 gas and can be used as the basis in detecting H_2O_2. Samples on plain glass substrates have greater capability in sensing and samples on Pt-coated have greater recovery after exposure. UV-Vis spectra show that the energy band gap are in the range of the documented values of ZnO [12,13], with ZnO samples on Pt-coated having less values than those of samples on plain substrate. XRD patterns and data further illustrate that ZnO signature peaks are present in the spectra. This suggests that ZnO was indeed really grown. This further shows that the ZnO nanorods from bath containing 3 M NH_4OH were wider than those from 1 M.

7. References

[1] Klapes N A and Vesley D 1990 *Applied and Environmental Microbiology* **56**502–3
[2] Webb B C 2001 *Journal of Pharmaceutical Science & Technology* **55**49–50
[3] Retrieved from http://chemistry.about.com/cs/howthingswork/a/aa101203a.htm
[4] Retrived from http://www.earthclinic.com/Remedies/hydrogen_peroxide_general.html
[5] Fox*et al.* 2002*Agency for Toxic Substances and Disease Registry***7722** 1–3
[6] Hart P W and RudieA W 2007 *TAPPI EPE Conference***3425**233–5

[7] Retrieved from http://www.newscientist.com/article/dn2600-torpedo-fuel-leak-sank-kursk.html
[8] Wehrum W L 1993 *Proceedings Safety Progress***12** 199–202
[9] Näther N*et al.* 2006 *Sensors***6** 308–17
[10] Kuwata Sand SadaokaY 2000 *Sensors and Actuators* B **65**325–6
[11] BenedetJ*et al.*2009*Anal. Bioanal. Chem.***395**371–6
[12] Janotti A and van de WalleC 2009 *Report on Progress in Physics***354**64–5
[13] ChoiY*et al.* 2010 *IEEE Transactions on Electron Devices***57**26–7
[14] Tan S T*et al.* 2005 *Journal of Applied Physics***98**1–5
[15] Kim J G*et al.* 2008 *Journal of the Korean Physical Society***53**271–5
[16] Ohshima T *et al.* 2003 *Surface and Coatings Technology***2564**517–20
[17] Dikovska A O *et al.*2005*Journal of Optoelectronics and Advanced Materials*71329–34
[18] Barankin MD*et al.* 2007 *Solar Energy Materials and Solar Cells***5648**924–30
[19] WangM*et al.* 2008 *Crystal Growth and Design***8**501–6
[20] MitraPand KhanJ 2006*Materials Chemistry and Physics***98**279–84
[21] LeeS*et al.* 2007 *National Kaohsiung University of Applied Sciences***3851**545–52
[22] MitraP and MukhopadhyayAK 2007 *Bulletin of the Polish Academy of Sciences: Technical Sciences***55**281–5
[23] SivapunniyamA*et al.*2011*Sensors and Actuators* B: *Chemical***101**645–50
[24] IevtushenkoA*et al.*2008*Proceedings XXXVII International School of Semiconducting Compounds***114** 1123–9
[25] AsifMH*et al.*2008 *Research Letters in Nanotechnology***2008**1–4
[26] MintangL 2010*Synthesis of ZnO Nanowires and Applications as Gas Sensors*Master Thesis, University of Saskatchewan, Saskatoon, Saskatchewan, Canada
[27] HahnY 2011 *Korean Journal of Chemical Engineering***28**1797–1813
[28] LeeS*et al.* 2007*National Kaohsiung University of Applied Sciences***5** 545–52
[29] LeeG J*et al.* 2011 *Journal of Nanoscience and Nanotechnology***11**511–7
[30] Joint Committee on Powder Diffraction Standards, Powder Diffraction File, card no: 36-1451

Acknowledgement

The authors would like to thank Prof. Rosemarie Terio of Physics Department of MSU-Main, and Mr. Alnasser Kasim of Turkish Chamber of Commerce Of the Philippines and Miss Christine Adelle Rico for helping the researchers in obtaining the XRD data.

The authors also acknowledge Prof. Bianca Rae Sambo, Prof. Majvell Odarve, Prof. Joy Cristy Piagola, Prof. Jess Gambe of the Physics Department of MSU-IIT for their assistance in this study.

IOP Conf. Series: Materials Science and Engineering **79** (2015) 012017 doi:10.1088/1757-899X/79/1/012017

Full-custom design of split-set data weighted averaging with output register for jitter suppression

MC Jubay[1] and OJ Gerasta

Mindanao-State University-Iligan Institute of Technology
Iligan City, Philippines

E-mail: mark.jubay@analog.com, olgajoy.labajo@msuiit.edu.ph

Abstract. A full-custom design of an element selection algorithm, named as Split-set Data Weighted Averaging (SDWA) is implemented in 90nm CMOS Technology Synopsys Library. SDWA is applied in seven unit elements (3-bit) using a thermometer-coded input. Split-set DWA is an improved DWA algorithm which caters the requirement for randomization along with long-term equal element usage. Randomization and equal element-usage improve the spectral response of the unit elements due to higher Spurious-free dynamic range (SFDR) and without significantly degrading signal-to-noise ratio (SNR). Since a full-custom, the design is brought to transistor-level and the chip custom layout is also provided, having a total area of 0.3mm2, a power consumption of 0.566 mW, and simulated at 50MHz clock frequency. On this implementation, SDWA is successfully derived and improved by introducing a register at the output that suppresses the jitter introduced at the final stage due to switching loops and successive delays.

1. Introduction

As the demand for the increase of performance of digital-to-analog converters (DAC's) continues to grow, several optimization schemes have been proposed. One factor worth considering is the problem regarding component mismatch, which is basically due to non-idealities introduced in the actual manufacturing process (e.g., inaccuracy and process variation) [1]. Matching effects gained importance as process dimension decreases and power supply lowers [2]. An existing method to cope with this problem is through static means (e.g., application of laser technology and some digital error correction techniques). These schemes have been effective in addressing the problem, but on the other hand can be impractical to someapplications [3]. Having these limitations of static matching, dynamic system became into an option [4]. Dynamic approach can only be applied in equally-weighted elements. This is because the weights are equal and thus they can be interchanged or used independently. This type employs equally-weighted current or voltage units and the number of the unit elements used directly comprises its resolution. Such implementation can potentially result to more circuit complexity, but on the other hand, it is still preferred on some applications due to its advantages (e.g. guaranteed monotonicity, reduced glitching noise and linearity). They are applied in ADC and DAC. The output of an ADC which is in thermometer bits can be fed directly to the input of the DAC unless conversions are necessary. Though this is possible, doing so would consequently

[1] To whom any correspondence should be addressed.

IOP Conf. Series: Materials Science and Engineering **79** (2015) 012017 doi:10.1088/1757-899X/79/1/012017

degrade the output of the DAC in the presence of element mismatch. Unequal and nonrandomized selection of the unit elements are the drawbacks of using directly a thermometer code. Due to these disadvantages, a basic DWA algorithm is commonly implemented for averaging. Basic DWA algorithm though provides averaging by cyclic usage of the unit elements in time. This, on the other hand, suffers the problem on some type of input since no randomization is done in the selection. Switching the elements equally would be necessary in the reconstruction of the discrete analog signal. This will make the errors spread out over the frequency spectrum resulting to nonlinearities averaged out over time [4]. Other than this, randomization breaks up the input-dependent tones generated by the selection process itself [1,4]. Considering these two criteria, distortion in the frequency domain is reduced in such a way that the SNR is not significantly affected or degraded. SDWA provides the averaging and randomization at the same time by using the DWA output and splitting the elements independently.

In this implementation, Split-set DWA is designed in full-custom. This allows more freedom to work to the lowest design level with the purpose of optimizing the layout area, minimizing glitches or delay problems, and lessening the overall power consumption. Timing limitations that result to glitches and jitter problems are compensated using a register.This register is used to block or filter out the glitches or jitter that occur at the output due to successive digital switching.

2. Experimental section

2.1. SDWA algorithm

SDWA operates by splitting the unit element set into subsets in a special way, and randomizing each subset independently. For N elements, SDWA is carried out in the following steps:

i. Apply DWA to the N-unit elements of DAC for (M-1) clock cycles;

ii. In clock cycle M (where M may be predetermined, or identified by a pseudorandom digital signal reaching a predetermined value), split the set of all unit elements into two subsets. Subset S_K contains elements 1 through k, where k is the highest unit element index used in clock cycle M: its complement S_{K_BAR} contains elements with indices k+1 through N;

iii. Rotate or scramble all elements of S_K within subset S_K, and similarly rearrange the elements of S_{K_BAR} internally within S_{K_BAR}.

iv. Return to step 1, starting with the unit element now occupying position k+1.

2.2. Design architecture

The implementation of SDWA algorithm is composed of a basic DWA block and somecomplementing blocks.

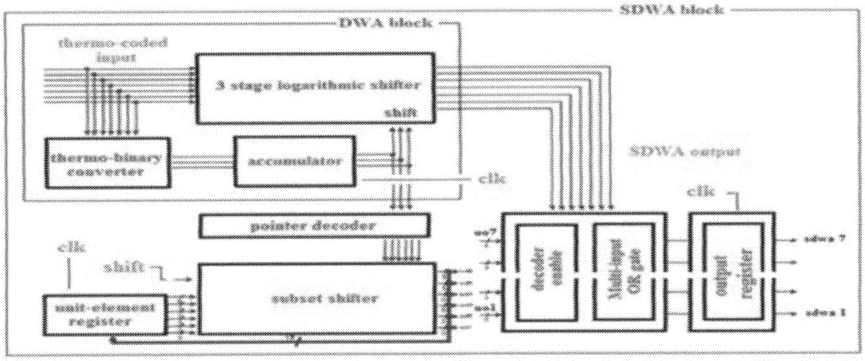

Figure 1. SDWA Circuit Block Diagram.

2.2.1. Data weighted averaging

The implementation of the Basic DWA uses logarithmic shifter which is controlled by the output of the accumulator block. Thermometer-to-binary converter provides the encoding of the input data to be able to generate the correct pointer or input value to the logarithmic shifter.

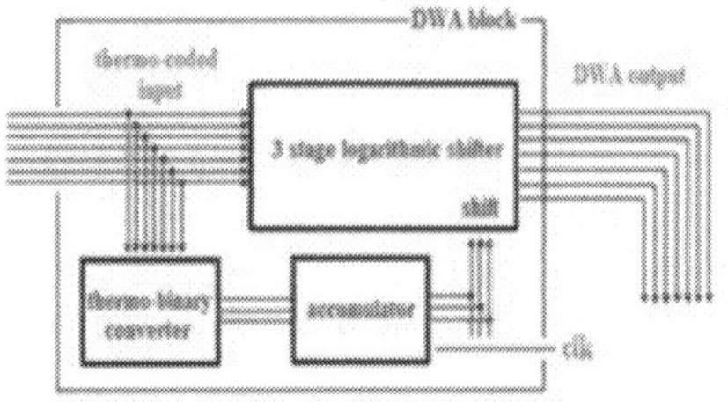

Figure 2.DWA circuitblock diagram.

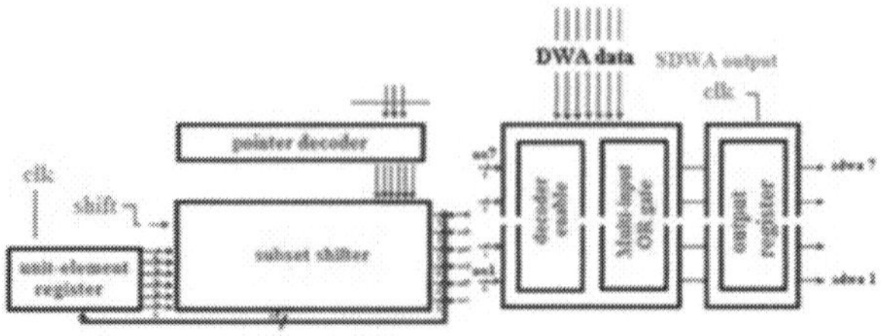

Figure 3. Complementing blocks.

The averaging characteristic of the basic DWA is induced in SDWA by using the DWA as the enabling data of the decoder-OR gate block. The splitting of the unit elements into subsets are provided by the subset shifter block at the control of the pointer decoder data. The updated subset shifter output data is controlled by the *shift* input which can be a predetermined scrambling-enable data. When this input data is not at enable state, the output of the SDWA block will just be a basic DWA. The unit element register block stores any update in the subset shifter output at every clock period.

3. Experimental section

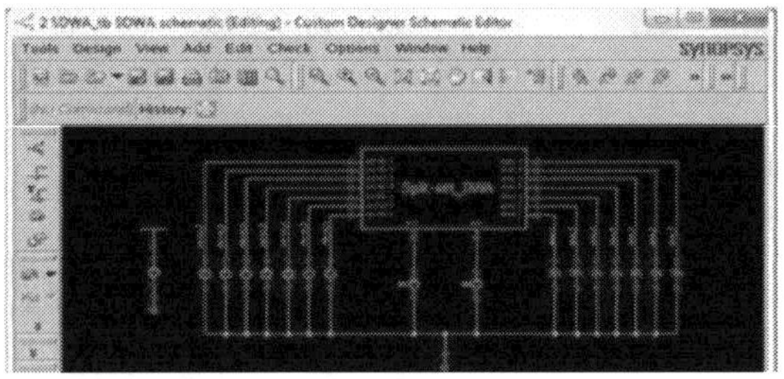

Figure 4. SDWA testbench setup in Synopsys.

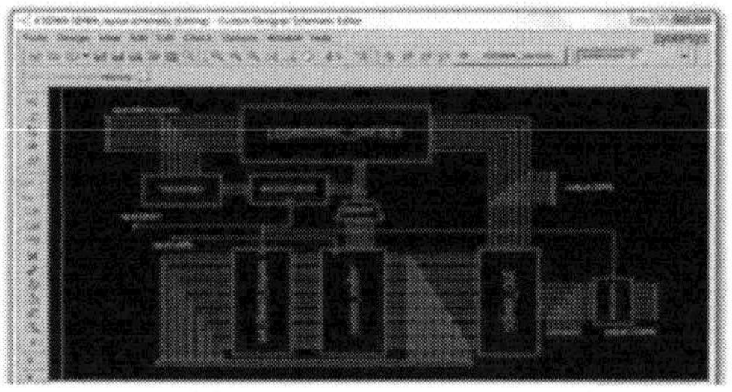

Figure 5. Split-set DWA inside blocks in Synopsys.

Fig. 4 above shows the SDWA testbench with 7-bit thermometer code as input. It also requires a clocking signal and a shift enable input. The shift input defines the SDWA scrambling frequency and which can be predefined or identified by a pseudorandom digital signal reaching a predetermined value [3]. For verification, we let the shift enable input to be equal as the clock. This means that SDWA is performed in every clock period. Shift input differentiates DWA to SDWA. When shift is enabled (high), the output is SDWA; when shift is low, output is DWA. All the following simulations are done in 10 pF output load, supply voltage of 1.2 volts having clock frequency at 50MHz.

3.1. SDWA simulation for DC input (constant 2 for 16 cycles)

A constant input sequence of two levels is maintained for 16 clock periods and is inputted to the SDWA circuit is shown in Fig. 6. In fig. 6, the first 2 less significant bits are used for 16 clock periods. This input can be fed directly to the unit elements for reconstruction but later on will provide significant problems in the presence of mismatch. Other than this, it is logical that the first two unit elements will be overused compared to the other bits (bits 3 to 7). This characteristic of thermometer code that prioritizes the less significant bit in usage can result to potential problems when non-idealities are being introduced during reconstruction.

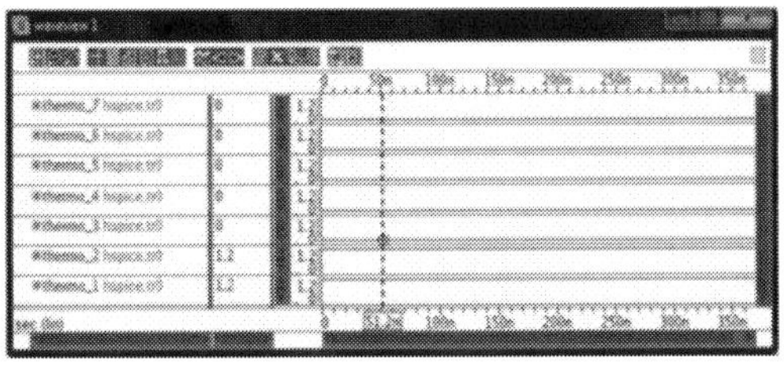

Figure 6. SDWA DC input.

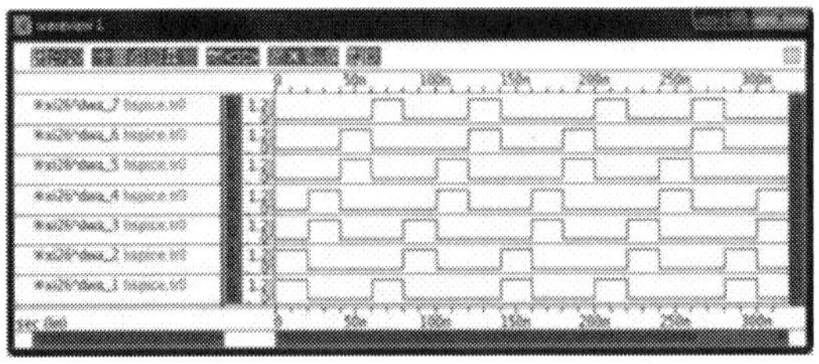

Figure 7. Generated DWA data for the DC input).

As a first major process in SDWA, it requires the input to undergo data weighted averaging (DWA). As shown in Fig. 7, DWA distributes the usage of the unit elements through cyclic assignment. Since the input has 2 bits in high for 16 clock periods, the output does the same; only that the assignment is different. Through cyclic assignment of DWA, each bit will be used almost equal as the others. This algorithm makes the long term average usage of the unit elements equal. Output is less susceptible to the effect of mismatch due to errors are averaged as well.

For the purpose of analysis, we provide the DWA block shown in Fig. 2 and Fig. 3. In this way we can picture out and trace the logic level at each different block.

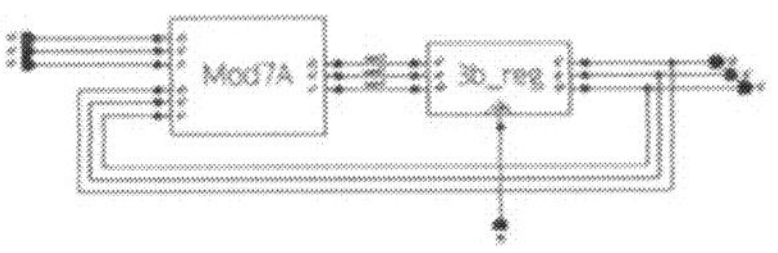

Figure 8. Accumulator.

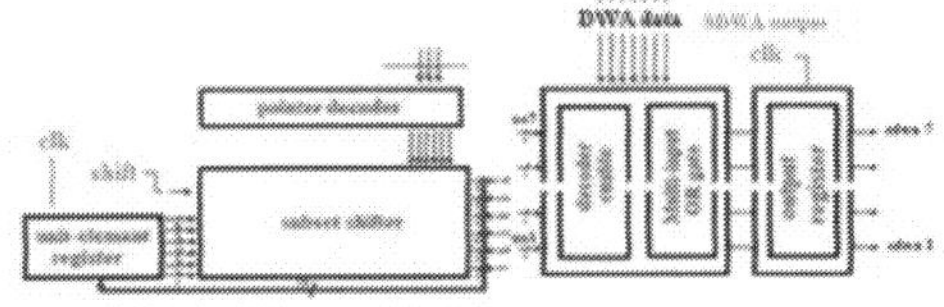

Figure 9. Complementing blocks.

The unit element registers inside the unit element register block are initialized to a particular position value defined by 3-bit data. What determines the position is the value of the 3-bit data itself. Initially, each 3-bit register is set to hold a distinct value from 001 to 111 at the starting time. As explained in the previous chapter, we implement this initialization by introducing delay from the supply to the preset or clear of that specific register. At the first thermometer input, the shift data of the subset shifter is not yet enabled. This does not provide any scrambling or rotation of the initialized position value; thus, the initial position is moved to the output and then is fed to the SDWA output block. Using the obtained DWA data at the first thermo input, un-shifted position data is now is used as the reference position of the SDWA out block at the input. At this period, DWA data is 0000011 and this enables the first two decoders. Since the first and second decoders have obtained its input from the un-shifted data of the unit element register which are 001 and 010 respectively.Using DWA data 0000011 will provide SDWA output of 0000011. It should be noted that at any clock period, each decoder at the input of SDWA out block must have a distinct position value (001 to 111). These position values will not reach the output unless enabled by the DWA data. When shift data of subset shifter is enabled, randomization is done while the SDWA output block maintains the long term equal element usage. We note that in every complete cycle of the thermo input, the subset shifter provides the randomization while the SDWA output block does the long term equal elements usage. For the rest of the clock cycles, we refer to Fig. 11.

At the initial condition, we do not allow any shifting in the logarithmic shifter. This can be realized by initializing the pointer register inside the accumulator at 000 (output). Having no shifting, the thermo input (0000011) is preserved at the output. It is to be noted that clocking will only start at the second period of the thermo input; so that the register will not yet store the output from the Modulo-7 Adder. Prior to the first clock cycle, the Modulo-7 adder has just read the output of the thermo-to-binary converter which is 010 (2 in decimal). At the first clock period, the output of the thermo-to-binary converter passes the modulo-7 adder having an addend of (000) and the output is loaded to the pointer register. Since this loaded value is equal to the first thermo input, the logarithmic shifter will provide shifting levels equal to the input. Doing this, the following bit/s will have a distinct assignment in position. In other words, it is not allowed to reuse the same bit position. Any incoming data will be accumulated but the output only rotates from 000-111. Thus any overflow will just be looped back to the input to maintain DWA. The system is just like playing with cards and distributing them in cyclic way to number of people.

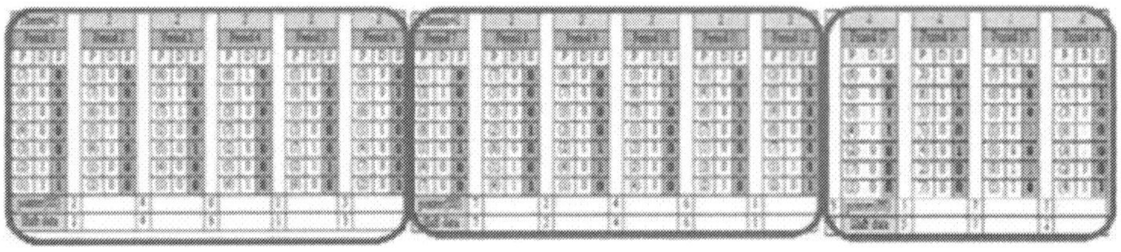

Figure 10. Output Verification (periods 1-16).

Logic level analysis is shown in Figure 11 for 16 periods of thermometer input. Note that P (position), D (DWA), and S (SDWA) are used as identifiers. Looking in the first period, position data are initialized from one to seven (001-111). At the same time DWA and SDWA are initialized to use the less significant bits. In period 2, position data is modified by enabling the subset shifter at level 2. This splits the position data from (1-2) and (3-7) and individually rotated in the sense shown in Fig. 12. The accumulator provides modulo-7 addition of the thermometer inputs to generate new pointer values. Pointer (SS)/shift data is a modulo-7 sum of all the thermometer inputs. Since our input is constant 2 thermometer levels, Pointer (SS)/shift data increases two each succeeding clock period but after reaching seven it loops back to one. The position value enabled by the DWA data becomes the SDWA output.

3.2. Result analysis 1

3.2.1. SDWA randomization (DC input – 16 periods)

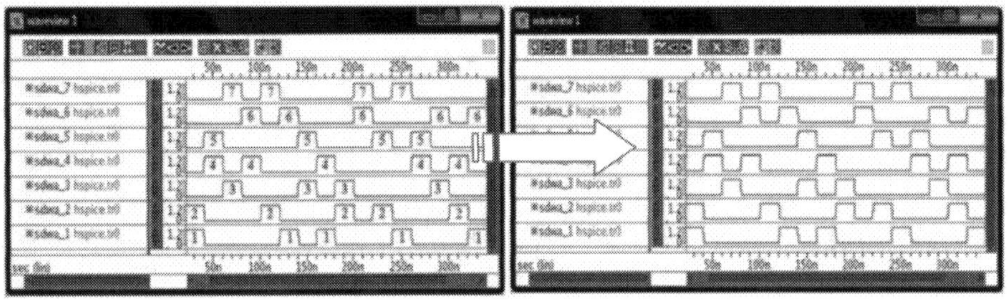

Figure 11. SDWA Randomization (DC input, 16 periods).

To show that SDWA provides randomization, we refer to Fig. 11. Looking at the figure, we label each bit in high with their respective bit placement. Collecting the numbers generated, we have: 12, 45, 37, 46, 27, 16, 35, 14, 23, 67, 25, 17, 45, 36, 24, 16. Given this 16-period sequence, it shows that elements are used without any particular pattern, or in other words, the usage is randomized.

3.2.2 SDWA Averaging (DC input – 16 periods)

To show that SDWA provides averaging, we refer to Fig.12 and Fig. 13 at 16 cycles having DC input.

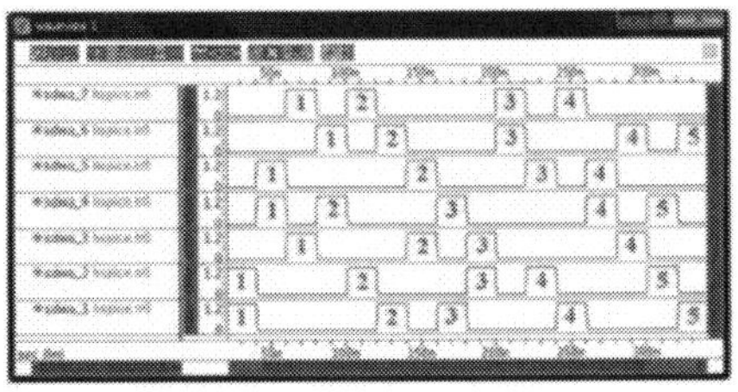

Figure 12. SDWA averaging (DC input).

	PERIOD															
BIT	1	2	3	4	5	6	7	8	9	10	11	12	13	14	15	16
[7]	0	0	1	1	2	2	2	2	2	3	3	4	4	4	4	4
[6]	0	0	0	1	1	2	2	2	2	3	3	3	3	4	4	5
[5]	0	1	1	1	1	1	2	2	2	2	3	3	4	4	4	4
[4]	0	1	1	2	2	2	2	3	3	3	3	3	4	4	5	5
[3]	0	0	1	1	1	1	2	2	3	3	3	3	3	4	4	4
[2]	1	1	1	1	2	2	2	2	3	3	4	4	4	4	5	5
[1]	1	1	1	1	1	2	2	3	3	3	3	4	4	4	4	5

Figure 13. SDWA averaging (DC input).

Figure 13 shows tabulated averaging obtained from Fig. 12,Generated Split-set DWA data (at 10pF load, 16 periods). At each period we count how many times the particular unit elements/bits are used for the 16-period sequence. As tabulated above, in every period, it is shown that the elements are equally used or differs only one bit from the other. A bit is not used twice unless all the others are used once. This characteristic is maintained throughout the whole period and thus for long term, we can say that it provides equal element usage.

3.2. SDWA simulation (time-varying thermometer input, 13 periods)

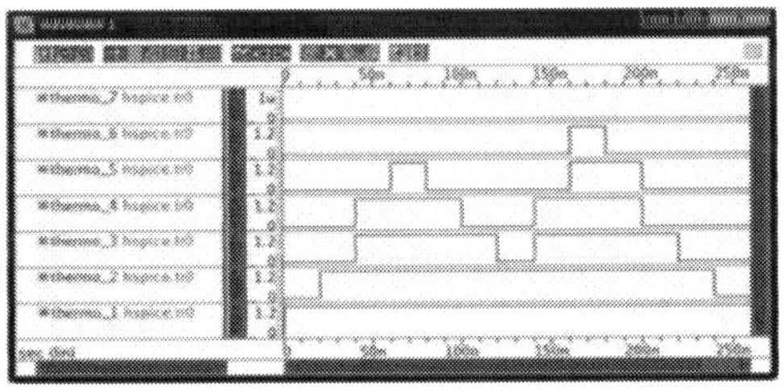

Figure 14. SDWA simulation (with time-varying thermometer input).

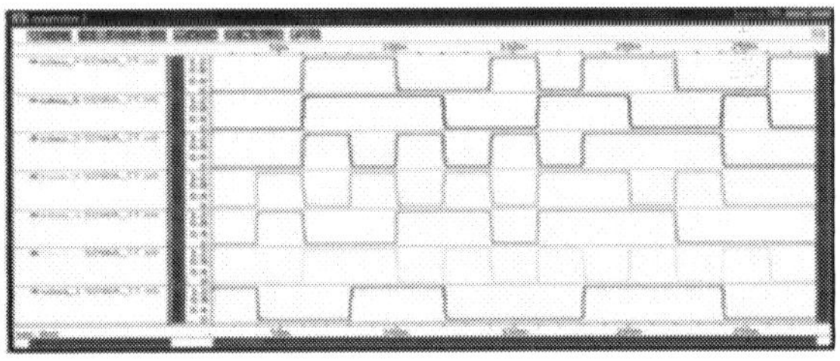

Figure 15. SDWA Output (with time-varying thermometer input).

Figure 14 shows a digital time-varying thermometer input for 13 clock periods. Having this digital input, SDWA output is shown in Figure 15. It can be observed that averaging is done since each unit element is used six times. On the other hand, randomization is also present since there is no particular pattern in element assignment.

3.3. SDWA simulation at three major corners (TT, FF, and SS) with time-varying input

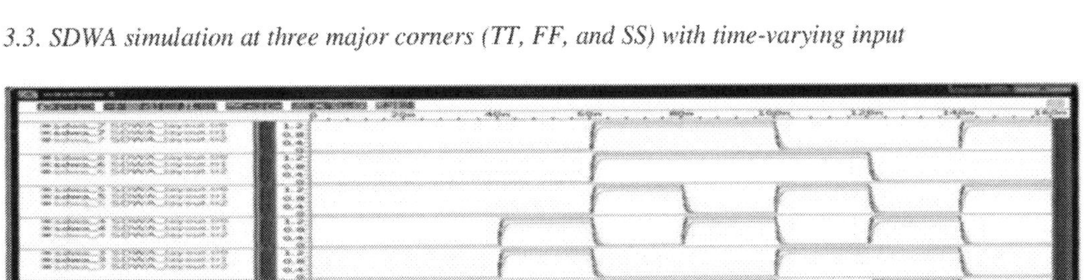

Figure16. Layout simulation at three major corners (TT, FF, SS) for 7 periods.

TT corner (red) is simulated at 25°C, with the normal supply level and using typical-typical Synopsys CMOS library. FF corner (blue) is simulated at 0°C, at the 1.1 of supply level and using fast-fast Synopsys CMOS library. SS corner (green) is simulated at 125°C, with 0.9 of the supply level and using the slow-slow Synopsys CMOS library. The three corner simulations provide the same output logic.

3.4.SDWA output simulation (schematic vs. layout)

Figure 17 shows the overlay mode of the schematic vs. layout output simulation. The graph verifies the logic equivalence of the schematic and the layout output. Graph colored in red is the schematic output simulation while the one in blue is the layout simulation. The pre-simulation and post-simulation as shown above verify the functionality of the design.

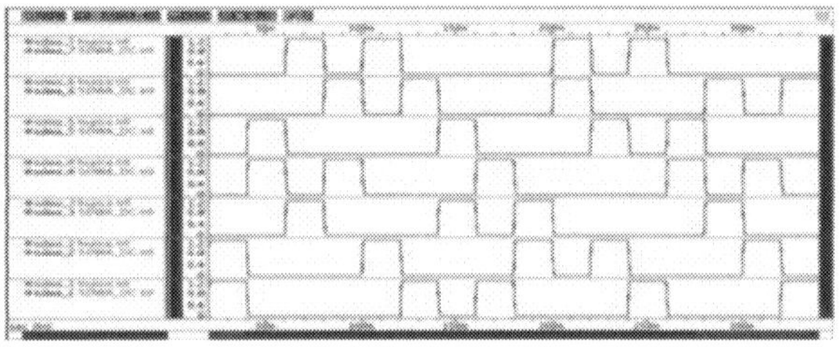

Figure 18. Output delay.

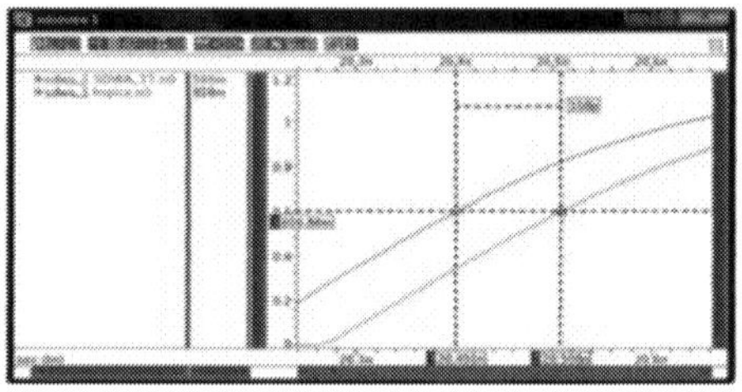

Figure 17. SDWA output simulation.

Figure 18shows the delay between the schematic and layout output with the time-varying discrete input. The layout output simulation is delayed by 108 picoseconds in reference with the schematic simulation.

3.5.SDWA output simulation (with and without the output register)

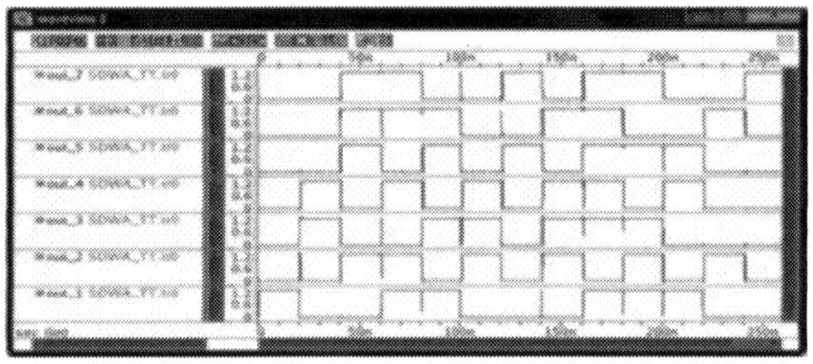

Figure 19.SDWA simulation (without output register block).

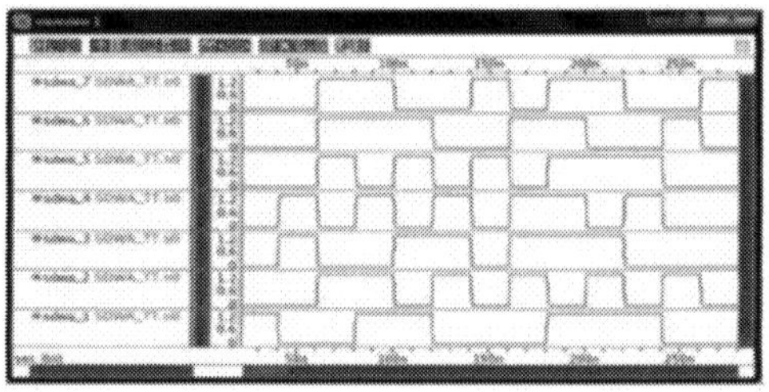

Figure 20. SDWA simulation (using output register).

Due to the randomization scheme of SDWA, unmatched delays are also randomized. These timing problems are clearly illustrated in Fig. 19. where the output are distorted by glitches and jitters. These are inevitable to some extent because randomization results to enequal or unparallelled delay between processes. The problem could extend more in the actual layout and which is very realistic. The researchers wish to filter or minimize these problems by introducing an output register.

The output register will store the data before it will be used in the next block. Doing this, the significant distortion is eliminated or atleast minimized significantly.In consequence of using an output register, the input to output response will be delayed at least one cycle of the clock. In Fig. 20, it is shown how the output is made better since of lesser distortions

3.6. SDWA output simulation (power consumption)

At typical NMOS-Typical PMOS corner (TT corner), the total power consumption is 566.348microwatts or approximately equal to 0 .566 milliwatts at 50 Megahertz clock frequency.

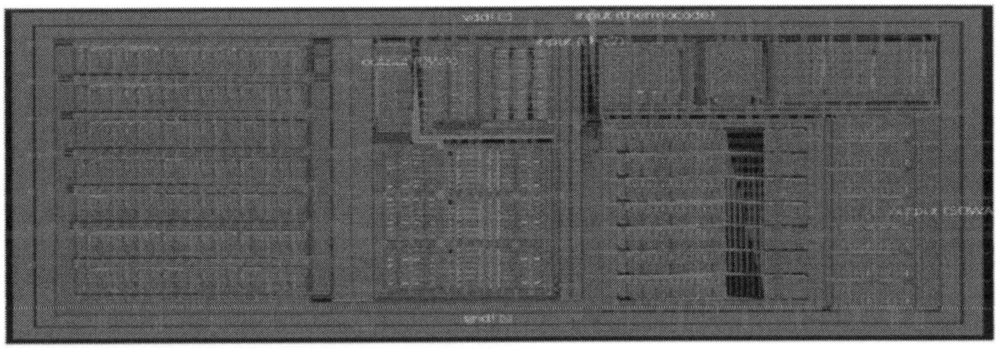

Figure 21. SDWA zoomed layout.

4. References

[1] Sall E, Andersson K O andVesterbacka M 2004 *Proceedings of the 6th Nordic Signal Processing Symposium* 137–40
[2] Pelgrom M J M, Tuinhout H P and Vertregt M 1998 *International Electron Devices Meeting* 915–8
[3] Wang R and Temes G C 2006 *Electronics Letters* **42** 248–9
[4] Bruce J W 2000 *Dynamic Element Matching Techniques for Data Converters* (Las Vegas: University of Nevada)
[5] Wang R 2006 *A Multi-Bit Delta Sigma Digital-to-Analog Converter* Dissertation, Oregon State University, Oregon, US

Acknowledgment

This work was supported by Department of Science and Technology – ERDT Fund, Philippines and MSU-IIT College of Engineering, Micro-Electronics Laboratory.

IOP Conf. Series: Materials Science and Engineering **79** (2015) 012018 doi:10.1088/1757-899X/79/1/012018

Characterization of POF for liquid level and concentration sensing applications

B G Lumanta[1], R T Candidato Jr and R L Reserva

Physics Department, College of Science and Mathematics, Mindanao State University-Iligan Institute of Technology,Iligan City, 9200, Philippines

Email: bridgetgwen@gmail.com

Abstract. Measuring liquid level and solution concentration play an important role in commercial and technological applications. For years, polymer optical fibers (POFs) have been very attractive for industrial applications because of their unique characteristics. In this work, we created simple, low cost and efficient set-up for sensing liquid level and solution concentration using POFs. We have calculated the acceptance angle of the POF to be 30° from numerical aperture (NA) measurements (NA≈0.500).Images of a single POF showed the presence of impurities within the fiber which can contribute to power loss of the transmitted light. Light leakage was also observed when the fiber was bent to a tight radius, i.e. beyond its minimum bend radius of 15 mm. The experimental results show that as liquid level increases, the output power decreases. Furthermore, when the liquid concentration was increased, its response showed a greater loss of optical power due to the light rays in the submerged region of the POF tend to be refracted out of the fiber instead of being totally internally reflected and transmitted when index of refraction of the surrounding liquid medium is increased.

1. Introduction

In the development of fiber optic technology, different researches had been conducted which are focused on suitable design of fibers. Some of the studies being done were directed to the designing of sensing systems with the use of these optical fibers, which led to the production of fiber-based sensing devices and components. At present, the development of sensors is remarkably responding to the needs and demands of consumers.

Nowadays, fiber optic technologies have been developed for application communication systems. However, the developments of high quality and competitively priced optoelectronic components and fibers have contributed to the growth of fiber optic technology in sensing applications [1,2]. In addition, low production cost and installation have contributed to the prevalent use of optical fibers in sensing. In recent years, optical fiber based sensors have attracted increasing interest owing to their inherent characteristics such as immunity to electrical noise, ease of miniaturization, the possibility of real time monitoring and remote sensing [3]. Additional advantages of fiber optic sensors in

[1] To whom any correspondence should be addressed.

comparison with conventional sensors are immunity to electromagnetic interference, multiplexing capability, distributive sensing and non-intrusiveness, lightweight, high sensitivity and spark-free [4]. These features make them a powerful tool for environmental and industrial process monitoring.

Liquid-level measurements in practice have been attracted by the splendid properties of optical fiber sensors. Level detection plays an important role in commercial and technological applications [5] such as monitoring of the water level in catchment areas, scrutinizing of illegal dumping of trash and pollution in canals, early warning of impending drought [4], monitoring of liquid level in oil tanks which is important in the petroleum and chemical industries [6]. Chemical industries widely require fiber-based liquid sensors since they are non-electrical and spark-free, and would not likely cause severe disasters when controlling volatile liquids.

Correspondingly important is the measuring and controlling of the concentration of solutions in chemical industries, sugar-manufacturing and pharmaceutics, to guarantee and improve product quality [7].The use of optical fibers in measuring solution concentration is a necessity to some industries since they are safe to be used in volatile liquids. Plastic optical fibers (POF) are also resistant to certain organic acids which allow them to be used to measure concentrations of some strong acids in water [8, 9].

In the present work, we characterized POFs made form Polymethyl Methacrylate (PMMA) core material and fluorinated polymer cladding material for liquid level sensing and concentration sensing applications. Single POF sensor system was constructed and utilized for the liquid-level sensing tests. Two single POFs were used to create the sensor arrangement—one was unstripped and the other fiber's cladding was stripped—to examine how would the stripping of cladding affects the response of the fiber sensor to the increasing level of liquid. Two fiber bundles were also made for the sensing of variation in concentration of glycerin solution. The prepared POF was then subsequently subjected for numerical aperture measurement and structure examination.

2. Experimental procedures

2.1. Numerical aperture measurement

Numerical aperture (NA) ofan optical fiber is the measure of its ability to gather light. NA was determined by allowing the laser beam pass into the fiber and measuring the angular dependence of the intensity of the transmitted light as shown in Figure 1. Measurement of the numerical aperture of POF was done four times using four varying lengths of single POF. The lengths of the fibers were 6inches, 12 inches, 18inches and 24 inches. The experiment was then carried out in a dark environment to assure that no stray lights aside from the source could contribute to the measurement to be obtained. Blue chromalase laser having 500mW power output operating at 488 nm served as the light source.

The laser beam was coupled into the single POF.The light which experiences total internal reflection was transmitted within the fiber until it reached the fiber end. This transmitted light was received by the sensor probe and the corresponding output power was then displayed in the screen of the Thorlabs Optical Power and Energy Meter detector. The far-end of the fiber was positioned at the center of the angular translator so that the distance between the far-end of the fiber to the sensor probe was equidistant at any angle. The single fiber was fastened to some wood blocks using an adhesive tape in order to ensure stability of the set-up. In this work, bending of the fibers along its length was avoided so that the light coming from the source travels in straight path and hence, maximum power was received by the detector and the intensity of transmitted light was measured at different angles. NA was calculated using this equation:

$$NA = \frac{1}{2}(x_2 - x_1) \qquad (2.1)$$

where the two values of x corresponds to the 5% intensity points. Numerical aperture was then determined using one half of the full angle between the two 5% intensity points.

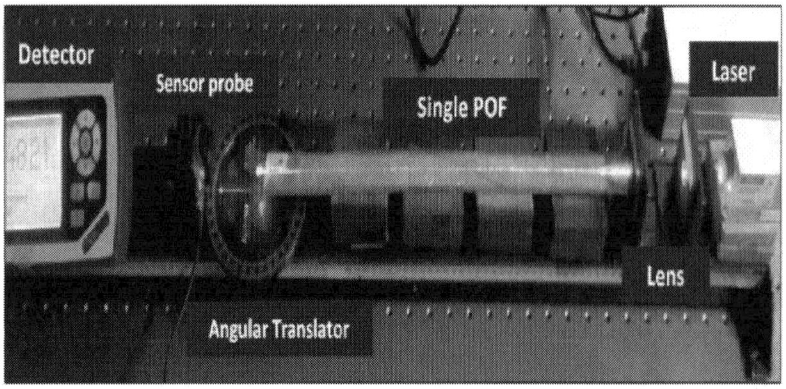

Figure 1. Experiment set-up for the measurement of the numerical aperture of the POF.

2.2.Assembly of the liquid-level sensor

The sensing capability of the liquid-level sensor was examined by immersing the POF in liquid and measuring the power transmitted through it as the liquid level was increased. The sensing performance of the concentration sensor system was evaluated through submerging it first, in distilled water, then in 20, 40, 60, 80, 100% v/V (v is the volume of glycerol and V is the total volume) concentration of glycerin solution. Glycerin is chosen in this study since it has important roles for various industrial applications.A design for liquid-level sensor is presented in this study as shown in Figure 2a. Figure 2b shows a simple sensor arrangement for liquid-level and concentration sensing and was constructed out of the characterized POF. This is based on the mechanism of total internal reflection in the optical fiber. The sensing performance of the fabricated fiber sensor system was installed and liquid level measurement was conducted at ambient room temperature.

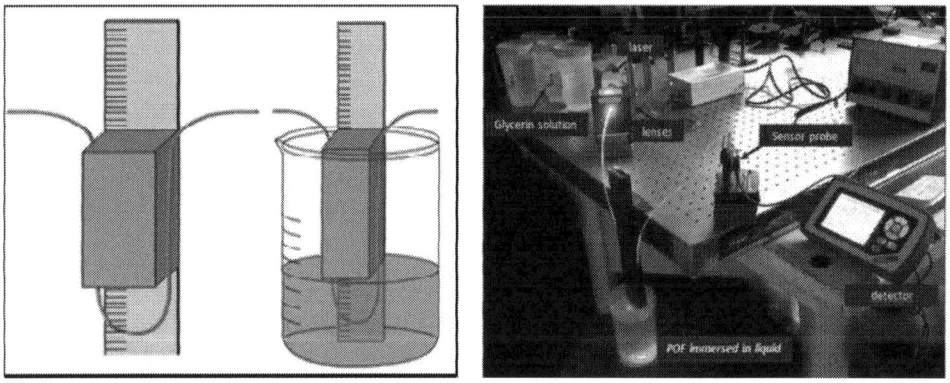

(a) (b)

Figure 2. (a) POF-liquid assembly while (b) is the whole set-up for liquid-level sensing and concentration measurements.

A container was filled with distilled water up to 100mm level. The light intensity transmitted through the fiber was first recorded before it was immersed in the water. The measurement of the intensity of the transmitted light was then recorded in every 10 mm level of the sensor. The set up for the measurement of the index of refraction of the liquid medium and the geometry of the coupling between the emitter and receiver was also installed. The bundled POFs are arranged parallel to each other and their tips are placed in front of a reflector (mirror). The light is launched into the transmitting fiber bundle and on emergence is then reflected by the reflector and is received by the receiving fiber.

The light intensity is sensed by a power meter to be displayed as output power in micro watts. The incident light on the reflector forms a cone of emittance from transmitting fiber. It is reflected back in the form of expanding cone of light. The cone diameter depends on refractive index '*n*' of the medium (liquid) and the separation between the reflector and the probe at any fixed position between probe and the reflector. If the space between the probe and the reflector is filled with a liquid of refractive index '*n*', then cone angle

$$\theta_\alpha = \sin^{-1}(\frac{NA}{n}) \tag{2.2}$$

here *NA* is the numerical aperture of the fiber and *n* is the index of refraction of the medium.

In evaluating the sensing performance of the concentration sensor, the sensor was first immersed in distilled water, then in different concentrations of glycerin solution, 20, 40, 60, 80 and 100% v/V, where 'v' is the volume of the soap and 'V' is the total volume of the solution. The power that was transmitted at the emitting fiber and received at the receiving fiber, then transmitted to the end of the receiving fiber was read by the power meter every time the solution concentration is varied. The separation distance between the tip of the probe and the reflector was varied to check how it will affect the light as read by the detector. The spacing between the tip of the probe and reflector is 10 mm and the readings were taken in successive steps of 5mm.

3. Results and discussions

3.1. Power output transmitted through POF

Figures 3a show the actual power of the transmitted light versus the sine of the angular position where the measurement was taken. The plot tells that the maximum power of light transmitted through the fiber was taken at the center orientation or where the fiber end and the sensor probe of the detector is in line with each other. It can be noticed that as the sensor probe was rotated in either clockwise or counterclockwise direction, the intensity of the transmitted light that was recorded decreases. Symmetry in the measurement was found with respect to the center.

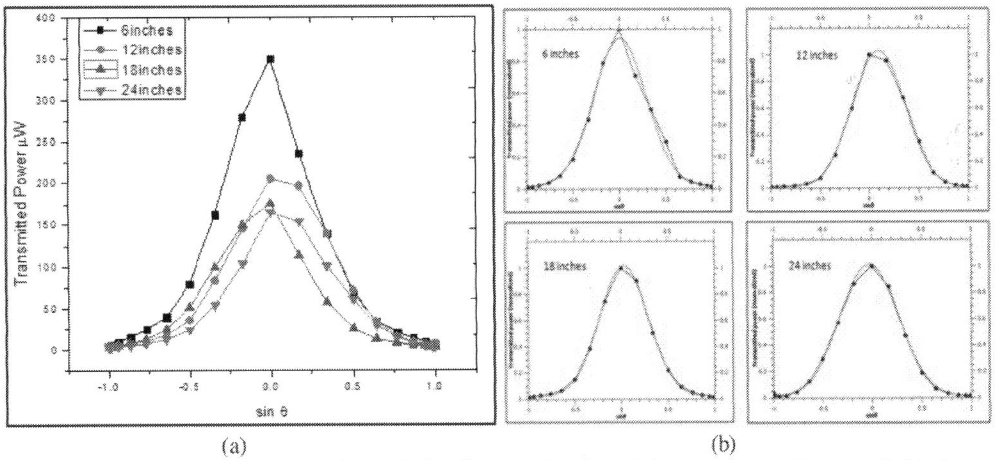

(a) (b)

Figure 3. (a)Transmitted power through the fiber vs the sine of the angular position and (b) shows the intensityof light (in normalized points) received by the detector as a function of the acceptance angle from the four single plastic optical fibers of different lengths.

It can be also observed that a decrease in the peak power taken at the center orientation as the length of the fiber is varied. It can be noticed that as the length of the fiber gets longer, the maximum

IOP Conf. Series: Materials Science and Engineering **79** (2015) 012018 doi:10.1088/1757-899X/79/1/012018

power of light transmitted through the fiber taken where the fiber end and the sensor probe of the detector is in line with each other, decreases. Since attenuation is directly proportional to the distance travelled by light, the longer the fiber, the farther the light has to travel, and the more the optical signal is loss. As light travels in the core, it interacts with the molecules in the core. This is attributed to the elastic collisions between the light wave and the molecules which result in Rayleigh scattering. If the light is scattered at an angle that does not support continued forward travel, the light is diverted out of the core and attenuation occurs. Optical power loss could also be because there is a reflection of minimal amounts of light in all directions as the signal is being transmitted through the fiber, some of the light goes out of the core, while some travels back to the source. Beer's Law, tells us that transmitted power decreases exponentially with propagation distance through the fiber. Hence, there is a greater decrease in the output power as the fiber's length is increased. The numerical aperture of the fiber is shown in Table 1.

Table 1. Calculated numerical aperture of the fibers.

	6" fiber	12" fiber	18" fiber	24" fiber
NA	0.561	0.505	0.583	0.584

3.2. Structure of the plastic optical fiber (POF)

Single POF was viewed at a closer look using a microscope. The image was then viewed in the computer through the Digital CCD Camera which is attached to the microscope. The structures of both bent and straight fibers were inspected. The fiber was first observed without bending while the light is passed through its length. Figure 4a shows the image, showing the presence of impurities and defects in the fiber. These impurities and defects cause light attenuation within the fiber. Attenuation occurs when there is absorption of the light that travels down the fiber which absorption is induced by the imperfections in the atomic structure of the optical fiber by the presence of oxygen defects or by the diffusion of hydrogen molecules into the fiber. Thus, in sensing, a decrease in output power can be accounted even without exposing the fiber to variations of the index of refraction of the surrounding medium. The output power read by the detector is lesser than that of the power emitted by the source as the light has been transmitted through the length of the fiber. Hence, it is necessary to measure the output power before the sensing tests be conducted so that the optical power loss that will be accounted during the sensing is just caused by the change of the refractive index of the medium where the fiber is being subjected.

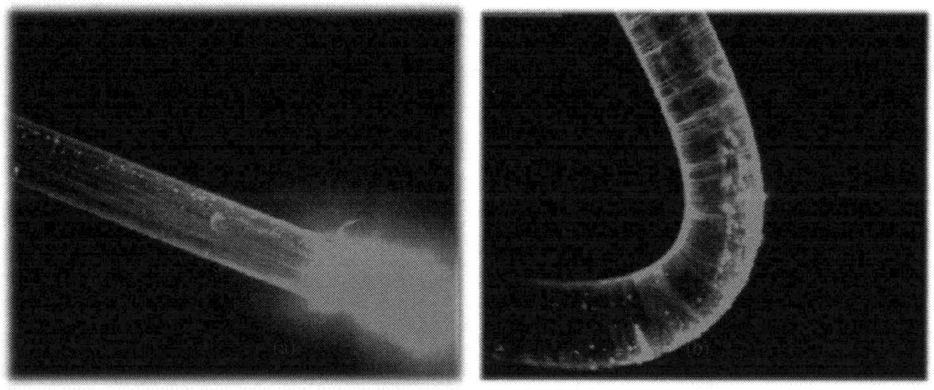

Figure 4.(a) Shows presence of impurities and defects and(b) shows leakage of light in the bent region of the fiber.

The plastic fiber was bent and the bent area was investigatedshown in Figure 4b. The defects are evident and are indicated by the presence of lines and dots. It can also be noticed that light rays tend to escape out of the fiber in the bent region. Thus we can say that there is loss of total internal reflection of light in the bent region since the bend curvature creates a very sharp angle for the light to be reflected back into the core, and some of it escapes through the cladding layer, causing attenuation. This attenuation rapidly increases as the radius is decreased to an inch or less. It is very important that in creating the sensor configuration, the fiber must not be bent beyond its minimum bending radius, which is 15 mm, so as not to cause core breakage and to avoid a large amount of power loss.

3.3. Liquid level sensor performance

Table 2 shows no apparent loss of total internal reflection (TIR) when the fiber was bent and stripped. Hence, it is fine to utilize these fibers in sensing since the loss of TIR that will be accounted in the sensing will just be due to the change of the index of refraction of the surrounding medium and not to the sensor's configuration. In this test, the sensor's response depends on the index of the core, the cladding layer, and the liquid in test. Several trials were done in this experiment and all show the same response.

Table 2. Power transmitted through the fiber *(**unstripped**: straight and bent; **stripped**: straight and bent).*

Fiber description	Output Power (µW)
Straight, unstripped fiber	317.1
Bent, unstripped fiber	316.9
Straight, stripped fiber	316.6
Bent stripped fiber	316.4

Figures 5 and 6 show the response of both the stripped and unstripped fiber sensors to increasing level of distilled water first, then to glycerin solution of different concentrations (20, 40, 60, 80, 100% v/V, where 'v' is the volume of the glycerin and 'V' is the total volume). All showed the same response, that is, the transmitted power accepted by the detector decreases as the level of liquid increases. The decrease in the transmitted power read by the detector at the output end of the fiber is caused by the loss of total internal reflection (TIR) as the liquid gradually covers the length of the fiber when its level was raised. When the fiber was immersed in the liquid and as the liquid level was increased by 10 mm increment, some of the light rays traveling through the fiber's length in the submerged region of the fiber were refracted, instead of being totally internally reflected.

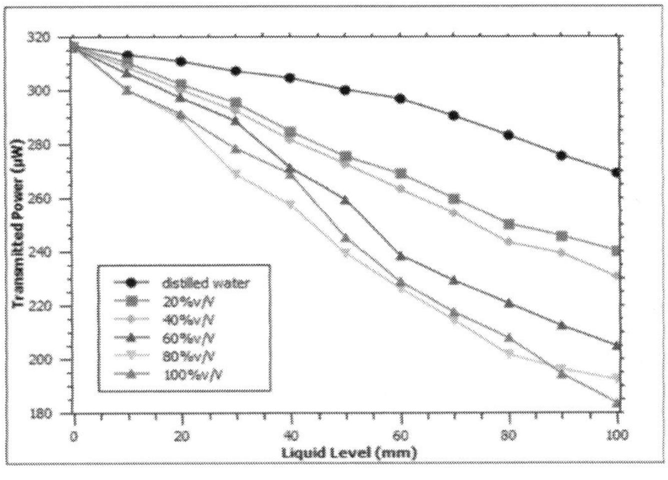

Figure 5. The response of single POF (*unstripped*) to increasing level of water and glycerine solution concentrations.

Table 3. The response of the single, unstripped fiber sensor in terms of the slope, ΔP/Δd.

Liquid in test	ΔP/ Δd (μW/mm)
Distilled water	-0.465
20% v/V	-0.803
40% v/V	-0.886
60% v/V	-1.195
80%v/V	-1.310
100%v/V	-1.366

Figure 5is the response of the unstripped fiber sensor to increasing level of liquid with different concentrations of glycerin solution.It can be deduced from the given plot above that as the concentration of the glycerine solution was increased, the transmitted power measured by the detector decreases as the liquid level increases. The decrease of the transmitted power per level per concentration is not proportionate. This might be because of the inhomogeneity of the solution. As presented in Table 3 is the unstrippedfiber's response to increasing level of liquid and increasing glycerine solution concentration in terms of the slope, ΔP/ Δd.The negative slope indicates that the output power measured by the detector at the fiber's end decreases as the liquid level increases. It can be noticed that as the concentration of glycerine solution was increased, the slope becomes more negative which means that as the level of liquid was increased with increasing concentration. There is a greater power loss experienced in the region of the fiber that was immersed in the liquid. This is because light rays tend to leak out from the fiber as the refractive index of the surrounding medium is increased.

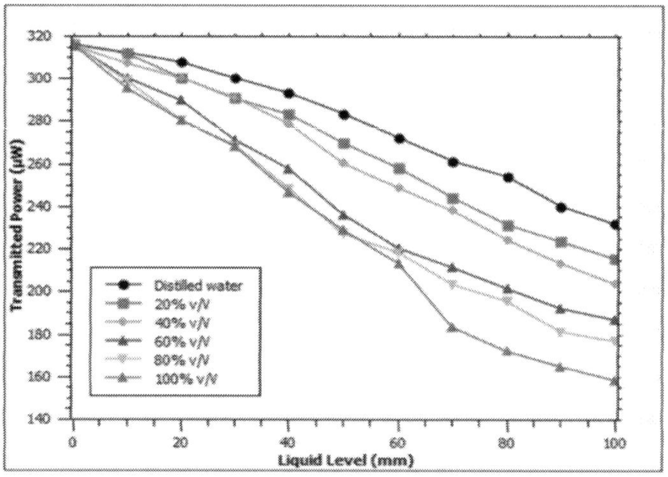

Figure 6. The response of single POF (*stripped*) to increasing level of water and glycerine solution concentrations.

In Figure 6, the response of the stripped fiber sensor to increasing level of liquid with increasing concentration of glycerin solutions is shown. There is also a decrease of the output power as the level of liquid increases. Furthermore, there is a greater power loss and hence a greater decrease of output power accepted by the detector as the solution concentration was increased. However, we can observe that compared to the unstripped fiber, the stripped fiber sensor has a better sensing capability since its response to increasing liquid level and increasing solution concentration in terms of the slope, as shown in Table 4, it is more negative than that of the unstripped fiber's response as shown in Table 3. This is because the bare fiber is more exposed to the change of the index of refraction of the

IOP Conf. Series: Materials Science and Engineering **79** (2015) 012018 doi:10.1088/1757-899X/79/1/012018

surrounding medium as its concentration varies. When the bare fiber is immersed in the liquid, light escapes out of the core in the submerged region since the liquid in test has a greater index of refraction than that of the fiber's core. Total internal reflection will only occur if the light strikes a denser medium and is approaching a less dense medium. In the stripped fiber, the air served as its cladding and TIR occurs since the air has a lower index of refraction than that of the core, that is $n_{air}=1<n_{core}=1.42$. But when the stripped fiber was immersed in the distilled water with n=1.33, greater than that of $n_{air}=1$ which served as the bare fiber's cladding, some of the light rays traveling along the fibers are refracted and escape out of the fiber when they pass on the submerged region. As the water was then replaced with glycerine solution of 20, 40, 60, 80, 100% v/V, where 'v' is the volume of the glycerin and 'V' is the total volume, the medium in which the fiber was submerged becomes denser, and hence higher and higher index of refraction. Thus, greater is the accounted optical power loss for higher concentration of glycerol, that is, the response of the single, stripped fiber sensor in terms of the slope, ΔP/Δd, becomes more and more negative as the solution concentration increases.

Table 4. The response of the single, stripped fiber sensor in terms of the slope, ΔP/ Δd.

Liquid in test	**ΔP/ Δd (μW/mm)**
Distilled water	-0.881
20% v/V	-1.073
40% v/V	-1.181
60% v/V	-1.363
80%v/V	-1.440
100%v/V	1.671

The response observed shows a linear response where the slope is negative; indicating that as there is an increase in the level of water, there is a corresponding decrease in the output power. The response of the fiber sensor is in accordance to the hypothesis. Theoretically and experimentally proven, the power received by the detector should decrease when liquid level rises, that is, if the index of refraction of the surrounding medium is greater than that of the cores.

3.4. POF as concentration sensor

A simple arrangement for sensing the change of solution concentration using the fiber bundles has been assembled. The ends of the fabricated fiber bundles were polished with a suitable grit (600-grit) to flatten and achieve uniform fiber bundle tips. We preferred to use fiber bundles so as to achieve an easy and better coupling of light into the fiber. One fiber bundle served as the light emitter, and the other as light receiver. As light from the source was emitted at the end of the emitter fiber, the power received by the receiving fiber, and the by the detector at its end is dependent on the refractive index '*n*' of the liquid. Thus, when concentration of the glycerol was varied, and since concentration and index of refraction are directly related, the transmitted power read by the detector also changes. Several trials were done in this experiment and all show the same responses.

When the probe was at close proximity of the reflector, a minimum output power is observed. The increase in separation distance between the probe and the reflector increases the light launched into the receiving fiber which results in rapid increase of output current and reaches to a maximum with steep front slope as presented in Figure 7. It can be deduced that at any separation the light reaching the receiving fiber is dependent on the concentration of solution. After reaching maximum, the light intensity starts decreasing for larger separation resulting in moderate back slope. The increase in the area of transmitting fiber cone with increasing separation and inverse square law caused the decrease in output power. The inverse square law states that the intensity of illumination is proportional to the inverse square of the distance from the light source. So, entire region of the receiving fiber tip is covered by the reflected beam at higher values of the displacement, and there is no change in the overlap area, then the fall in output power follows the inverse square law. Emitting fiber gets narrower, and hence the decrease in output power. But the response is opposite. This reverse response of the fiber bundle might be because of the arrangement of the fibers bundled together. It is possible that the light as it travels within the fiber bundle with the presence of the liquid surrounding it might just travel from one fiber to another fiber. As *n* of the solution increases with concentration, it might be that the light rays being refracted by one fiber is absorbed by the other fiber, and hence the increase in output power.

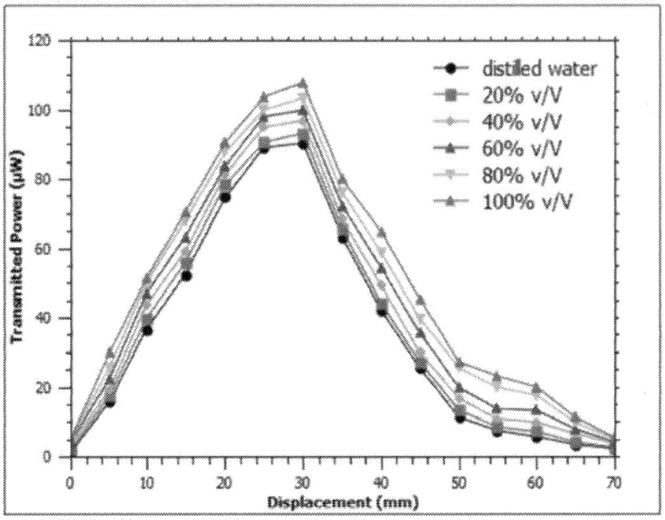

Figure 7. The output power vs. the displacement (distance between probe and reflector).

4. Conclusions

In this work, we have characterized plastic optical fibers of different lengths. We have successfully design sensor arrangements in liquid level and concentration sensing purposes and the performance were evaluated using distilled water and glycerin solutions of varying concentration (20, 40, 60, 80, 100% v/V).The calculated numerical aperture of the fiber is 0.5 and the fiber's acceptance angle is measured 30°allowing light to be transmitted through its length at light's entrance angle equal or less than 30° measure from the fiber axis of propagation. The micrograph images have revealed the presence of defects and impurities in the fiber. It was also found that the stripped fiber sensor has a better sensing capability than that of the unstripped fiber since it has higher response to increasing liquid level and increasing solution concentration. These results can be accounted because the bare fiber is more exposed to the change of the index of refraction of the surrounding medium as its concentration was varied. When the glycerol concentration was increased, it was found that the transmitted output power also increases as shown in the increment of peak power. Results revealed that an increase in index of refraction gavea corresponding increase of the transmitted output power.

5. References

1] Cigada, A. *Fiber-Optic Sensing*. http://www.ni.com/opticalsensing/.
[2] Lee, B. *Review of the present status of optical fiber sensors.* Opt. Fiber Technol. 9 (2003) 57-79.
[3] Vijavan, A., Fuke, M., Hawaldar, R., Kulkami, M., Amalnerkar, D., Aiyer, R.C. *Optical fibre based humidity sensor using Co-polyaniline clad.* Sensors and Actuators B 129 (2008) 106-112.
[4] Kuang, K.S.C., Quek, S.T., Maalej, M. *Remote flood monitoring system based on plastic optical fibers and wireless motes.* Sensors and Actuators A 147 (2008) 449-455.

[5] Musayev, E., Karlik, S.E. *A novel liquid level detection and its implementation*. Sensors and Actuators B 129 (2008) 106-112.
[6] Gao, J.Z., Zhao2, Y.L., Jiang, Z.D. *Design and performance characterization of a fiber optical sesnsor for liquid level monitoring*. Journal of Physics: Conference Series 13 (2005) 77-80
[7] D. Sengupta, M. Sai Shankar, P. Saidi Reddy, R. L. N Sai Prasad, K. SrimannarayaNA. *Performance of the reflected concentration sensor using plastic fibers.*Optoelectronics and Advanced Material–Rapid Communications Vol. 4, No. 8, August 2010, p. 1066 - 1069
[8] Institute of Microelectronics and Optoelectronics, Warsaw University of Technology, 75 Koszykowa Str. 00-662 Warsaw, Poland.*Intelligent Fiber* 69 *Optic Sensor for Estimating the Concentration of a Mixture-Design and Working Principle.*Sensors 2007, *7*, 384-399.
[9] Lomer,M., Echevarria, J., Zubia, J., Lopez-Higuera, J.M.*In Situ Measurement of Hydrofluoric and Hydrochloric Acid Concentrations using Plastic Optical Fibers.*
[10] Wang, A., Gunther, M.F., Murphy, K.A., Claus, R.O. *Fiber-optic liquid- level sensor.* Sensors and Actuators A 35 (1992) 161-164. [11] Sharma, M., Brooks, R.E. *Fiber Optics Sensing in cryogenic environments.* In: Proceedings SPIE Conference on Fiber Optics for Communications and Control, vol. 224, Washington, DC, USA, April 8-9, 1980, pg.46-52. [12] Spenner K., Singh, M.D., Schult, H., Boehnel, H.J. *Experimental investigation of fiber optic liquid level sensors and refractometers.* In: Proceedings OFS 1, vol. 1, London, UK, 1983, pg. 96-99.
[13] James, S.W., Khaliq, S., Tatam, R.P. *A long period grating liquid level sensor,* in: Proceedings 15th Optical Fiber Sensors Conference Technical Digest, vol. 1, Portland, OR, USA, May 6-10, 2002, pg. 127-130.
[14] Raatikainen, P., Kassamakov, I., Kakanakov, R., Luukkala, M. *Fiber-optic liquid level sensor.* Sensors and Actuators A 58 (1997) 93-97.
[15] Botaccini, M., Burani, N., Foroni, M., Poli, F., Selleri, S. *All-plastic optical-fiber level sensor.* Microw. Opt. Technol. Lett. 46 (2005) 520-522. 70
[16] Khijwania, S.K., Gupta, B.D. *Maximum achievable sensitivity of the fiber optic evanescent field absorption sensor based on the U-shaped probe.* Opt. Commun. 175 (2000) 135-137.

Acknowledgments
Heartfelt thanks to the members of Photonics Laboratory, MSU-IIT, Philippinesfor the great motivations of this work and to DOST-SEI for the scholarship grant to the first author.

IOP Conf. Series: Materials Science and Engineering **79** (2015) 012019 doi:10.1088/1757-899X/79/1/012019

Verification on the use of the Inoue method for precisely determining glomerular filtration rate in Philippine pediatrics

MJ D J Magcase[1], A Q Duyan[2], J Carpio[3], CA Carbonell[2,3] and J D Trono[2,4]

[1]St. Luke's College of Medicine - WHQM, Cathedral Heights, Sta. Ignaciana St., QC, Philippines

[2]Physics Department, De La Salle University, 2401 Taft Avenue, Manila, Philippines

[3]Nuclear Medicine Department, St. Luke's Medical Center, Cathedral Heights, Sta. Ignaciana St., QC, Philippines

Email: jade.trono@dlsu.edu.ph

Abstract. The objective of this study is to validate the Inoue method so that it would be the preferential choice in determining glomerular filtration rate (GFR) in Philippine pediatrics. The study consisted of 36 patients ranging from ages 2 months to 19 years old. The subjects used were those who were previously subjected to in-vitro method. The scintigrams of the invitro method was obtained and processed for split percentage uptake and for parameters needed to obtain Inoue GFR. The result of this paper correlates the Inoue GFR and In-vitro method (r = 0.926). Thus, Inoue method is a viable, simple, and practical technique in determining GFR in pediatric patients.

1. Introduction

Kidney disease causes more mortalities annually than cancers of the breast and prostate, it is currently the ninth most common cause of mortality in the United States and Asians are among the races which are at an increased risk for kidney disease [1]. Chronic Kidney Disease (CKD) is the principal cause of sickness and mortality in children worldwide [2]. Renal cancer among one of them is one of the most common causes of renal dysfunction for pediatric patients, along with nephritides and infections [3]. In alleviating these ailments, it is essential to have an efficient, safe, and practical diagnostic method due to the nature of the disease and the fragility of pediatric patients [4]. Such that, earlier diagnosis and suitable treatment will prevent further progression of renal disease and mortality [1].

Currently, the ninth leading cause of death in the Philippines is due to kidney failure, killing one Filipino hourly. Approximately 120 Filipinos per million populations suffer from failure of the kidneys, increasing the need for kidney transplants and dialysis [5].

Glomerular Filtration Rate (GFR) is the parameter used to determine renal function and to assess any problems in the renal system [6]. It is mainly determined by blood sampling, but urinalysis and imaging testing have gain distinction since 2009 [2].

[4] To whom any correspondence should be addressed.

There have been four prominent methods used in determining GFR, each employing either the use of endogenous or exogenous markers. These methods are: creatinine clearance, inulin clearance, in-vitro method and the camera-based method. Creatinine clearance being the method primarily used for hospital diagnosis [7]. In spite of this, this method is not the ideal indicator for estimating GFR [8].

Until recently, exogenous radioisotope markers have been used in nuclear medicine for determining GFR measurements. One such marker, Inulin, bears the distinction of an ideal marker [8].The inulin clearance was traditionally considered the gold standard for measuring GFR, but it is expensive, cumbersome and is only typically used for research rather than in clinical diagnostic purposes [8, 9]. The radiopharmaceutical 99mTc-DTPA has been found to have similar, if not equal qualities as of Inulin; qualities such as volume distribution, transit time, and plasma clearance which are the reasons why Inulin is considered an ideal marker [10]. Currently, the plasma disappearance method, or simply the In-vitro method, uses such a marker and has been recognized as the valid gold standard for nuclear GFR by the Pediatric Committee of the European Association of Nuclear Medicine [3]. Bloodletting in two to three separate occasions is necessary in the evaluation of nuclear GFR through the use of this method. This could be very traumatic, most especially for pediatric patients [11, 12].

Camera based methods, such as the Gates method and the Inoue method, combine the use of 99mTc-DTPA and a gamma camera to assess renal function, which may alleviate the dilemma of children foregoing bloodletting, and the ensuing trauma [11]. This has also been found to be a precise method for the evaluation of GFR independently of inulin and creatinine clearance, which are the methods primarily used for measuring GFR [9]. The forerunner for this camera-based method is typically that of the Gates' method, but inaccuracies have been found in measuring GFR in pediatric patients. Inaccuracies which have been resolved by Yusuke Inoue's camera-based method [11].

In differentiating the two camera-based methods; patient positioning, renal depth imaging tool andequations vary in acquiring renal depth and the process in acquiring the gamma counts from the scintigrams make them unique from one another. In acquiring renal depth, the Gates' Method makes use of the formulas by Tonnesen *et al.* paired with an ultrasound while the patient is in a seated position and as for the Inoue Method, formulas by Raynaud et al. for pediatric patients and Taylor *et al.* for adult patients using a CT scan in a supine-positioned patient [11]. CT scan use anatomical markers for reference and yields better measurements and renal depth estimation. The Region of Interests (ROI) is vital to image processing of the scintigram and before the ROI's are to be contoured it is important to distinguish the time the ROIs should be drawn, the shape of the kidney ROI, and the shape of the kidney background ROI. The Gates' method makes use of the renal scan from 5 minutes onwards, a time where the tracer has reached the renal pelvis, the non-functional part of the kidneys [11]. The C-shape for Gates' Method is drawn in order to compensate for the renal pelvis. It is notable that the C-shape omits a certain part of the functional part of the kidney this potentially underestimated GFR. The Inoue Method however, makes use of the 2.0-2.5 minute duration which is a functional phase where the radioisotope has not reached the renal pelvis. This time difference is mainly done to remove the ROI's compensatory C-shape by the Gate's method making the ROI shape fit a more natural contour of a bean since the 99mTc-DTPA has not yet reach the non-functional region of the kidneys.

The ROI shape to account for background correction is also different for both methods, asemilunar contour for Gates and perirenal contour for Inoue. A perirenal background correction shape provides a better correction for attenuated gamma rays that where inadvertently collected [11].

The objective of this study is to validate the Inoue method that it would be the preferential choice in determining GFR in Philippine paediatrics.

2. Materials and methodology

The measurement of In-vitro GFR was conducted by the Nuclear Medicine department of St. Luke's Medical Center, Quezon City, accordingly any form of patient interaction and handling of nuclear materials such as ^{99m}Tc-DTPA will not be of concern. Any form of consent is not necessary, since the consent form of St. Luke's Medical Center, Quezon City applies to any protocols that the patient will performing. The figurative data gathered from these procedures are considered to be thatof the hospitals property, thus the data could possibly be used for research or statistical purposes. Furthermore, amongst the data gathered names and any form of labelling that will single out or put at risk the patient's identity will not be included in any way or form.

In this study, thirty six (36) pediatric patients subjected to in-vitro GFR determination from 2011to 2012 were used in the study. Patient's scintigrams and general information (age, weight and height) were collected and analyzed. The subject's ages range from 2 months to 19 years of age with a mean age of 7.65 ± 6.98. Subjects have varying degrees of renal function in a 4.5 month data gathering period. Patient numbers 3, 6, 26 and 29 were excluded in the statistical determination. Since, Patient 3 and 6 have incomplete parameters necessary for the calculation of the Inoue GFR; missing left kidney and missing injection site, respectively. Moreover, patient 26 and 29 are missing their In-vitro GFR measurements, thus plotting them would be impractical. Data obtained will be used to determine if the measured results for in-vitro method will also hold true for the Inoue equation for estimation of GFR.

Retrospective in-vitro GFR measurements and respective general information such as age, weight and height was collected to be used in renal depth estimation and correlation. The scintigrams of the pre-injection syringe, post-injection syringe, injection site (1 min), injection site background (1 min), and kidneys was retrieved to be able to process the images for gamma (γ-) counts.

The eNTEGRA Workstation, a software which interprets the electrical signals from the gammacamera to digital scintigrams, was used to process the images into viable forms of data. Areas of the scintigrams mentioned above are measured for their γ-counts, either by whole area counting or selective area (ROI) counting. Whole area counting is done for the pre-injection syringe and post-injected syringe, thus the need to draw ROIs is not necessary. Then, the γ-counts were recorded to be used for the equations to estimate Inoue GFR. As for, selective area counting the scintigrams of the injection site and kidneys were used employing the parameters of Inoue. After adjusting for adequate intensity and contrast to properly distinguish the borders of the kidneys, the ROIs are drawn manually in contour to the borders of the kidneys. The shape of the kidney ROI (bean shape) and kidney background ROI (perirenal) using the compound image from 2.0-2.5 min are followed as indicated in the Inoue Method.[11] With the necessary shaping done, γ-counts and area of each kidney ROI are recorded and will be used to estimate GFR.

The retrospective data from previously gathered procedures of ^{99m}Tc-DTPA via in-vitro method for nuclear GFR measurement of pediatric patients was used as the reference values to correlate with the Inoue method and to validate the hypothesis that percent uptakes of the kidney are proportional to In-vitro GFR measurements. This study will implement the parameters and equations proposed by Yusuke Inoue [11]. Linear regression and t-test were used to determine the statistical significance of the two values correlation to each other. The r-values should approximate to +1 in order to substantiate a strong correlation and the p-value should reject the null hypothesis, thus should be greater than 0.05.

3. Results and discussion

The pre-estimated in-vitro GFR measurements were used as the reference value to estimate the relation of the calculated ^{99m}Tc-DTPA % uptakes of the kidneys. If proven to have adequate correlation, it can be assumed that the total % uptakes of the kidneys correspond to the gamma counts measured in the blood samples extracted from the in-vitro method.

The correlation of the measured total percent uptake and in-vitro GFR was plotted on a scatter plotand determined of its relation to each other. The r-value calculated is 0.916, indicating a strong correlation between total percent uptakes and In-vitro GFR (fig. 1). Thus, estimation of GFR with the use of scintigrams can be done with adequate precision.

Since the results in Figure 1 agree with hypothesis that % uptakes are proportional to the gamma counts taken from the blood samples from the in-vitro GFR method, the method and general equations of Inoue can be applied to this study. The Inoue GFR was calculated for each patient and was correlated using linear regression analysis with that of the reference method, in-vitro GFR. It was found that there is a strong correlation between the two GFR values having an r-value of 0.926 (fig. 2).

Furthermore, using t-test we assume that the null hypothesis (Ho) has no significant difference between the In-vitro method and the Inoue method. The p-value calculated is 0.9037, which is a large value for p (>0.05), which means that there is not enough evidence to reject the null hypothesis. In conclusion, there is no statistical difference between the In-vitro method and the Inoue method.

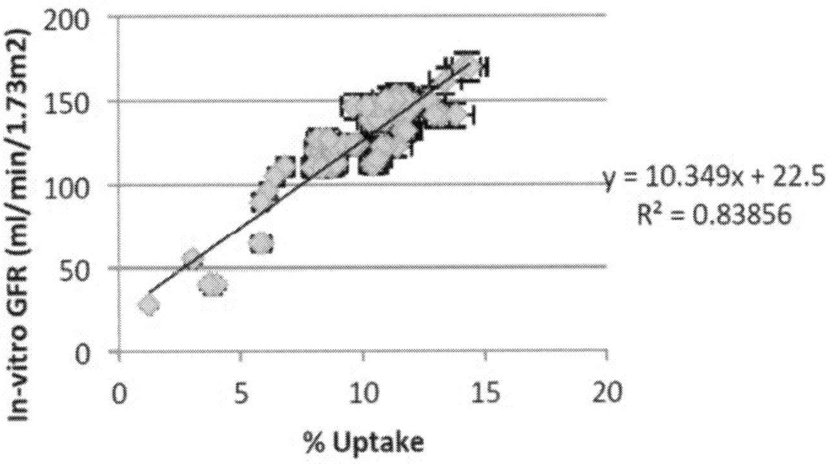

Figure 1. Relationship between total % uptake of the kidneys and Normalized In-vitro GFR measured by reference method (r = 0.916)

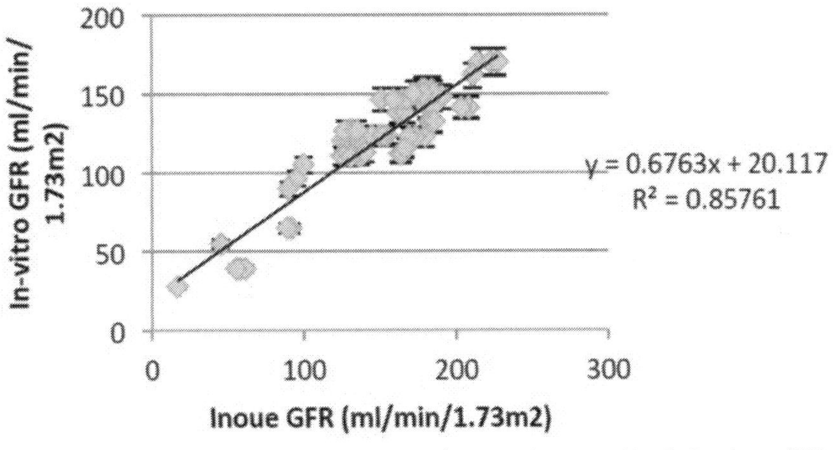

Figure 2. Relationship between Inoue GFR and Normalized In-vitro GFR measured by reference method (r = 0.926).

4. Conclusion

This study confirmed the hypothesis that total percent uptakes of the kidneys taken from the scintigrams are proportional to that of normalized In-vitro GFR with an r-value of 0.916, such that we can assume that GFR can be determined with the use of only the scintigrams via renal percent uptake, thus removing the need to withdraw blood. Estimation of GFR via camera based method in children has been validated due to the strong correlation (r = 0.926) and a large p-value (p = 0.9037) of the statistically calculated Inoue GFR with the reference value (In-vitro GFR). The strong correlation yields a precise value of the clinical gold standard. It is highly encouraged to use the camera based method by Inoue for pediatric patients, given its benefits from other GFR acquisition methods. The

Inoue method has a relatively quick processing time which is void of laboratory processing, has a lesstraumatic procedure where it is void of bloodletting, and is not subjected to deviation based on patients with lesser body surface area (BSA). This paper ascertains that the Inoue method is a viable method that satisfies the gold standard and is a reliable diagnostic tool in determining GFR in pediatric patients.

References

[1] 2013 *Fast Facts* Retrieved from http://www.kidney.org/news/newsroom/factsheets/FastFacts.cfm
[2] Kanitkar M 2009 *Medical Journal Armed Forces India* **65** 45–9
[3] Mattman A *et al.*2006 *Journal of the American Society of Nephrology* **17** 487–96
[4] Simon J, Amde M and Poggio E D 2001 *Cleve. Clin. J. Med.* **78** 189–95
[5] 2012 *Rise in kidney failure cause alarms health experts*Retrieved from http://www.interaksyon.com/article/%2033699/%20rise-in-kidney-failurecases-alarms-health-experts
[6] BrownS 2012 *International Journalof Biological and Life Sciences* **8** 248–53
[7] Sugihartono T, Nusi I A, Boedi S P, Kuumobroto H O and Umar K 2010 *Folia Medica Indonesiana* **46** 167–71
[8] Soares A A, Eyff T F, Campani R B, Ritter L, Camargo J L and Silveiro S P 2009 *Clin. Chem. Lab. Med.* **47** 1023–32
[9] ChoiJ D *et al.* 2010 *Eur. Urol.* **58** 900–5
[10] Macunluoglu B *et al.*2011*Int. Urol. Nephrol.* **43** 257–63
[11] Inoue Y, Ohtake T, Homma Y, Yoshikawa K, Nishikawa J and Sasaki Y 1998*J. Nucl. Med.* **39** 1784–8
[12] Inoue Y *et al.*2000 *J. Nucl. Med.* **41** 823–9

IOP Conf. Series: Materials Science and Engineering **79** (2015) 012020 doi:10.1088/1757-899X/79/1/012020

Polymer electrolyte system based on carrageenan-poly(3,4-ethylenedioxythiophene) (PEDOT) composite for dye sensitized solar cell

C A Ng and D H Camacho[1]

Chemistry Department, De La Salle University
2401 Taft Avenue Manila 1004, Philippines

E-mail: drexel.camacho@dlsu.edu.ph

Abstract. Poly(3,4-ethylenedioxythiophene) (PEDOT)-κ-carregeenan polymer electrolyte blend was prepared and incorporated as the electrolyte system in dye-sensitized solar cells (DSSC). Polymer blends prepared with different κ-carrageenan concentrations and molecular weights were investigated. It was found that the conductivity of the polymer blend increases with higher κ-carrageenan concentration, and lowers with degraded κ-carregeenan. The polymer blend was incorporated in a DSSC and yielded a solar cell with efficiency (η) of 0.421%.

1. Introduction

Dye-sensitized solar cells (DSSC) as next generation solar cells are promising alternatives to the conventional silicon-based solar cells owing to the low energy production requirement, low cost of raw materials, high configurational tunability and relatively good energy conversion efficiency [1].

The Gratzel configuration of the DSSC consists of a dye adsorbed on the surface of a nanocrystalline semiconductor oxide, an electrolyte solution and a platinum counter electrode all sandwiched between conductive glasses. This solar cell converts energy from sunlight through the photon excitation of the dye molecule ejecting electrons into the semiconductor oxide and to the external load. The ground state configuration of the dye molecule is regained by the donation of electrons from the electrolyte system through a redox process catalyzed by the platinum coating from the counter electrode [2].

The most efficient design of DSSCwith 11% energy conversion employs an electrolyte system composed of the redox couple I^-/I_3^- dissolved in volatile acetonitrile solvent [3,4]. However, its long term stability and performance is limited due to the use of low boiling liquid in the electrolyte system.To further improve DSSC for practical applications and commercial exploitation, studies have been done to replace the liquid system with other materials. Various materials studied for use in DSSC as alternatives to the liquid system includes room temperature ionic liquids [5-7] molten salts, organic and inorganic hole materials, and polymer electrolytes [8-13]. Among these alternatives, polymer electrolytes are attractive due to their unique hybrid network structure, negligible vapor pressure, high fabrication flexibility and thermal stability [14-17].Moreover, using polymer electrolytes allow better

[1] To whom any correspondence should be addressed.

electrode penetration, improving the long-term performance of the solar cell. Polymer electrolytes are composed of alkaline metal salts in a high molecular weight polymer with polar entities like amide, ether and ester linkages [18,19].These electrolytes have been studied and employed in various devices such as separation membranes, solid-state batteries and solar cells [1].

One conductive polymer widely studied today for applications in various electronic devices is poly(3,4-ethylenedioxythiophene) (PEDOT). It is a very promising material due to its high conductivity, good stability, good transparency, low oxidation potential and low band gap. The ion diffusion properties of PEDOT are better compared to other conductive polymers. It has found applications as antistatic coating, electrical coating, supercapacitor, organic light emitting diode, biosensor and in photovoltaic applications [20,21]. The inclusion of other negatively charged polymers as dopant to the PEDOT matrix has been shown to improve its conductive properties as well as the electrochemical, optical, mechanical and transport properties of the PEDOT composite. The quality of the film was also observed to be better with polyelectrolyte dopants [22,23]. This study explores the use of naturally occurring κ-carrageenan as the negatively-charged polymer dopant to PEDOT for application in DSSC.

2. Materials and methods

2.1. Materials
EDOT (3,4-ethylenedioxythiophene) monomer was obtained from Aldrich, pure κ-carrageenan was obtained from Shemberg Corporation. Standard DSSC materials including FTO coated glass (TCO22-7), ruthenium dye-sensitizer (Ruthenium 535-bis TBA), TiO_2 semiconductor oxide (Ti-Nanoxide D11111) and platinum solution (Platisol 41121) were bought from Solaronix.

2.2. Preparation of PEDOT and PEDOT-κ-carrageenan blends
PEDOT was prepared using 0.01M EDOT in acetonitrile, and polymerizing the monomer with the addition of 0.02M $FeCl_3$ oxidant. PEDOT-carrageenan blends were prepared with the monomer EDOT, κ-carrageenan solution and $FeCl_3$ oxidant. Different concentrations of κ-carrageenan solution (0.5%(*w/v*), 0.2%(*w/v*) and 0.1%(*w/v*)) were used to investigate the effects of κ-carrageenan on the conductivity of the resulting PEDOT blends. To investigate the effect of the molecular weight of the dopant carrageenan on PEDOT, blends with degraded carrageenan were also prepared. The degraded carrageenan was prepared by dissolving 2g of κ-carrageenan in 100mL deionized water and was heated to 80°C. The pH of the carrageenan solution was adjusted to 12 from an original pH of 8.7 in the addition of 2M NaOH. The degradation process was then initiated by adding 20mL 30% H_2O_2 and left to degrade for 4 hours[24]. The resulting degraded carrageenan was collected by precipitating it with ethanol and drying in an oven. The average molecular weight of the resulting carrageenan was measured using GPC.

2.3. Polymer blend measurements
Cyclic voltammetry (CV) characterization was performed using a Powerlab/4SP Potentiostat to study the electrochemical behavior of the PEDOT-carrageenan blends. This was done by electrodeposition of the polymer on FTO glass at 1500mV for 210s in a three electrode chamber with the FTO glass as the working electrode, Ag/AgCl reference electrode and platinum counter electrode.

The electrodeposited polymer on FTO glass was then ran through the potentiostat to measure the CV properties of the polymers. A three electrode setup was also used with the Ag/AgCl reference electrode, platinum counter electrode and the deposited polymer as the working electrode. Measurements were obtained by running through CV with a scan rate of 100mV/s in a 4mM $K_3[Fe(CN)_6]$ solution saturated in nitrogen.

Comparison of the PEDOT-carrageenan blend polymers with pure PEDOT was done by obtaining the infrared spectra of the polymers. This was carried out in a Nicolet 6700 FT-IR using KBr disc method. Differential scanning calorimetry (DSC), Perkin Elmer Differential Scanning Calorimeter

DTA7 with TAC7/DX Thermal Analysis Controller, was performed to record the glass transition temperature of PEDOT and the PEDOT-carrageenan blends. Around 10mg-20mg of samples were used for the measurements and were run from 35°C to 300°C. Conductivity measurements of the polymer blends were performed through the 4-point probe technique. The polymer deposited on the FTO glass was used for the tests. Samples were connected to the 4-point probe apparatus by wires attached to the surface of the sample.

2.4. DSSC fabrication and analysis

DSSC fabrication followed the traditional fabrication method. TiO_2 paste was placed and spread throughout the active area and was sintered at 450°C for 30 minutes. The photo anode was completed by immersing the FTO glass with the sintered TiO_2 in $3x10^{-4}$ M ruthenium (Ruthenium 535-bis TBA), solution in ethanol to allow the dye molecules to adsorb to the surface of the TiO_2 layer. The counter electrode was prepared byspreading a layer of Platisol throughout the surface of the FTO glass. The counter electrode was then fired at 450°C for 10 minutes. The solar cell was completed by sealing the two electrodes and the electrolyte system was introduced into the solar cell. DSSCs prepared were evaluated under Abet Technologies AM1.5 solar simulator to determine the efficiency of the solar cell. The solar cells were illuminated under $100mV/cm^2$ to get the voltage-current curve of the solar cell. From the I-V curve, the open circuit voltage (V_{oc}), short circuit current (I_{sc}) and fill factor (*ff*) were obtained to calculate the efficiency of the solar cell.

3. Results and discussion

3.1. Preparation of the PEDOT-carrageenan blends

The presence of counter ion dopant (e.g. PSS) for PEDOT establishesstability on the polymer. K-carrageenan with its well-spaced sulphate ester groups are potential polyelectrolyte dopant for PEDOT. Exploration of the blend between PEDOT and κ-carrageenan has not been well studied.

To investigate the effect of the κ-carrageenan molecular weight on the properties of PEDOT, the commercial carrageenan (MW_{ave} = 6.4 x 10^5) was degraded through oxidative degradation with hydrogen peroxide in alkaline condition. The hydroxyl radical attacks the glucosidic linkages in the carrageenan chain to afford the low *MW* carrageenan sample (MW_{ave} =1.5 x 10^4). The FT-IR analysis for the degraded κ-carrageenan revealed peaks similar to the original κ-carrageenan attributed to the absorbances of O-H (3427 cm^{-1}) and C-H (2955 cm^{-1}) bonds, polymer bound water (1644 cm^{-1}), sulphate stretching (1381 cm^{-1}), C-O-C bonds (1156 cm^{-1}) and C-O-S bonds (848 cm^{-1}) [25-26]. A pronounced reduction in the intensities of the glycosidic bonds (1072 cm^{-1}) can be observed owing to the degradation of the original κ-carrageenan.

The original and the degraded carrageenans were blended separately with PEDOTthrough the oxidation of EDOT monomer in the carrageenan matrix to afford the composites. FTIR spectra of the resulting PEDOT-κ-carrageenan blends were compared to pure PEDOT and pure carrageenan. PEDOT being an electrically conducting polymer showed no significant peaks owing to its conjugated nature undergoing geometric configuration [27].Comparison of the spectra of the composite with the pure κ-carrageenan shows shifts in peaks particularly the O-H stretching in carrageenan (3432 cm^{-1}) shifted to 3418 cm^{-1} for the PEDOT-carrageenan blend. Notable differences in the C-O-C signals for both spectra can be observed. The C-O-C cyclic ester asymmetric stretching at the 1184 cm^{-1} and the C-O-C $\alpha(1,3)$ linkage stretching at 700 cm^{-1} for pure carrageenan were observed at a shorter frequency in the spectrum of the PEDOT-0.5% carrageenan blend. The bathochromic or red shift can be consistently observed for other PEDOT-carrageenan blends (PEDOT-0.2% carrageenan and 0.1% carrageenan), regardless of the carrageenan concentration. These results show the two polymer chains interacting as evidenced by the red shift of the O-H and C-O-C peak absorptions due of the increased solvation of the polymeric oxygen lone electron pairs thus forming extensive H-bonding in the presence of H_2O allowing better interaction between the PEDOT and κ-carrageenan chains.

The FT-IR spectra of the PEDOT-degraded carrageenan blends were also compared with the spectrum of the degraded pure carrageenan to confirm the PEDOT blend formation. Similar observations were noted comparing the pure degraded k-carrageenan with the blue-colored composites of PEDOT containing varying % of degraded carrageenan (0.5%, 0.2%, 0.1%) indicating the interactions of the polymer chains in the formation of the blend. These observed bathochromic shifts of the OH and C-O-C absorptions are due to the interactions between the PEDOT chains and degraded carrageenan blends suggesting physical entanglements between polymer chains.

3.2. Surface morphology and elemental analysis
Surface morphology characterization of the chemically and electrochemically prepared samples was determined through SEM analysis (Figure 1).

Figure 1.SEM image of pure PEDOT film prepared (a) chemically and (b) electrochemically deposited on FTO glass.

The surface of chemically prepared PEDOT (Fig. 1a) showed defined shapes with microparticle deposits while PEDOT electrodeposited on FTO glass at 1500mV (Fig 1b) showed a formation of blob with thread-like deposits of PEDOT. The observed difference in the surface morphology is due to the different modes of polymerization resulting to their differences in morphologies.Figure 2 shows the SEM images of carrageenan and the PEDOT-carrageenan blends prepared electrochemically.

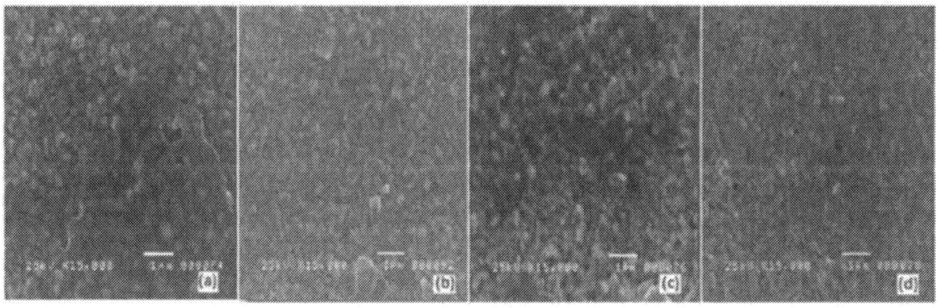

Figure 2.SEM Images of (a) pure carrageenan, (b) PEDOT-0.1% carrageenan, (c) PEDOT-0.2% carrageenan and (d) PEDOT-0.5% carrageenan electrochemically deposited on FTO glass.

The SEM image of pure carrageenan (Fig. 2a) shows a smoother surface compared to the electrochemically prepared PEDOT (Fig 1a). The PEDOT-carrageenan blends (Fig. 2-b-d) showed more similar surface morphology with pure κ-carrageenan than the pure PEDOT, where smaller, grain-like deposits can be observed. Although the surface morphologies of the blends are similar to pure carrageenan, the presence of the blue color of the blends confirm the presence of PEDOT in the carrageenan matrix.The SEM images of the chemically prepared PEDOT-carrageenan blends (Fig. 3.7) were observed to have similar surface morphologies with each other, having fine particles on the surface. Comparing both methods, the electrochemical method yielded a film with smoother surface because of the controlled conditions applied equally throughout the film.

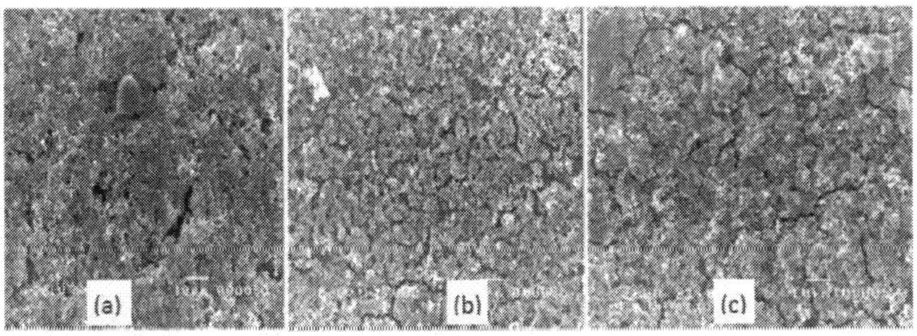

Figure 3.SEM Images of (a) PEDOT-0.1% carrageenan, (b) PEDOT-0.2% carrageenan and (c) PEDOT-0.5% carrageenan prepared chemically.

The SEM images of the composite consisting of PEDOT and degraded carrageenan in 0.1%, 0.2% and 0.5% concentrations (Fig. 4) were observed to have similar surface morphologies with the PEDOT-carrageenan blends, indicating that the incorporation of κ-carrageenan, whether degraded or not degraded, alters the film formation of PEDOT. The effect of adding whether the original or degraded carrageenan yielded a PEDOT-carrageenan blend with the same surface morphology.

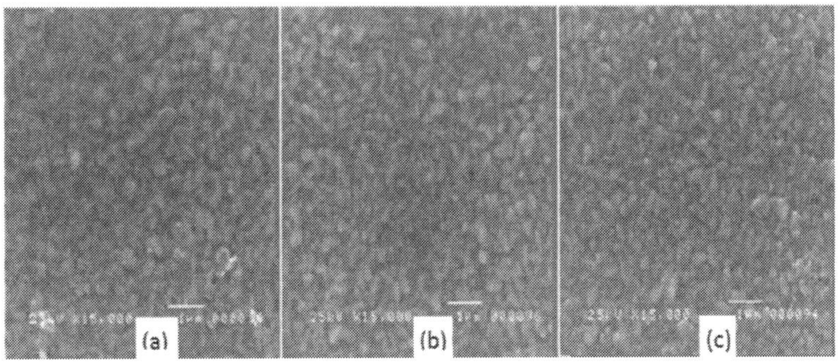

Figure 4. SEM Images of (a) PEDOT-0.1% degraded carrageenan, (b) PEDOT-0.2% degraded carrageenan and (c) PEDOT-0.5% degraded carrageenan deposited on FTO glass.

EDX analysis was also performed on the electrodeposited polymer samples to look at the elements present in the polymer samples to confirm the deposit of polymer. The EDX analysis of both the PEDOT-carrageenan and PEDOT-degraded carrageenan blends yielded the expected C, O and S elements.

3.3. Conductivity measurements

Conductivity of the polymer blends were tested to identify which blend would be the most conductive for incorporation into a dye-sensitized solar cell. Conductivity was determined through the four-point probe Van der Pauw method and through cyclic voltammetry (CV). Through the Van der Pauw method, the conductivity of the polymers was determined by measuring the voltage at constant current of 12 mA and measuring the thickness of the polymer film through SEM. The resulting conductivity was obtained by first calculating the resistivity of the film using equation (1) then theconductivity was calculated using equation (2). Table 1 shows the conductivity results.

$$s = \frac{\pi d}{ln2}\left(\frac{V_1}{I_1} + \frac{V_2}{I_2}\right) f(\frac{R_1}{R_2}) \quad (1)$$

$$\rho = \frac{1}{s} \quad (2)$$

Table 1. Conductivity of PEDOT and PEDOT-carrageenan blends.

Sample	V_1 (V)	I_1 (mA)	V_2 (V)	I_2 (mA)	Thickness (cm)	Conductivity (S/cm)
PEDOT	2.414	12.11	1.595	12.14	6.50E-05	10.42
PEDOT-0.1% carrageenan	5.439	12.21	1.255	12.20	5.14E-05	9.24
PEDOT-0.2% carrageenan	6.139	12.18	1.380	12.13	3.10E-05	13.61
PEDOT-0.5% carrageenan	2.163	12.11	1.524	12.11	4.51E-05	16.23
PEDOT-0.1% degraded carrageenan	7.030	12.08	1.580	12.10	5.38E-05	6.85
PEDOT-0.2% degraded carrageenan	4.319	12.03	3.486	12.05	4.18E-05	8.20
PEDOT-0.5% degraded carrageenan	3.555	12.10	2.298	12.10	5.04E-05	9.17
k-carrageenan	-	-	-	-	-	3.149

PEDOT was observed to have a conductivity of 10.42 S/cm, consistent to literature conductivity [28].Generally, the conductivities of the PEDOT-carrageenan blends were observed to be higher than the PEDOT-degraded carrageenan blends. The most conductive blend is the PEDOT-0.5% carrageenan composite exhibiting 16.23S/cm conductivity, which was higher than the conductivity of the electrically conducting pure PEDOT and pure κ-carrageenan. The conductivity of the polymer blends was seen to be related to the concentration of carrageenan, where by increasing the concentration of carrageenan increases the conductivity of the polymer blend.

CV measurements were also performed to look at the redox behavior of the polymers (Figure 5).

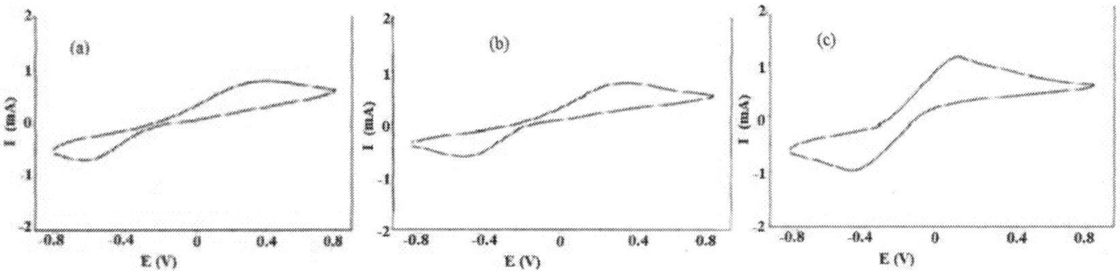

Figure 5.CV cycles of (a) PEDOT-0.1% carrageenan, (b) PEDOT-0.2% carrageenan and (c) PEDOT-0.5% carrageenan from -800mV to +800mV.

The CV cycles of PEDOT-carrageenan blends showed the reversible property occurring between the -800mV to +800mV range. The peak currents of the cycles were also seen to increase as the concentration of carrageenan increased, with the PEDOT-0.5% carrageenan blend having a cycle with the highest $I_{pa}$and I_{pc} values. This trend was observed to be consistent with the conductivities measured by the four-point probe method, since the peak currents of the PEDOT-carrageenan blends increased as concentration increased showing the PEDOT-0.5% carrageenan blend as the most conductive.

The CV profile of the PEDOT-degraded carrageenan blends on the other hand showed very narrow cycles and only the PEDOT-0.5% degraded carrageenan sample had an observable redox cycle. The differences in the conductivities and redox behavior of the PEDOT-carrageenan blends and the PEDOT-degraded carrageenan blends could be attributed to the ability of the long carrageenan chains to keep excess solvent in its matrix increasing its ability to transport ions. It has been observed that gelation of κ-carrageenan does not occurbelow a certain critical molecular weight[29].The concentration of the polysaccharide also affects the conductivity and redox behavior of the blend. The relationship between the concentrations of the polysaccharide;increasing the concentration of the conducting polysaccharide increases conductivity,was consistent with the literature [30].

3.4. *Thermal analysis*

Thermal analysis of PEDOT and the PEDOT-carrageenan blends(Table2) showed a general observation that incorporation of carrageenan to PEDOT increase the glass transition temperature (T_g). This is due to the incorporation of a high molecular weight polymer chain to PEDOT, which stiffens the PEDOT chain causing the $T_g$to increase.

Table 2. Glass transition temperatures (T_g) of PEDOT and PEDOT-carrageenan blends

Sample	T_g (°C)
PEDOT	64.431
PEDOT + 0.1% carrageenan	85.207
PEDOT + 0.2% carrageenan	77.683
PEDOT + 0.5% carrageenan	72.748
PEDOT + 0.1% degraded carrageenan	79.475
PEDOT + 0.2% degraded carrageenan	83.116
PEDOT + 0.5% degraded carrageenan	93.618

The decreasing trend of the observed T_g of the PEDOT-carrageenan blends as the concentration of κ-carrageenan increased can be attributed to the ability of the long chainκ-carrageenan polymerstotrapwater molecules into its matrices to form a gel, making the chains more fluid. The T_g of the PEDOT-degraded carrageenan composites, on the other hand, had an opposite trend, which can be attributed to the molecular weight of the degraded carrageenan. A study previously done by Souza, H. *et al.* [29]showed the relationship of the thermal properties of κ-carrageenan with the molecular weight. They were able to show that the gel forming property of κ-carrageenan is dependent on the molecular weight of the polysaccharide; below a certain critical molecular weight (MW = $3.0x10^5$), no κ-carrageenan gels are formed. Due to the limited ability of degraded carrageenan to form gels, itwould not be able to keep water in its matrix efficiently resulting to a more rigid polymer system, higher T_g.

3.5. *DSSC measurements*

Considering the conductivity and CV observations, the best candidate for the polymer electrolyte blend to be incorporated in DSSC is the PEDOT-0.5%carrageenan blend. The fabrication of the DSSC follows the protocol described by Solaronix [31].The PEDOT-0.5%carrageenan blend was introduced as a gel and sandwiched with the counter electrode and the anode assembly. This is referred in this study as unsealed DSSC. Another protocol, where the sides of the cells were sealed with the 100μm thick film sealant and the PEDOT-0.5%carrageenan gel was introduced through a hole previously made on the counter electrode. The hole was further sealed with the thermoplastic sealing material and a piece of glass to afford a sealed DSSC.

Efficiency of the solar cells employing the electrolyte system was evaluatedby irradiating under100mW/cm^2 AM 1.5G light source. A DSSC with the traditional iodide/triiodide liquid system was likewise prepared as a standard. TheI-V curves of the DSSCs are shown in Figure 6.

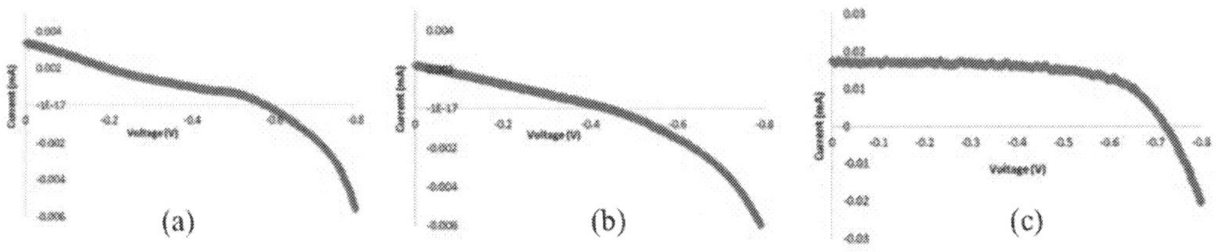

Figure 6.I-V curve of theDSSC employing PEDOT-0.5%carrageenan electrolyte system in (a) sealed configuration, (b) unsealed configuration; compared to the (c) control using traditional I^-/I_3^- liquid electrolyte.

From the I-V curves, V_{oc}, $I_{sc}$and fill factor were determined (Eqn. 3) and the efficiencies of the solar cells were calculated (Eqn. 4) (Table 4).

$$ff = \frac{V_{max} \times I_{max}}{V_{oc} \times I_{sc}} \quad (3) \tag{3}$$

$$\eta = \frac{V_{oc} \times I_{sc} \times ff}{P_{in}} \quad (4) \tag{4}$$

Table 4: DSSC efficiency data.

Electrolyte System	V_{oc} (V)	I_{sc}(mA)	*ff*	η (%)
PEDOT-carrageenan (Sealed)	0.565	3.37	0.221	0.421
PEDOT-carrageenan (Unsealed)	0.439	2.19	0.259	0.249
Iodide-triiodide (Sealed) Control	0.724	17.36	0.631	7.93

The DSSC employing the traditional liquid electrolyte system had an efficiency of 7.93% which is consistent with the literature (efficiency range between 7-11%). The DSSC with the polymer blend as electrolyte system was observed to have a very low efficiency of less than 1%. This low efficiency can be attributed to the very stiff configuration of PEDOT in the electrolyte system. Because of the stiffness of the PEDOT chains, electron transport is not as efficient compared to the liquid system, as seen from the I_{sc}values, where the cell with the liquid electrolyte recorded greater I_{sc} values than the PEDOT-carrageenan cell. The fill factors of the PEDOT-0.5%carrageenan DSSC were also lower than the I^-/I_3^-system. The low fill factor of the DSSC shows the high resistance of the system leading to internal loss of the current produced by the solar cell. Comparing the DSSC with literature employing various polymer electrolytes yielded DSSC with varying efficiencies ranging from PEO system with 0.01% efficiency [32] toPEPI-EO with an efficiency of 5.3%[33].

The electrolyte activity of the PEDOT-carrageenan system is based on the redox action of PEDOT, which in the presence of the carrageenan polymer matrix, undergoes charge hopping mechanism by the self-exchange of charges between the redox couple. However, basing from the CV data and molecular modeling, there is an observed preference for the reduced form of PEDOT such that the process is not 100% reversible. The CV peak current ratio in the PEDOT-0.5% carrageenan blend is slightly less than unity indicating that only some fraction of the reduced PEDOT in the forward scan are available for reoxidation on the reverse scan. There is in effect a kind of barrier that prevents efficient charge hopping mechanism. Inside the DSSC, various reactions are going on in the pico-second and millisecond scale that interfereswith the forward process of electron flow. Changing the electrolyte system specifically affects the dye regeneration process and the redox process. A stiff polymer electrolyte system although conductive, limits the ionic mobility,lowersthe rate at which the electrons could be used by the excited dye and decreases overall conversion in the solar cell.

4. Conclusion

Poly(3,4-ethylenedioxythiophene) (PEDOT) is a highly conductive polymer electrolyte utilized for many purposes; and conductivity improved with the addition of dopants. In this study, the natural polysaccharide κ-carrageenan was used as dopant and incorporated into PEDOT producing a more conductive polymer material. Increasing the concentration of κ-carrageenan resulted to an increase in the conductivity of the PEDOT-carrageenan blend. The utilization of a lower molecular weight carrageenan and using it as the dopant electrolyte causes the conductivity to decrease due to its inability to efficiently trap water in the polysaccharide matrix, resulting to a highly crystalline polymer material. The PEDOT-carrageenan blends, as polymer electrolyte systems in dye-sensitized solar cells show low efficiency in converting solar energy to electrical energy.

5. References

[1] Kong F T, Dai S Y and Wang K J 2007 *Adv. Optoelectron.* **2007** (75384) 1–13
[2] Wang B, Kang B, Wang P and Qui Y 2006 *Sol. Ener. Mat. Sol. Cells* **90** 549–73
[3] Chiba Y, Islam A, Watanabe Y, Komiya R, Koide N and Han L 2006 *Jap. J. Appl. Phys.* **45** L638–40
[4] Kuang D, Klein C, Ito S, Moser J E, Humphry-Baker R, Zakeeruddin S M and Grätzel M 2007*Adv. Func. Mater.* **154**
[5] Kubo W, Kambe S, Nakade S, Kitamura T, Hanabusa K, Wada Y and Yanagida S 2003 *J. Phys. Chem.* B **107** 4374–81
[6] Sakaguchi S, Ueki H, Kato T, Kado T, Shiratuchi R, Takashima W, Kaneto K and Hayase S 2004*J.Photochem. Photobiol.* A **164** 117–22
[7] Lee K M, Chen P Y, Lee C P and Ho K C 2009 *J. Power Sources* **190** 573–7
[8] Nogueira V C, Longo C, Nogueira A F, Soto-Oveiedo M A and De Paoli M A 2006*J. Photochem. Photobiol.* A: *Chem.***181** 226–32
[9] Kang M S, Kim J H, Won J and Kang Y S 2006*J. Photochem. Photobiol.*A: *Chem.***183** 15–21
[10] Li M, Feng S, Fang S, Xiao X, Li X, Zhou X and Lin Y 2007 *Electrochim. Acta* **52** 4858–63
[11] Huo Z, Dai S, Wang K, Kong F, Zhang C, Pan X and Fang X 2007 *Sol.Energ. Mat. Sol. Cells* **91** 1959–65
[12] Stathos E, Chen Y and Dionysiou D D2008 *Sol.Energ. Mat. Sol. Cells* **92** 1358–65
[13] Chatzivasiloglou E, Stergiopoulos T, Kontos A G, Alexis N, Prodromidis M and Falaras P 2007 *J. Photochem. Photobiol.* A: *Chem.* **192** 49–55
[14] Katsaros G, Stergiopoulos T, Arabatzis I M, Papadokostaki K G and Falaras P 2002 *J. Photochem. Photobiol.* A: *Chem.* **149** 191–8
[15] Karim M A, Song M, Park J S, Kim Y H, Lee M J, Lee J W, Lee C W, Cho Y R, Gal Y S, Lee J

H and Jin S H 2010 *Dyes Pigments* **86** 259–65
[16] Chen D, Zhang Q, Wang G, Zhang H and Li J 2007*Electrochem. Commun.* **9** 2755–9
[17] Yamaguchi T, Tobe N, Matsumoto D, Nagai T and Arakawa H 2010*Sol.Energ. Mat. Sol. Cells* **94** 812–6
[18] Ganesan S, Muthuraaman B, Matthew V, Madhavan, J, Maruthamuthu P and Suthanthiraraj S A 2008*Sol.Energ. Mat. Sol. Cells* **92** 1718–22
[19] De Freitas J N, Longo C, Nogueira A F and De Paoli M A2008*Sol.Energ. Mat. Sol. Cells* **92** 1110–4
[20] Hunholz D, MacDiarmid A G, Sarno D M and Jones W E2001*Chem. Commun.* 2444–5
[21] Du X and Wang Z 2003*Electrochim. Acta* **48** 1712–7
[22] Sonmez G, Schottland P and Reynolds J R 2005*Synth. Met.* **155** 130–7
[23] Döbbelin M, Marcilla R, Pozo-Gonzalo C, andMecerreyes D 2010*J. Mat. Chem.* **20** 7613–22
[24] Sun T, Tao H, Xie J, Zhang S and Xu X 2009*J. Appl. Polym. Sci.* **117** 194–9
[25] Fan L, Wang L, Gao S, Wa P, Li M, Xie W, Liu S and Wang W 2011*Carbohydr. Polym.* **86** 1167–74
[26] Tranquilan-Aranilla C, Nagasawa N, Bayquen A and Dela Rosa A 2012*Carbohydr. Polym.* **87** 1810–6
[27] Zhan L, Song Z, Zhang J, Tang J, Zhan H, Zhou Y and Zhan C 2008*Electrochim. Acta* **53** 8319–23
[28] Fabretto M, Hall C, Vaithianathan T, Innis P, Mazurkiewicz J, Wallace G and Murphy P 2008 *Thin Solid Films* **516** 7828–35
[29] Souza H, Hilliou L, Bastos M and Goncalves M 2011*Carbohydr. Polym.* **85** 429–38
[30] Ueno H, Endo Y, Kaburagi Y and Kaneko M 2004 *J. Electroanal. Chem.* **570** 95–100
[31] *Solaronix: Dye-Sensitized Solar Cells Assembly*Retrieved from http://www.solaronix.com/documents/dye_solar_cells_for_real.pdf
[32] Kang M, Kim J, Kim Y, Won J, Park N and Kang Y 2005 *Chem. Commun.* 889–91
[33] Benedetti J E, Goncalves A D, Formiga A L B, De Paoli M A, Li X, Durrant J R and Nogueira A F2010*J. Power Sources* **195** 1246–55

Acknowledgment
This work was supported by the De La Salle University Science Foundation, the University Research Coordinating Office (Grant # 50 S/U 311) andDOST-PCIEERD-GIA. The authors would also like to thank the Solid-State Physics Laboratory of De La Salle University particularly Dr. Gil Nonato Santos and Mr. Rey Correa.

IOP Conf. Series: Materials Science and Engineering **79** (2015) 012021 doi:10.1088/1757-899X/79/1/012021

Utilization of *nata de coco* as a matrix for preparation of thin film containing spin crossover iron (II) complexes

D Onggo[1], O K Putri and M Aminah

Inorganic and Physical Chemistry Research Group, Faculty of Mathematics and Natural Sciences, Institut Teknologi Bandung, Jl. Ganesha 10 Bandung 40132 Indonesia

Email: djulia@chem.itb.ac.id

Abstract. Spin crosover iron(II) complexes have potential applications as smart materials since the complexes show reversible transition between diamagnetic low-spin(LS) state and a paramagnetic high spin(HS) state under the application of temperature change, pressure or light irradiation. The complexes generally prepared as a powder compound isolated from direct reaction between aqueous iron(II) solutions with ligand in ethanol or methanol solution. For application as electronic molecular devices, the complex was prepared as a thin film using several matrixes derived from nafion, silica and other synthetic polymer. In this work, *nata de coco*, a natural bacterial cellulose polymer, has been utilized as a matrix for preparation spin crossover triazole iron(II) complexes. The morphology of the complex and the composition of elements on the surface of *nata de coco* have been explored using SEM-EDX analysis.

1. Introduction

Nata de coco is a jelly-like food product originates from the Philippines, produced by fermentation of coconut water using *Acetobacter xylinum*. It consists of natural cellulose fibers and commonly known as Bacterial Cellulose (BC). The BC fiber has excellent mechanical properties and high water holding capacity, high tensile strength and biodegradable; make it potentially applied in a wide range of polymer science researches. BC has also been modified using many organic polymer to improve toughness and water absoprtivity and it is known popularly as a biodegradable and environment-friendly 'green' composite [1-3].

The cellulose fibers in *nata de coco* form a unique three dimensional network with nano-size cavities that suitable for *in situ* metal deposition like Platinum, Palladium and silver nano particles synthesized from their complex solutions. The thin film material containing nano metal particles have been evaluated as fuel cell materials [4-5]. The ability of *nata de coco* as a matrix for preparation nano metal particles from their aqueous complex solution, inspired us to utilize it as a matrix for preparation transition metal complexes to produce a thin film.

Spin crossover iron(II) complexes are smart materials that undergo transition between paramagnetic high spin(HS) and diamagnetic low spin(LS) states that can be tune by external perturbation such as temperature, pressure or light irradiation[6]. The transition between the two states

[1] To whom any correspondence should be addressed.

is abrupt, reversible and generally accompanied by distinct color changed. This property offers promising opportunities for application in molecular sensors and electronic devices [7].

Several iron(II) complexes have been synthesized from iron(II) precursors such as iron(II) sulphate heptahydrate, iron(II) perchlorate hexahydrate and iron(II) tetrafluoroborate hexahydrate. Most of thesalts are very soluble in water or ethanol. The reaction of iron(II) salts with 1,2,4 triazole (H-trz) andamino triazole(NH_2-trz), the very well known ligands, produces solid complexes in powder form. To improve its functional devices properties, the complexes need to be synthesized in nanometer scale and prepared as a thin film.

Numerous matrixes have been investigated to prepare thin film materials: Nafion, a fluorinated polymer has been used to prepare Fe(H-trz)-nafion and Fe-(NH_2-trz)-nafion. At room temperature, the film containing the complexes is colorless and change to lilac at liquid nitrogen temperature. Nafionis anionic resin, therefore no other anion variation can be prepared using Nafion matrix [8]. A thin film of [Fe(H-trz)2(trz)](BF_4) in a silica matrix has been prepared and it has been reported that the original optical and magnetic properties of spin crossover materials are still observed in the film [9]. The utilization of *nata de coco* as a matrix to prepare spin crossover iron(II) complexes was considered in order to demonstrate that the natural polymer with unique molecular structure has a scientific contribution in preparation of thin film smart materials[10].

2. Experimental methods

The following chemicals: $Fe(BF_4)_2.6H_2O$; 1,2,4 triazole (H-trz = $C_2H_3N_3$); 4-amino, 1,2,4 triazole(NH_2-trz= $C_2H_4N_4$), ascorbic acid and argon gas were used without further purification. *Nata de coco*has been prepared using local coconut water and *Acetobacter xylinum* was obtained from local laboratory in ITB.

The preparation of *nata de coco* sheet was adopted from the work by Radiman and Yuliani[11]. *Nata de coco* sheet was cut into squares of 3cm x 3cm, and soaked for one hour inaqueous $Fe(BF_4)_2$ solution (30 mL, 0.1M) containing ascorbic acid (0.1g) to prevent oxidationof iron (II). After that, it was rinsed with water and methanol before it was soaked in amethanol solution of H-trz (30 mL, 0.3M) for another one hour. The sheet containing H-trziron(II) tetrafluoroborate complex was rinsed using methanol and dried overnight at room temperature to form a thin film. Similar procedures have been applied for the preparation of the film containing NH_2-trziron (II) tetrafluoroborate complex.

The magnetic properties were measured using Magnetic Susceptibility Balance Sherwood Scientific Ltd, operated at room temperature. Scanning Electron Microscope (SEM) JEOL JSM-6400 has been used to analyze the complex morphology. The atomic ratio has been carried out using SEM-Energy Dispersive X-ray analysis (EDX) JEOL-JSM 6510LA.

3. Results and discussion

A white sheet of *nata de coco* in 20 cm long, 20 cm wide and 4-5 cm thick can be simply obtained after 6 days fermentation process of the coconut water. At this moment, the amount of water inside the *nata de coco* is still extremely high. Therefore the polymer sheet needs to be pressed and dried atroom temperature to obtain *nata de coco* in form of a thin film. The SEM image of the film shows that*nata de coco* consists of cellulose fibers with average diameter of about 50 nm with its length of several micrometers. The fibers produce three dimensional structures as shown in Figure 1 with varied sizes cavity or holes. These holes are suitable for matrix in the preparation complexes.

When *nata de coco* sheets soaked in aqueous iron(II) solution followed by imersion in methanol solution of H-trz ligand, no change in color is observed. However this does not mean that the complex has not been formed inside the cavity of cellulose fibers. A distinct lilac color was observed on the drying process of the *nata de coco* sheet to produce a thin film. SEM image of the film containing this complex shows that the complex shape is long rod with average diameter around 100 nm and with various length from 0,5 up to 1 μm, as shown in Figure 2. The counter anion (BF_4^-) was detected from EDX data and the ratio of Fe:F was found to be 1:4. This value is in agreement with the complex formula reported as Fe(H-trz)$_2$(trz)] BF_4[9]. From this data it is known that one H-trz molecule in the complex has been deprotonated and became anionic trz.

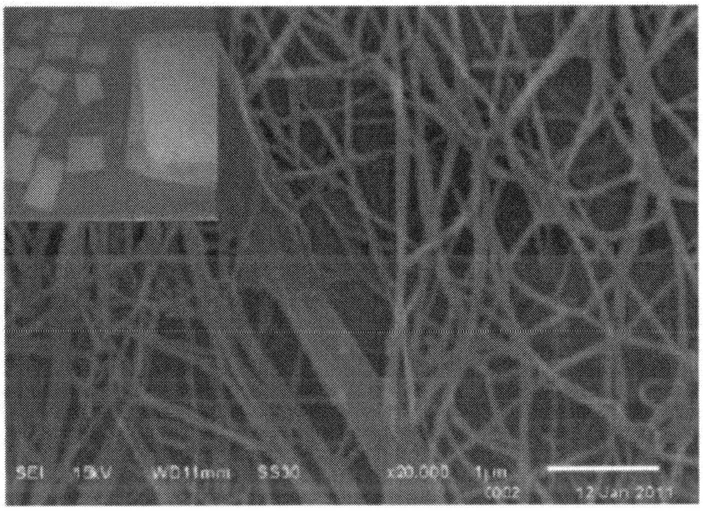

Figure 1. SEM image of *nata de coco* sheets, (inset) sheets form.

The preparation of *nata de coco* film containing NH_2-trziron(II) tetrafluoroborate complex produces a transparent film, which indicates that the iron(II) is in the high spin states. When the filmis immersed in liquid nitrogen, its color became lilac, as shown in Figure 3. The color changing of the complex from colorless to lilac demonstrates the reversible transition from high spin iron(II) to lowspin iron(II) in the complex film.

Figure 2. SEM image *nata de coco* film containing Fe(H-trz)$_2$(trz)]BF_4, (inset) thin film.

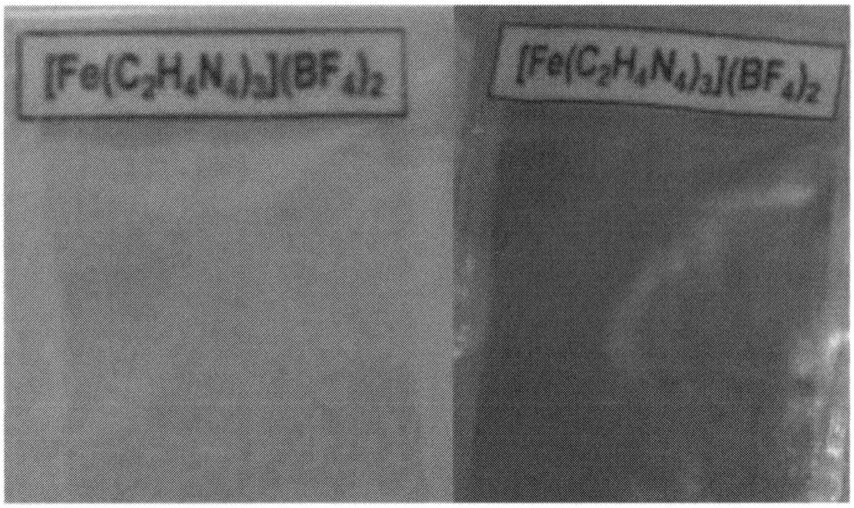

Figure 3.*Nata de coco* film containaing NH_2-trziron(II) complex, (a) at room temperature and (b) at low temperature.

The preparation of the complex from direct reaction of aqueous solution of iron(II) salt and H-trzligand without using *nata de coco* as a matrix, the solid product was obtained as a powder compound. The magnetic moment of this complex is 1 BM, which indicates that the iron(II) in the complex is in the low spin states. The SEM image of this complex showed that in the solid powder form, the long-rod shapes are overlapped each other to form irregular clusters as shown in Figure 4a.

When the H-trz ligand was replaced with NH_2-trz, the white powder of complex was obtained, indicating that the iron(II) in the complex was in the high spin sates. Magnetic properties of this compound was found to be 5 BM resulting from 4 unpaired electrons of iron(II). The molecular shapes of the complex are long rods with large slices between the rods as shown in the SEM image of the complex in Figure 4b. These larger shapes might be due to the amino substituen in triazole ligands.

(a) (b)

Figure 4. SEM image of iron(II) complexes with (a) H-trz and (b) NH_2-trz(inset) their powder form.

4. Conclusion

Based on our current works, it is found that *nata de coco* is a good matrix for the preparation of iron(II) spin crossover complexes. The hydrophylic properties of *nata de coco* make it easy to diffuse aqueous iron(II) solution inside cellulose fibers and anchored it via hydrogen bonding. The mostcrucial part of the preparation complex inside *nata de coco* is in the dehydrating process of *nata de coco* to form a thin film. When the drying is too fast the thin film will shrink and the film surface will not be smooth, because the water content in *nata de coco* is extremely high. The slow drying process will improve the film quality in terms of surface film. However the other unwanted chemical reaction like oxidation reaction of iron(II) might occur and reduce the quality of thin film.

5. References

[1] Shoda M and Sugano Y 2005*Biotechnology and Bioprocess Engineering* **10** 1–8
[2] Svensson A, Nicklasson E, Harrah T, Panilaitis B, Kaplan D I, Brittberg M and Gatenholm P2005 *Biomaterials* **26** 419
[3] Chen P, Cho A Y and Jin H J 2010*Macromolecular Research* **18** 309–20
[4] Yang J, Sun D, Li J, Yang X, Yu J, Hao Q, Liu W, Liu J, Zou Z and Gu J 2009*Electrochimica Acta* **54** 6300–5
[5] Evans B R, O'Neill H M, Malyvanh V P, Lee I and Woodward J 2003*Bisensors and Bioelectronics* **18** 917–23
[6] Gütlich P and Goodwin H A 2004 *Spin Crossover in Transition Metal Compounds, Topics inCurrent Chemistry*(Berlin: Springer-Verlag) pp 233–235
[7] Kahn O and Martinez C J 1998*Science* **279** 44–8
[8] Nakamoto A, Kojima N, Jun L X, Moritomo Y and Nakamura A 2005*Polyhedron* **24** 2909–12
[9] Faulmann C, Chahine J, Malfant I, de Caro D, Cormary B and Valade L2011 *Dalton Trans.* **40** 2480
[10] Onggo D, Real J A, Mulyani I, Syahbanu I and Aminah M 2011 *The Study of Thermal spincrossover tris amino triazole iron(II) complex in natural biopolymer nata de coco*Monograph of the XXIII International Conference on Coordination and BioinorganicChemistry;*New Trends in Coordination, Bioinorganic and Applied InorganicChemistry*ed M Melník, P Segľa and M Tatarko (Bratislava: Press of Slovak University ofTechnology)
[11] Radiman C and Yuliani G 2008 *Polym Int.* **57** 502–8

Acknowledgement

I would like to thank Henry Aritonang for the *nata de coco*preparation. This work has been supported by Riset Kompetensi 2009-2011 and Riset Inovasi ITB No. 254/I.1.C01/PL/2011.

The effects of precursor concentration and thermal annealing on the growth of zinc oxide nanostructures grown on silicon substrate

H M DPaculba[1], A C Alguno and R M Vequizo

Materials Science Laboratory, Physics Department, Mindanao State University - Iligan Institute of Technology, A. Bonifacio Avenue, Tibanga, 9200 Iligan City, Philippines

E-mail: paculba.hazielmarie@gmail.com

Abstract. This study focuses on the growth of Zinc Oxide (ZnO) nanostructures on SiO_2/Si(100) substrate via chemical bath deposition (CBD) with varying NH_4OH concentration and annealing temperature. The grown ZnO nanostructures were characterized via SEM-EDS for the surface morphology and elemental composition and UV-Vis spectroscopy for the reflectance measurement. Increasing the concentration of NH_4OH produced denser ZnO nanostructures composed of rods having smaller diameter. It is believed that at higher concentration of NH_4OH, more $Zn(OH)_2$ seed will act as nucleation site for ZnO formation which suggests higher probability of ZnO growth. Thermal annealing increased the average diameter of ZnO nanorods. Annealing provided enough energy for unstable atoms to rearrange into a more suitable position. This would result to larger rods that have been formed in expense of the smaller rods. Furthermore, it is confirmed in the UV-Vis spectroscopy results that ZnO nanostructures were successfully grown on SiO_2/Si(100) substrate. This successful growth of ZnO nanostructures is a promising material for solar cell technology.

1. Introduction

Man's continuous search for the advancement of science and technology has brought about the production of materials in the nanometer scale.These miniaturized materials have paved the way for significant enhancements in various fields such as biology and medicine, agriculture and business and the electronics industry. In the electronics industry, nanomaterial addressed tremendous issues such as increasing the capabilities of electronic devices while reducing their weight and power consumption. Nanostructured materials possess enhanced optical and photoluminescence properties that are not present in their bulk counterpart.

Among the renowned semiconductor in the research industry is zinc oxide (ZnO). It has received considerable attention owing to its capability to be formed into various morphologies such as nanowires[1] and nanorods[2]. It is a wurtzite-type II-VI compound semiconductor with a relatively large band gap of 3.37 eV at room temperature (RT) [3] and has a large exciton binding energy of approximately 60meV[4] which makes it suitable for wide range of electronic and optoelectronic

[1] To whom any correspondence should be addressed.

device applications such as light emitting diodes [5] and lasers [6]. Besides, its high electron mobility and high thermal conductivity are also two attributes of ZnO that adds to its promising applications in the semiconductor industry [7].

The crystal quality of ZnO nanostructures is highly dependent on the growth conditions, the growth techniques and the substrates used [8]. For instance, Mitra and Khan [9] reported that growth rate and quality of the ZnO film produced were found to be sensitive to the pH of the zincate bath. Moreover, Kim *et al.*[10] found out that the large lattice mismatch (40.1%) and large difference in the thermal coefficients between ZnO film and Si substrate [8,11] will result in the introduction of dislocations and built-in residual stress on the deposited film. However, the unparalleled dominance of Si in the semiconductor industry and the potentials of ZnO as a semiconductor paved the way to the fabrication of ZnO on Si(100) substrate. To address this problem, we utilized amorphous SiO_2 to act as buffer layer between ZnO and Si.

In this study, ZnO nanostructures were deposited on SiO_2/Si(100) substratesthrough chemical bath deposition. As mentioned earlier, the large lattice mismatch and large difference in thermal expansion coefficients between ZnO film and Si substrate would cause unfavorable defects. This inherent problem of direct growth of ZnO on Si substrate can be solved by introducing an amorphous SiO_2buffer layer. This SiO_2 layer, which is readily incorporated on the Si substrate, will serve as a buffer wherein ZnO particles can easily relax during the growth, thus reducing the strain on ZnO/Si interface and avoiding the introduction of interfacial dislocation. Thermal annealing of the grown samples is carried out to eradicate the impurities incorporated during deposition. The reflectance measurement of the samples is measured using UV-Vis spectroscopy and the surface morphology is examined by scanning electron microscopy-energy dispersive x-rayspectroscopy (SEM-EDX).

2. Experimental procedures

In this study, the substrate used is SiO_2/Si(100). The 4-μm thick SiO_2 layer is readily incorporated in the Si(100) substrate. The pre-treatment of the substrate strongly affects the depositionof ZnO. The SiO_2/Si(100) were first cut accordingly into its desired size. The substrates are then cleaned in distilled water, acetone and ethanol. Zinc oxide were deposited by chemical bath deposition. ZnO growth was carried out by suspending the SiO_2/Si(100) substrate in an aqueous solution of zinc sulphate heptahydrate ($ZnSO_4 \cdot 7H_2O$) and ammonium hydroxide (NH_4OH). The concentration of NH_4OH was varied to two concentrations (2M and 3 M). First, 60mL of zinc acetate solution and 60mL of NH_4OH solution were mixed in a beaker. The mixture is stirred at 360 rpm for 30 minutes. Immediately after stirring, the pH of the bath is determined using digital pH meter. The pH is controlled by dropwise addition of HCl to the bath. The cleaned substrates were immersed vertically inside the bath using improvised holders made of glass. Heat is then applied to the ammonium zincate bath and the temperature is maintained at 70°C by keeping it inside a water bath. When the bath temperature reaches to 70°C, the deposition of ZnO in the substrates will start. The deposition time is 5 hours of constant stirring and application of heat. Using distilled water, the removed substrates are washed and then dried in air for at least 2 hours. Samples are annealed in a furnace at 400° to improve the crystallinity.

3. Results and discussion

This chapter presents the results and discussion of the experiments. Thorough discussions on the growth mechanisms of the synthesized samples, their surface morphology and elemental composition as well as their optical properties are elucidated in this chapter.

3.1. NH_4OH concentration dependence

To know the effect of NH_4OH concentration of the deposited structures ontoSiO_2/Si(100) substrate, the concentration of NH_4OH in the solution was varied while controlling the pH of the zincate bath. Figure 1 shows the SEM images of ZnO prepared from 2M and 3M NH_4OH concentration at different magnification. Figure 1 (a) shows lesser density of ZnO nanostructures grown on the surface of the substrate. Higher magnification is shown in the inset photo. It is observed that the deposited ZnO nanostructures exhibit individual rod-like formation of the nanostructures with scattered sea-urchin like structures. The sizes of these structures vary from 3 to 10μm. It is observed that a single sea-urchin-like structure is composed of several numbers of nanorods. Each nanorod has well-defined hexagonal surfaces which is a characteristic of ZnO. The average diameter of ZnO nanorods is measured to be approximately 510nm with corresponding length ranging from 3 to 5μm. It can be noticed that smaller urchin-like structures accumulate on top of the larger ones.

Meanwhile, higher concentration of about 3M NH_4OH has been prepared to grow ZnO nanostructures. Figure 1 (b) depicts the SEM images of ZnO sample obtained from 3M NH_4OH concentration at various magnifications. It is observed that the presence of sea-urchin-like structure is very apparent. Denser ZnO nanostructures are obtained for this concentration. It is noticed that more numbers of sea urchin-like structures are found while lesser numbers of individual rods are present. The sizes of the sea urchin-like structures vary from 3 to 8μm. It is noticed that appreciable amount of flakes covered the nanorod structures (not shown). These flakes are believed to be impurities which could be eliminated via thermal annealing. It is clearly shown in the inset photo that the sea urchin-like structures consist of nanorods with an average diameter of approximately 470nm which is smaller compared to the rods obtained using 2M NH_4OH concentration. All nanorods exhibit hexagonal cross section.

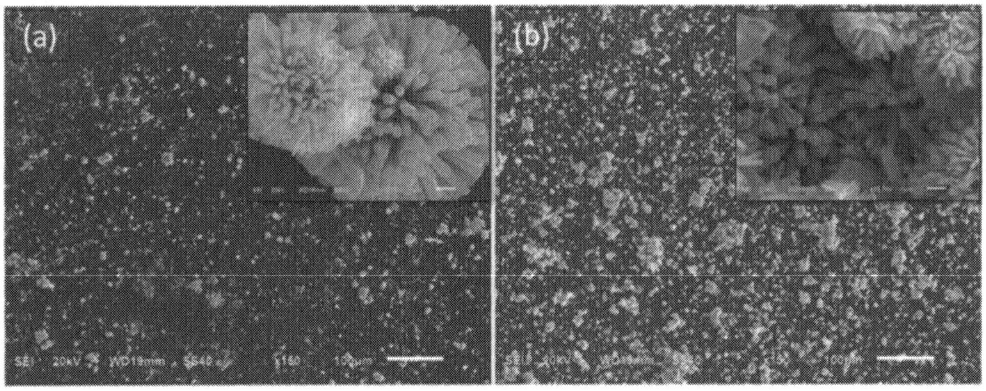

Figure 1. SEM micrographs of ZnO grown on SiO2/Si(100) substrate from 0.03M ZnSO4 and (a) 2M NH4OH and (b) 3M NH4OH with controlled zincate bath pH (11.2) at lower magnification and higher magnification (inset).

For 2M NH_4OH concentration, there are two possible ways in which ZnO will form as shown in the illustration presented in Figure 2. It is suggested that ion-by-ion growth might occur wherein ZnO will be formed through subsequent ionic reaction. The second growth process is the hydroxide cluster mechanism in which $Zn(OH)_2$ will adhere into the substrate and will attract ZnO forming sea-urchin like structure. In this concentration, individual rod formations and sea urchin-like structures co-exist as observed in the SEM images. During the addition of the ammonium zincate solution, $Zn(OH)_2$ precipitation could possibly reoccur leading to sea urchin-like structure formations. The presence of the small sea urchin-like structures that accumulate on top of the larger ones can be explained by the addition of ammonium-zincate solution after 3 hours.

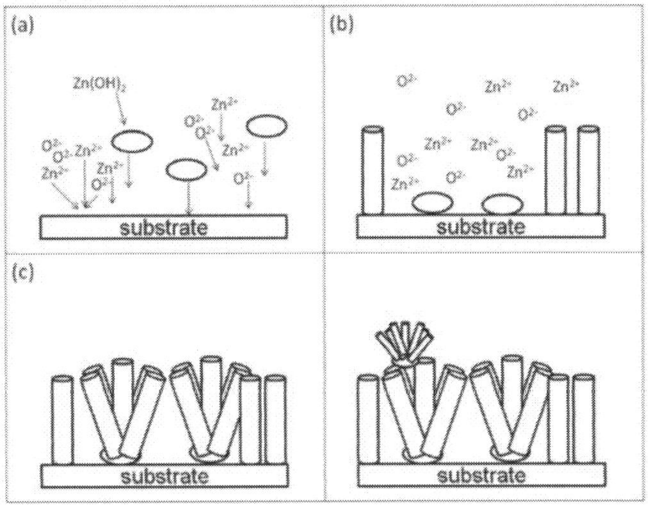

Figure 2. Growth mechanism of ZnO formation for bath with 2M NH_4OH concentration:(a) initial precipitation of $Zn(OH)_2$ ions in the solution, (b) $Zn(OH)_2$ adheres into the substrate and individual rod formation, (c) sea urchin-like formation with the nanorods and (d) accumulation of smaller sea urchin-like structure on top of the larger ones.

Meanwhile, for 3M NH_4OH concentration, more $Zn(OH)_2$ clusters are precipitated in the solution. Schematic illustration of the growth is presented in Figure3. It is believed that some clusters will adhere into the substrate while other clusters will dissociate to form Zn^{2+} and O^{2-} ions. $Zn(OH)_2$ clusters adhering into the substrate will attract ZnO atoms forming ZnO nanorods. Since large amounts of $Zn(OH)_2$ are attracting ZnO atoms, it will be distributed densely into the substrate.As a consequence, smaller rods will be formed. Again, the addition of the second solution after 3 hours might have caused the accumulation of smaller sea-urchin like structures over the larger nanostructures.

Different areas of the grown ZnO samples were considered and measured using EDS, i.e., an area with sea urchin-like structures and an area with no sea urchin-like structures. The EDS result reveals that the sea urchin-like structures are composed mainly of Zn and O which verifies that the produced samples are most likely ZnO. However, the area with no sea urchin-like structures has high peaks of Si and O. This can be explained by the presence of voids in the surface.

Figure 4 shows the reflectance of the samples measured through UV-Vis spectroscopy. Identical curves and peaks were observed for both concentration of NH_4OH with significant variation in the values of the reflectance between 200 to 400 nm. ZnO obtained from 3M NH_4OH concentration (labeled ZnO 3M) has lesser reflectance compared to ZnO sample obtained from 2M NH_4OH concentration (labeled ZnO 2M). Quantitatively, the intensity of the reflectance between 200 to 247.5nm reduces by about 5% and lessen to 3% in the 247.5 to 300nm range. However, at 300 to 380 nm, only 2% reduction can be observed. The decrease in the intensity of reflectance might be due to large amount of ZnO deposited unto the substrate at higher NH_4OH concentration which verifies that ZnO 3M is denser compared to ZnO 2M.

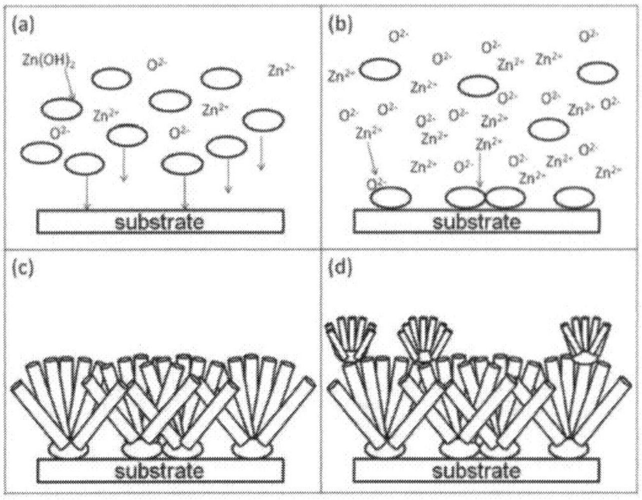

Figure 3. Growth mechanism of ZnO formation for bath with 3M NH_4OH concentration: (a) initial precipitation of $Zn(OH)_2$ ions in the solution, (b) $Zn(OH)_2$ adheres into the substrate and generation of free radicals into the solution, (c) sea urchin-like structure formation and (d) accumulation of smaller sea urchin-like structure on top of the larger ones.

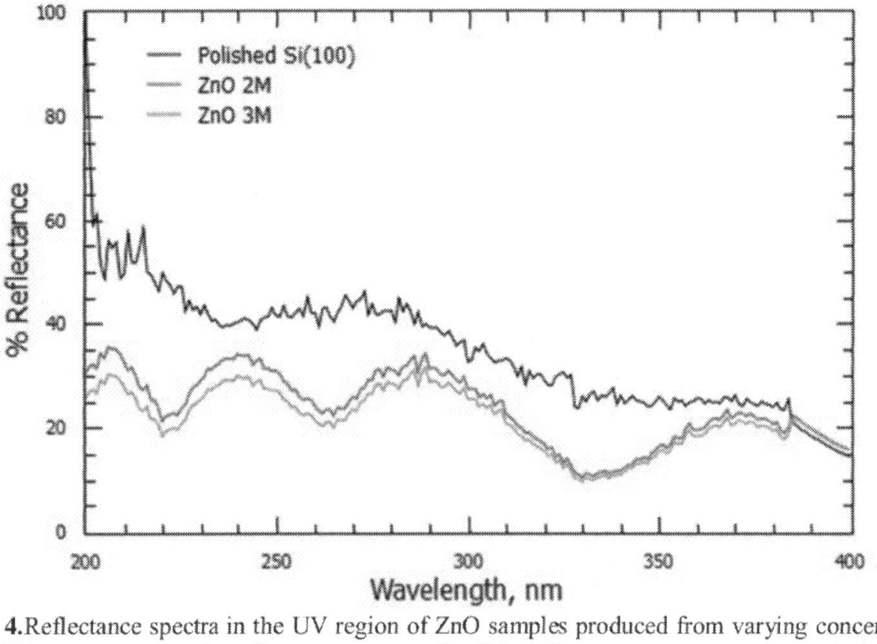

Figure 4. Reflectance spectra in the UV region of ZnO samples produced from varying concentration of NH_4OH as compared to the reflectance spectra of polished Si.

The density of the nanostructures deposited in the surface of the substrate will lessen the reflectance of the material since large area of SiO_2/Si(100) substrate which is more reflective than ZnO has been covered. Moreover, the reflectances of ZnO obtained are lesser compared to polished silicon. The difference in the reflectance intensity varies in the different region of the spectrum. At 206 nm,

20% reflectance difference is observed while at 220nm the reflectance differs by 28% as compared to the reflectance spectra of ZnO 2M. When compared to the reflectance spectra ofZnO 3M, the reflectance differs by about 25% at 206nm and about 31% at 220 nm. However, at 243 and 289 nm, the reflectance difference is only 6% while at at 265 nm, 282nm and 332 nm, the reflectance is approximately 17 %. The significant difference of the reflectance between the obtained sample and polished silicon is an indication that ZnO can be a good material for solar cell applications.

3.2. Effects of thermal annealing

Thermal annealing treatment is the most widely used method to improve the crystal quality of a material. It can affect the crystal structure, grain size and surface quality of thin films because it gives sufficient energy to allow the movement of atoms to occupy the correct site in the crystal lattice [12]. Moreover, annealing is done to eliminate the impurities present in the samples. The annealing time done in our samples is 2 hours and the annealing temperature is 400°C because at around 150°C the suspected impurities present in the samples such as water and $Zn(OH)_2$ will be decomposed [9]. To examine the effects of thermal annealing treatment, the sample is subjected to SEM-EDS and UV-vis spectroscopy.

Figure 5 shows the high magnification SEM images of ZnO deposited on SiO_2/Si(100) substrate from 0.03M $ZnSO_4$ and 2M NH_4OH annealed at 400°C for 2 hours as compared to an as-deposited sample. After annealing, it can be observed that nanorod formations and sea urchin-like structures are still present as shown in Figures 5 (b). The average diameter of the rods increases up to 783nm which is larger compared to the average diameter of the as-grown sample. After the heat treatment, the diameter of the rods is almost uniform compared to the as-grown sample shown in Figure 5 (a) wherein the diameter of individual rods is not uniform. These observations are thoroughly illustrated in Figure 6. It can be observed in Figure 5 (b) that rods tend to coalesce each other to form bigger rods. It is expected that extending the annealing time to longer period might cause the sea urchin-like structures to disintegrate because complete decomposition of the $Zn(OH)_2$ seed will take place.

Figure 5.SEM micrographs of ZnO grown on SiO_2/Si(100) substrate from 0.03M$ZnSO_4$ and 2M NH_4OH (a) as-grown and (b) annealed at 400°C for 2 hours.

Figure 6 (a) represents an as-grown ZnO sample before subjected to heat treatment. It is composed of irregularly-sized sea urchin-like structures and nanorods. It is believed that the smaller sea urchin-like structures and nanorods are unstable and will coalesce to other nanorods to form bigger-sized nanorods when applied by high annealing temperature as illustrated in Figure 6 (b). As mentioned above, this high-temperature annealing will provide enough amount of energy to allow the dissociated ZnO atoms to form into a more stable structure. This causes the uniformity and the increase of the diameter of the nanorods. This phenomenon is known as Ostwald ripening [13, 14]. Moreover, coalescence has taken place where two or more nanorods stick together. Figure 6 (c) is an illustration of the sample after annealing. It is assumed to be composed of thermodynamically stabe and pure ZnO rods with nanorod diameter larger by about 53% compared to the as-grown ZnO.

Figure 7 shows the reflectance spectra obtained from as-deposited and annealed ZnO samples shown in Figure 5. It can be observed that similar peaks are present. Furthermore, between 200 to 300 nm, it can be seen that there is a significant difference in their reflectances. The annealed ZnO shows lesser reflectance compared to the as-deposited one. Specifically, 5% reflectance reduction is observed between 200 to 245nm range. It decreases to 3% between 245 to 290nm but minimized to only 1% reduction between 290 to 400 nm. Thermal treatment at 400°C might have caused the decrease in the intensity of reflectance since some of the impurities such as H_2O and $Zn(OH)_2$ has decomposed. This suggests that the sample is purified after the heat treatment. Moreover, at 206 nm, 20% reflectance difference is observed between polished Si and the as-grown sample while at 220nm the reflectance differs by 25% as compared to the reflectance spectra of the as-grown sample. When compared to the reflectance spectra of the annealed sample, the reflectance differs by about 24% at 206nm and about 28% at 220 nm. However, at 243 and 287 nm, the reflectance difference is only 6% and 10%, respectively, while at 265nm and 333 nm, the reflectance is approximately 17% with respect to the as-grown sample. It is widely known that pure ZnO has lesser reflectance compared to Si as it can be used as window layer for solar cells [15].

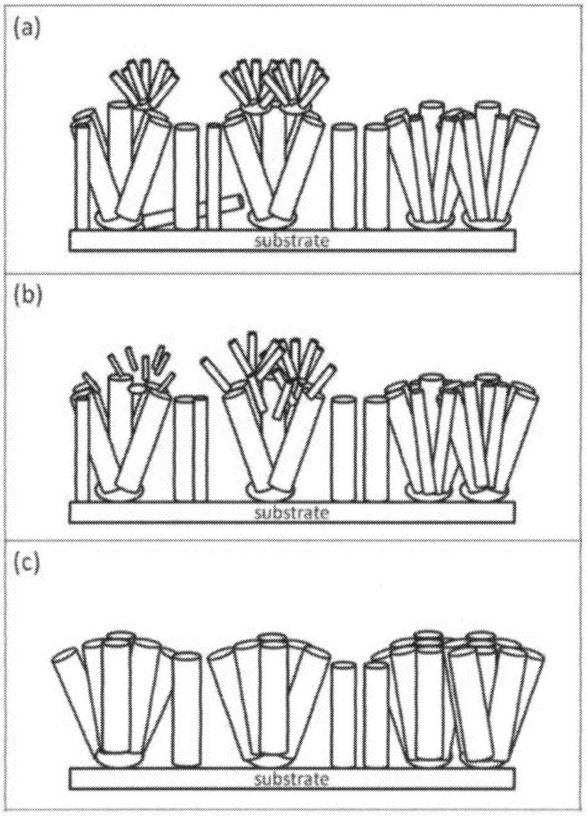

Figure 6. Formation mechanism of ZnO sample obtained from 2M NH_4OH annealed at 400°C: (a) before annealing, (b) during annealing where smaller, thermodynamically unstable clusters will coalesce spontaneously to allow growth of larger, thermodynamically stable rods, (c) thermodynamically stable and pure ZnO rods.

Meanwhile, Figure 8shows the high magnification SEM images of ZnO deposited on SiO_2/Si(100) substrate from 0.03M $ZnSO_4$ and 3M NH_4OH annealed at 400°C for 2 hours as compared with an as-grown sample. It can be observed that after the annealing process, dense sea urchin-like structures are still present as shown in Figure 8 (b). The average diameter of the rods ismeasured to be approximately 533 nm. This is larger compared to the average diameter of the as-grown sample. However, the presence of flakes covering the sea urchin-like structures is still visible but minimal compared to that of the as-grown sample. Unlike for the annealed ZnO sample obtained from 2M NH_4OH concentration, it can also be observed that the sizes of the smaller sea urchin-like structures accumulating on top of the larger ones have become larger instead of coalescence to other nanorods. After annealing, it shows that the diameters of the rods have become larger and the sizes of the urchins have become bigger. This phenomenon is illustrated in Figure 9.

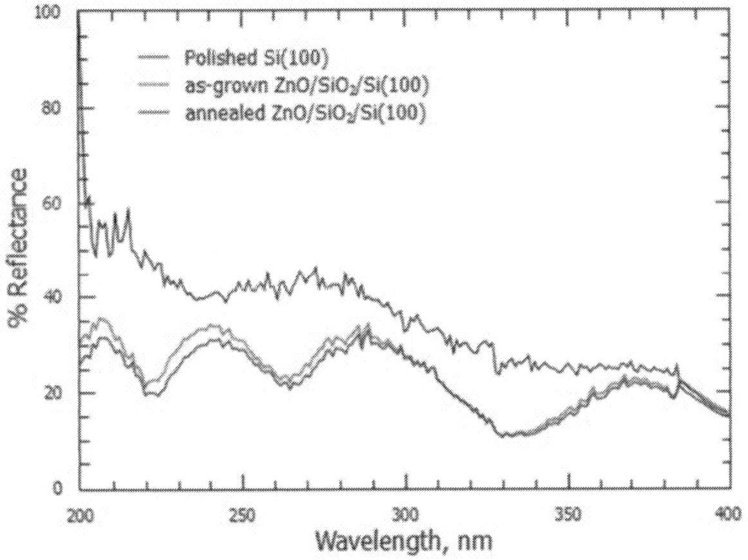

Figure 7. Reflectance spectra in the UV region of as-grown (red) and annealed (blue) ZnO samples obtained from 2M NH_4OH concentration.

Figure 8.SEM micrographs of ZnO grown on SiO_2/Si(100) substrate from 0.03M $ZnSO_4$ and 3M NH_4OH (a) as-grown and (b) annealed at 400°Cfor 2 hours.

Figure 9 (a) shows an as-grown ZnO. It is characterized by densely-packed sea urchin-like structures. On top of the larger sea urchin-like structures are smaller sea urchin-like structures that are covered with appreciable amount of flakes which are believed to be impurities. Upon annealing, Ostwald ripening will take place which might cause the diameter of the rods to increase and the amount of flakes to be minimized as shown in Figure 9 (b). The flakes might have contributed to the enlargement of the diameter of the nanorods. At high temperature (400°C), unstable molecules will dissociate. The dissociated atoms will eventually associate to other existing ZnO nanorods to form a more stable structure. Furthermore, the smaller sea urchin-like structures have become larger. Figure 9 (c) illustrates the sample after subjected to annealing treatment. It is assumed to be comprised of thermodynamically stable and partially purified ZnO rods with 13.40% increase in nanorod diameter and bigger sea urchin-like structures as compared to the as-deposited sample. The dominant elements present are Zn, O and Si. The Zn to O ratio for the annealed sample is close to the stoichiometric ratio of ZnO.

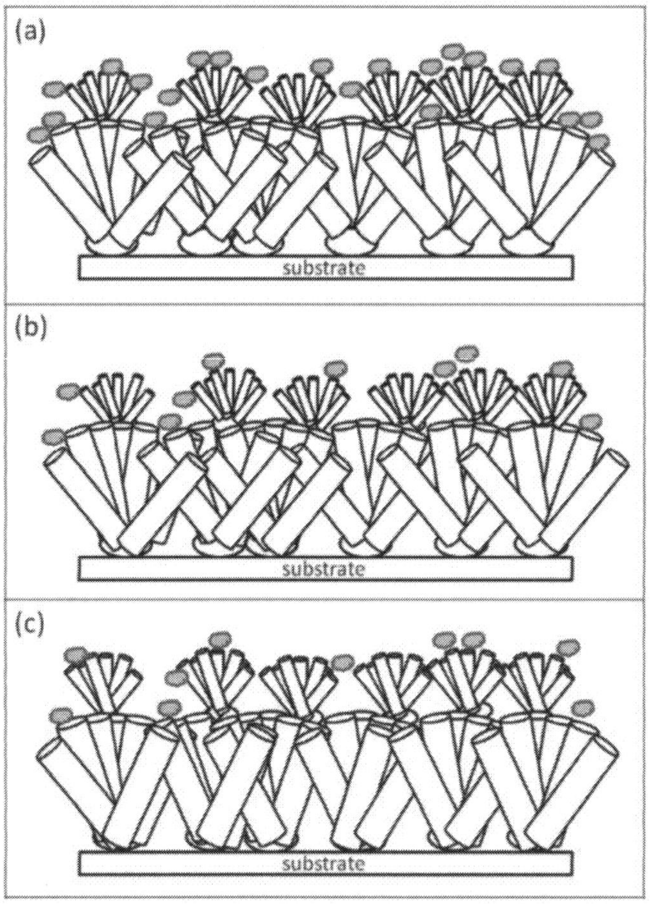

Figure 9. Formation mechanism of ZnO sample obtained from 3M NH_4OH annealed at 400°C: (a) before annealing, (b) during annealing where flakes and other thermodynamically unstable clusters will dissolve spontaneously to allow growth oflarger, thermodynamically stable rods, (c)thermodynamically stable and partially purified ZnO rods.

Shown in Figure 10 are the reflectance spectra of the annealed and unannealed ZnO sample obtained from 3M NH_4OH concentration in comparison with that of polished silicon. It can be noticed that the unannealed and annealed samples have similar peaks and the intensity of their reflectance differs to only about 2% with the annealed sample having lower reflectance. This implies that the sample is not completely purified. This can be explained by the incomplete decomposition of the impurities (i.e., flakes). In addition to that, the ZnO samples have lower reflectance compared to polished silicon. This observation is similar to what is observed in the previous samples. At 206 nm, 26% reflectance difference is observed while at 220nm the reflectance differs by 32% as compared to the reflectance spectra of the as-grown sample. When compared to the reflectance spectra of the annealed sample, the reflectance differs by about 27% at 206nm and about 32% at 220 nm. However, at 243 and 289 nm, the reflectance difference is only 10% while at at 265 nm, 282nm and 332 nm, the reflectance is approximately 18% both for the annealed and as-grown samples.

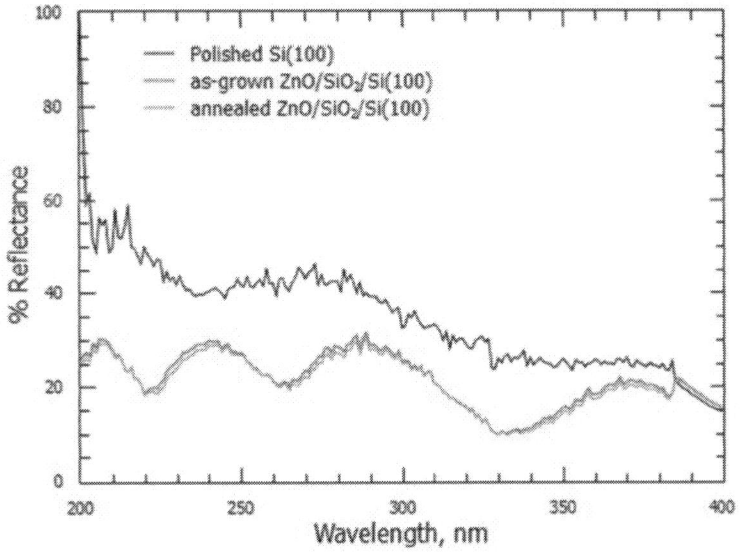

Figure 10. Reflectance spectra in the UV region of as-grown (red) and annealed (green) ZnO samplesobtained from 3M NH_4OH concentration.

4. Summary and conclusion

ZnO nanostructures were successfully produced on SiO_2/Si(100) using chemical bath deposition method. The effects of concentration of NH_4OH and thermal annealing were investigated. It was observed that ZnO grown from varying concentration of NH_4OH shows variation in the density of the nanostructures. The density increases as the concentration of NH_4OH is increased. However, the diameter of the rods decreases as the concentration of NH_4OH is increased. The intensity of the reflectance in the UV region decreases as the concentration of NH_4OH is increased. It can be further observed that there is a competition of the ion-by-ion growth mechanism and hydroxide cluster mechanism in bath with 2M NH_4OH concentration while bath with 3M NH_4OH concentration is dominated by the hydroxide cluster growth. SEM images of the thermally annealed ZnO samples reveal that the diameter of the rods increases after annealing the sample at 400°C due to Ostwald ripening. The reflectance shows that annealing reduces the reflectance of the sample since at this time pure ZnO is obtained. Only oxygen, zinc, and silicon were detected. The silicon element comes from the substrate. This confirms that the nanorods are primarily ZnO. Moreover, the growth mechanisms of ZnO formation for different experimental conditions have been illustrated.

IOP Conf. Series: Materials Science and Engineering **79** (2015) 012022 doi:10.1088/1757-899X/79/1/012022

5. References

[1] Lyu S C, Zhang Y and Lee C J 2003 *Chem. Mater.* **15** 3294–9
[2] Jin M *et al.* 2009 *Journal of Nanoscience and Nanotechnology* **9** 1–4
[3] ZaleszczykW *et al.* 2007 *Acta Physica Polonica* A **112** 357
[4] Huang J S and Lin C F 2008*J. Appl. Phys.***103** 014304
[5] Wang X, Cole J, Dabiran A Mand Jacobs H O 2007 *NSTI-Nanotech.* **4** 526
[6] ZnaidiL, Soler IlliaG J A A, BenhayiaS, Sanchez C and Kanaev A V 2003*Thin Solid Films* **428** 257
[7] Janotti A and Van de Walle C G 2009 *Rep. Prog. Phys.***72** 1
[8] Lee Y, Hu S, Water W, Huang Y,Yang M, Sheng J, Tiong K and Huang C 2007 *Solid State Communications* **143** 250
[9] Mitra P and Khan J 2006*Mater. Chem. and Phys.* **98** 279–84
[10] Kim M S *et al.* 2012 *Current Applied Physics* **12** S94–8
[11] Ortega-Lopez M, Avila-Garcia A, Albor-Aguilera M L, Sanchez-Resendiz V M 2003 *Materials Research Bulletin* **38** 1241
[12] Zhao X Q *et al.*2009 *Applied Surface Science* **255** 4461
[13] Madras G and McCoy B J2002 *Chemical Engineering Science* **57** 3809
[14] Madras G and McCoy B J2003 *Chemical Engineering Science* **58** 2903
[15] Pern F J, To B, DeHart C, Li X, Noufi R and Glick S H 2008 *Conference Paper NREL/CP-520-42792* 1–12

IOP Conf. Series: Materials Science and Engineering **79** (2015) 012023 doi:10.1088/1757-899X/79/1/012023

GSM &web-based flood monitoring system

J C Pagatpat[1], AC Arellano and OJ Gerasta

Mindanao-State University-Iligan Institute of Technology
Iligan City, Philippines

Email: jaime_pagatpat2000@yahoo.com,aileenchris.arellano@gmail.com, olgajoy.labajo@g.msuiit.edu.ph

Abstract. The purpose of this project is to develop a local real-time river flood monitoring and warning system for the selected communities near Mandulog River. This study focuses only on the detection and early warning alert system (via website and/or cell phone text messages) that alerts local subscribers of potential flood events. Furthermore, this system is interactive wherein all non-registered subscribers could inquire the actual water level of the desired area location they want to monitor. An estimated time a particular river waterway will overflow is also included in the analyses. The hardware used in the design is split into several parts namely: the water level detector, GSM module, and microcontroller development board.

1. Introduction

Residents from flat areas beside the mouth of Mandulog River were greatly affected by typhoonSendong(2011) and Pablo(2012). This river had a waterway more than 50 kilometers long which serves as the main drain of the mountainous area which covers about 70,000 hectares of land as shown in Figure 1. The watershed covers hundreds of smaller rivers connected to Mandulog River and form a network of channels similar to the branches of a tree. It is the main trunk where most of the storm water collected by the mountain watershed passes through before reaching the sea (Iligan Bay).

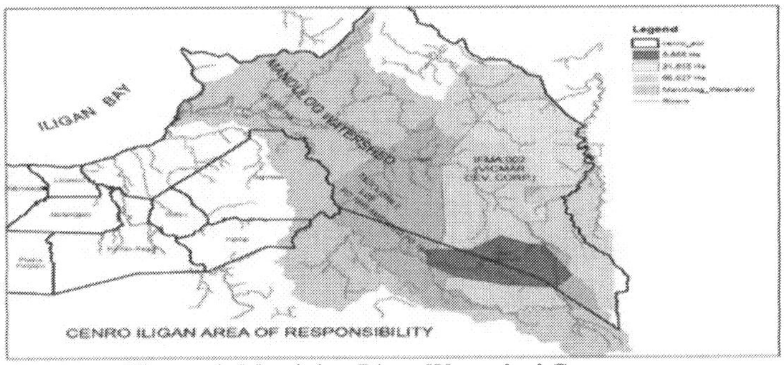

Figure 1. Mandulog River Watershed Coverage.

[1] To whom any correspondence should be addressed.

During periods of heavy rainfall, the stream network in the uplands collects rainwater and swellMandulog River, generating a flash flood which sweeps away everything in its path. The river channels swell and form an overbank flow that encroaches on adjacent fields. During the typhoon Sendong, the river of torrent carried tons of trees and eroded debris cascaded toward the communities of Bayug, Upper Hinaplanon, Hinaplanon and Santiago in Iligan. Floodwaters 7 to 10 meters high rampaged beyond the banks of the river and wiped out villages with incomprehensible force.

This special project develops flood detection, emergency alert and monitoring system which will warn the residents of a particular community for a possible flooding along the Mandulog river banks. River watersheds are mounted with a water level detection device which has a GSM kit that will send signals to the server node. If system detected a warning status, the server will automatically send a text message to those recognized individual subscriber, and will continue to update until the water level detected returns to normal. Moreover, a monitoring device will automatically upload the current status to the internet through a website and other social media sites. Individual subscribers could also inquire the latest water level status of the particular area they wanted to monitor by texting a keyword which directs the text message to a specific node where the device is mounted. Additionally, a specific area could be monitored and predicted when to overflow depending on the volume of the watershed and amount of rainfall. This could be a great help to nearby areas in order to be prepared for a possible flush floods.

2. Other flood monitoring system

These last decades, lots of flooding risk technologies have been developed to minimize the danger of flood in inhabited areas. The current government development is the Project NOAH of the Department of Science and Technology's (DOST). Automated rain gauges (ARG) and water level monitoring stations (WLMS) was installed along the country's major river basins (RBs). This project will come up with computer models for the critical RBs through LiDAR technology, automate the process of data gathering, modeling and information output, and release flood forecasts.

Most of the technologies being developed commonly apply flood detection and monitoring system. They usually use a sensing device, flood modeling using software, internet and mobile technology. However, the applications are usually for one way communication only. In order to get information, the individual needs to access the website. However, in order to access the information, it requires mobile phone that has an internet feature, and most people could hardly afford to purchase one. In addition to that, individuals are busy doing their daily routine, to a point that monitoring activity cannot be their priority. These are the reasons why communities are blinded with the current status of the nearby river watershed. The unawareness led to the overflow of the watercourses of Mandulog River waterway and the subsequent inundation of various localities causing extensive damages to properties and human life.

As compared to other works, the proposed technology is personalized and designed for individuals who are directly involved within Iligan City and nearby communities along theMandulog river waterway. This system is also a two-way topology where residents can interact and inquire the current status of the waterway.

3. Proposed research

This project builds a prototype that will detect the current water level across the watershed of Mandulog River and its surrounding areas through sensors. The geographical area of the river was sub-divided into areas where sensors were installed. Each sensor signifies a warning level. Once a sensor is triggered, an output signal will be relayed to a microcontroller which serves as a switch that triggers the connected GSM modem to send alert SMS message to the server. Then, the server will automatically send a text message to the numbers stored in the database. Also, the computer will then automatically relay the alert signal by uploading a warning post on a website or to social media sites like Facebook and Twitter. The process repeats as the water level continues to rise and triggers another sensor. Once the water level reaches its critical point, it will relay a message giving warning to the nearby areas. Furthermore, communities could inquire by a sending message thru keywords.

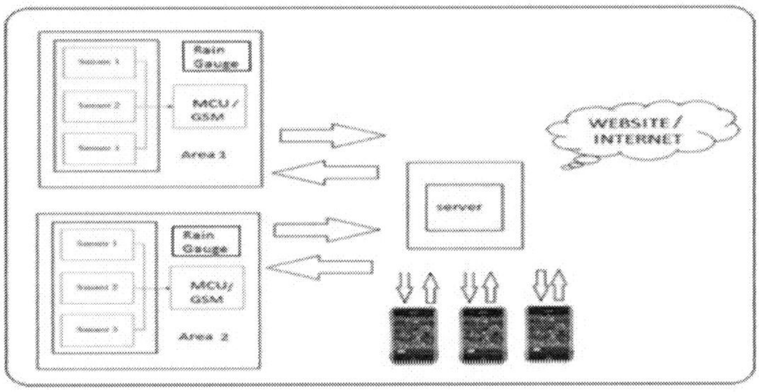

Figure 2. Overall System Design.

4. Project elements

The basic hardware components used in the implementation of this project are as follows: float switch for water level detector, inverter, rain gauge, GSM module, and microcontroller development board. GSM is controlled using a water level sensor.

4.1. Sensors

For a specific watershed area, water level sensors are mounted depending on desired number of levels. The sensors used in the study are of float switch type. The purpose of a float switch is for it to open or close a circuit as the level of a liquid rises or falls. The float encases a sealed magnet which moves up and down the length of the stem as a fluid level rises and falls. As the magnet passes by the contacts in the encased reed switch, they touch and complete a circuit between the two lead wires.

4.2. GSM

The GSM module is used to transmit the data from sensors to the PC System through wireless transmission. In this project, Model 900 GSM/GPRS Modem Kit is used which is a physical add-on and usually mounted to the top of Arduino. It comes with a library to send/receive SMS and voice calls, and establish TCP communication over the broadly spread GPRS network. To upload sketches to the board, connect it to computer with a USB cable and upload sketch with the Arduino IDE. Once the sketch has been uploaded, you can disconnect the board from your computer and power it with an external power supply. Then, the GSM library handles communication between Arduino and the GSM shield. The majority of functions are for managing data, voice, and SMS communication.

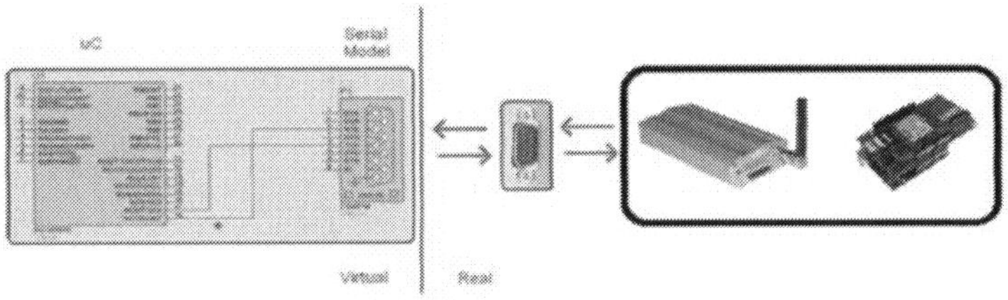

Figure 3. GSM Module Interface.

To configure a GSM Modem, a HyperTerminal is used. RS232 Cable is also used to connect to the modem, and to the PC's serial port as shown in Figure 3. A HyperTerminal utility should be installed in your PC. Additionally, Micro C is the tool used in PIC micros for the development of embedded systems without compromising its performance and control.

4.3. Rain gauge

The same concept is applied when the level of water droplets in the rain gauge increases - there is detection of change in the level of rain fall volume and this will trigger the GSM module to send an SMS. The rain gauge uses a measuring jug container calibrated in milliliters (ml) to measure the amount of rainwater collected. The measured amount is divided by the area of the funnel used in millimeters. This gives the amount of rain in millimeters that has fallen. Measured average rainfall volumeis tabulated in Table 1.The formula used is:

$$\text{Rain fall} = \frac{\text{Amount of rainfall in jug (mL)}}{\text{Area of the funnel (mm}^3\text{)}}$$

Table 1. Rain fall volume.

Level	Volume in mL	Area of funnel	Rain Fall
1	50	7850 mm	6.36 mm
2	100	7850 mm	12.74 mm
3	150	7850 mm	19.11 mm
4	200	7850 mm	25.48 mm

4.4. Control unit

The basic operation of control unit is controlling GSM thru microcontrollers which are defined by the particular program. GSM module is connected to an output pin of microcontrollers by way of a switch circuit which is connected to a transistor. The collector of this transistor is connected to the switch circuit and the emitter is grounded. In the switch circuit, one diode is used for sending signal in one direction and one inductor is for opposing the change of current flow. The output of switch circuit is connected with GSM module. Control unit would perform depending on the water level detected by the microcontroller. When the microcontroller sends 0/1 volt to the base of the transistor then it becomes off/on and its emitter and collector becomes open/closed conditions. Then the GSM will get a closed path giving an OFF/ON condition, idle (no message sent) or sending status.

5. Operation description and complete circuit diagram

The operation of the main river water level detector is described through the block diagram shown in Figure 4. Other parts needed in the design are PIC16F9877A microcontroller, crystal oscillator, 2 capacitors with a capacitance 22 pF, LED, water basin, water level sensor, transistor, and some capacitor. The crystal oscillator is composed of two 22 pF capacitors which perform as an external clock generator to execute the instructions of the program.

The microcontroller is embedded with a program that will trigger the output with given input signals from the sensors. The power supply is regulated to 5 Volts. This is enough supply for the microcontroller to operate. The sensors that are placed into the bank of the river are composed of stranded wires, one wire for each level. One of the wires is connected to ground of the microcontroller which is in pin 12 or pin 31. That indicates Level 0. As shown in the Figure 4, the wire for the levels 1, 2, 3 and 4 are connected to the microcontroller through pins RE1, RA0, RA1 and RA2, respectively. Pins were assigned according to level defined as follows: Pins RA0 (Critical), RA1 (Warning), and RA2(Normal). On open connection or level 1 of the main river water level detector is not filled with water, the resistance is 100kΩ. A voltage drop can be measured across the resistor connected parallel to the wire for level 1. The voltage drop is about +4V-+5V. The microcontroller is programed that if the input voltage is greater than +4V, then the rain drops collected has not reached such level. When the water level rises and immerses the wire for level 1, due to water conductivity, the wire for level 1 or of any level becomes connected with wire connected to ground. Then the microcontroller gets ground signal (0v) to that pin of that level. Once the level 1 is grounded or 0V, this indicates that the water level in the river has reached level 1, then the microcontroller sets the LED for level 1 to ON and the information of what are the input voltages of the pins RA1, RE0, RE1 and RE2 are is passed on to the Arduino Mega 2560 through the transmitter of the microcontroller which is the pin RC6 and processed to be sent to the GSM module for wireless transmission. The same goes for levels 2, 3 and 4. But on the 3rd and 4th level, the alarm will be triggered for warning. Every change in the level of the raindrops collected will be read by the microcontroller and goes on continuously. From the rain gauge set-up, the same set up is applied in the main river water level detectors for the crystal oscillator, power supply and ADC converter has been used.

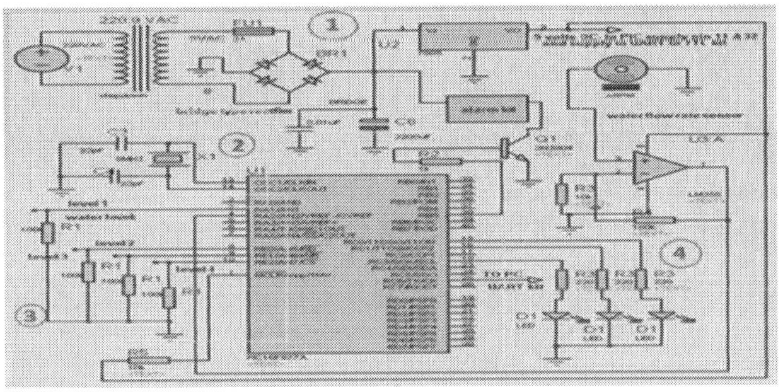

Figure 4. Circuit Design.

Table 2. Experimental Result of Water Level Sensing Unit.

Water Basin Input	RA0(LED 1)	RA1(LED 2)	RA2(LED 3)	Water Level	River Conditions
000	OFF	OFF	OFF	(0)EMPTY	LOW LEVEL
001	OFF	OFF	OFF	(1)50%* H	NORMAL
011	OFF	ON	ON	(2) 75%*H	WARNING
111	ON	ON	ON	(3)90%*H	CRITICAL

Each pins output will be display on the LED depending on the inverted signal from the microcontroller. The water level could be varied depending on the physical characteristics of the basin. The position or the height of the sensors where it is mounted will define the water level condition. In this experiment, we defined Level as 50%, Level 2 as 75%, and Level 3 as 90% of basin's height(H).

Likewise, RE1 pin of the microcontroller is used to detect the signal from the server. When pin gets high, the LED's will be turned ON which means there is a high signal received. This signal is a text inquiry from the subscriber about the current water level condition. If high signal is received, it sends a signal to the controls of the whole circuit and sends back a message to the subscriber thru the GSM module of the current level condition of the water.

6. Mandulog river prototype and measurements

The prototype replicating the Mandulog River is developed as shown in Figure 5. To simulate the behaviour of the Mandulog River, main riverand branching waterway were built. Initially, the amount of water filled is just below level 1 for the main river and also for the branching rivers. Upon simulation, River 1 is filled with water as an assumption that water from the rain added volume to the water in the river. The float switch rises up and there has been detection for level one. As the water level continues to rise up and it reaches level 2, there is detection for level 2 and so on for level 3. The same scenario occurs in river 2. On the main river, besides from the captured rainfall, water level also rises as water from rivers 1 and 2 are continually rising. As a result, there is an increase in the water level in the main river and that will trigger the water level detector. For every change in the water level, the GSM module will send an SMS of the change in water level.

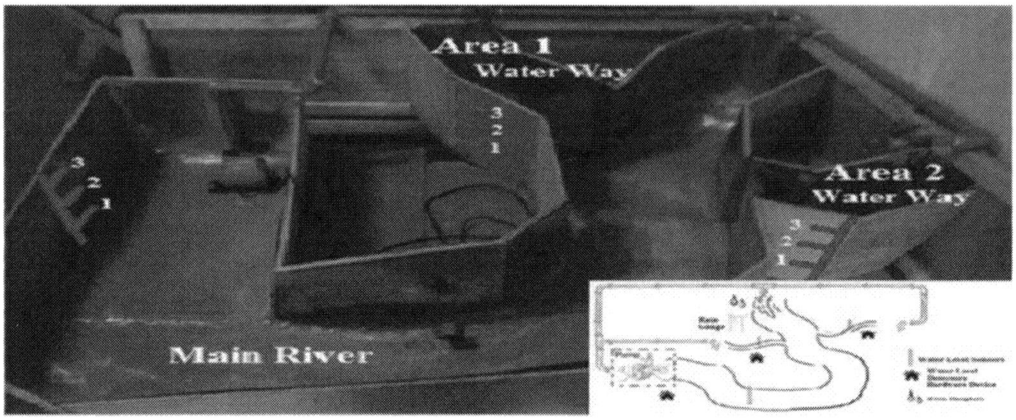

Figure 5. Mandulog River Prototype.

To calculate the amount of time each area will overflow, the basic formula in getting the volume of a rectangular solid, which is $V = LWH$, is used. By differentiating both sides of the equation, we can come up with a general equation for computing the speed of change of water level height at any given amount of pressure or flow rate$\left(\frac{dV}{dT}\right)$ and length and width of a particular terrain, that is:

$$\frac{dV}{dT} = L \times W \times \frac{dH}{dT}$$

From the basic equation, the time could be computed since the following parameters are available: $\frac{dV}{dT}$, which represents the flow rate measurements from the design rain gauge;H (height), L(Length), and W(Width) of a particular area. Moreover, take note that for the main river, the computation for branching waterway should be included.

Actual time in hours of testing is performed in order to analyze the time variation at each point of time. The tabulated time delay of operation shows the actual time of sending the generated commands from the mobile phone until loads are activated.

Table 3. Time delay response of the device.

Trial	Main River	Water Way 1	Water Way 2	Rain Gauge
1	1.7s	2.4s	3.4s	1.7s
2	3.3s	1.8s	2.7s	3.2s
3	2.5s	2.9s	2.3s	2.4s
Average	1.83s	2.36s	2.8s	2.43s

7. Website

The website has different interactive GUI as shown in Figure 6. You can view where sensors are mounted and its height for the water level indications. It was linked to a Google map, to locate specific area where the device is mounted. FacebookandTwitter users can also follow in order to monitor and get updated with the current water level status of Mandulog River. A 7-day weather forecast for Iligan City is also included.The current status of all areas monitored and mounted with the device is displayed in the server webpage as shown in Figure 7. Additionally, the histories from the previous dates of the real-time captured data of the water level activity are saved. Thus, you can still review the events and use this data for any purpose.

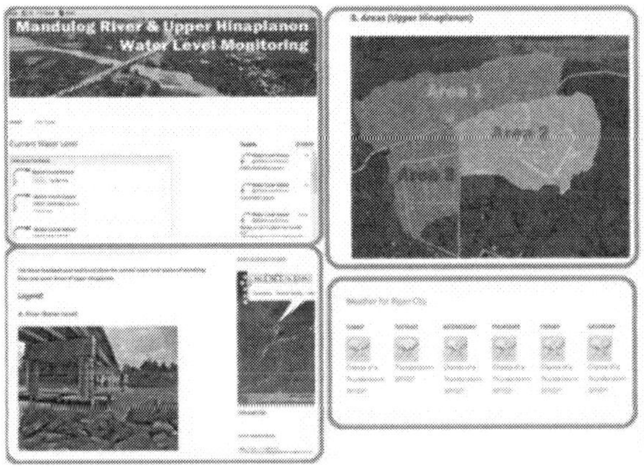

Figure 6. Web Page Design.

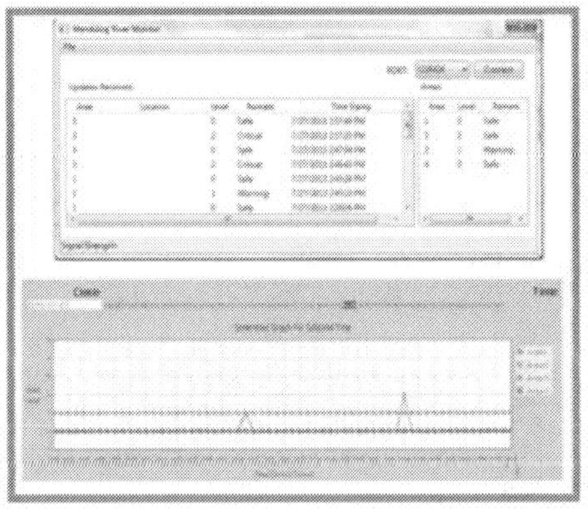

Figure 7. Server GUI.

8. Results and discussion

The data from the water level monitoring hardware were successfully transmitted using GSM sim900 and Arduino Mega 2560 and was successfully received by the computer system using GSM USB Dongle. The researchers have achieved the objectives of the study, which is to integrate different water level detectors from remote areas into one system. It is quite efficient for it hasa wide area coverage since it uses the cellular network that is used in mobile phones. The wireless communication set up that the researchers used is already web-`based, given that there will be a dedicated IP Address. Since dedicated address is only issued to commercial entities, this study will just be limited to the use a local host to present the data received from the hardware. The local host is used as the server and if given a dedicated IP Address, the data in the system can be broadcasted live to anyone in the internet. One of the best examples of establishment that have a dedicated IP address is MSU-IIT (http://my.iit.edu.ph). Moreover, the system made has the ability to handle more data provided that the hardware that transmits the data works well and it is within the coverage of the module. Since cellular networks are used in module, the system could cover the whole country. Data can be obtained from the devices located anywhere in the Philippines. The prototype was successfully designed to have a simulation which is the same as in the real situation in the Mandulogriver.

In terms of efficiency, a test was conducted by recording the time delay of the detection of the water level to be transmitted and recorded or updated in the website. The record time tabulated in 3 shows that it is efficient and reliable enough to operate via wireless. Through this system, the information could be available to anyone who could access the internet once the website will be given a domain and can be broadcasted live in the internet. Aside from the people near the river who would be alarmed once the water level will be of critical level, those who are away will also be informed of the current situation. With that, necessary preparations and safety measures can be done. It could be a help to prevent or lessen the damages that flood may bring.

[1] "Project Noah." [online]. Available from: <http://noah.dost.gov.ph/#about> [Accessed 28 February 2013].
[2] Cracknell, A.P. and Hayes, L. (2007). "Introduction to Remote Sensing." 2nd ed. London: Taylor and Francis.
[3] UrbanFlood Consortium, (2012) "The Story." *UrbanFlood* [online]. Available from: <http://www.urbanflood.eu/Pages/Publications.aspx> [Accessed 2 March 2013]

IOP Conf. Series: Materials Science and Engineering **79** (2015) 012024 doi:10.1088/1757-899X/79/1/012024

Surface characteristics and structural properties of sol-gel prepared ZnO-SiO_2 nanocomposite powders

E G Pantohan[1,3], **R T Candidato Jr**[2] **and R M Vequizo**[2]

[1]Natural Science and Mathematics Department, Arts and Sciences Program, Father Saturnino Urios University, Butuan City, 8600, Philippines

[2]Physics Department, College of Science and Mathematics, Mindanao State University-Iligan Institute of Technology, Iligan City, 9200, Philippines

Email: edmarpantohan@yahoo.com

Abstract. Zinc oxide-silica (ZnO-SiO2) composite powders with different surface morphologies were synthesized via sol-gel process using rice husk ash as silica source. Micrograph results revealed the surface modification of bare ZnO from hexagonal tips to hemispherical tips after incorporation of SiO2 to ZnO. After annealing at 400 and 500°C, the hemispherical tips became more apparent due to the removal of the flaky particles associated to the complexes of zinc. Annealing the sample powders to 600°C sharp tips was observed on the morphology of ZnO-SiO2. Infrared spectroscopy further support that SiO2 was incorporated to ZnO due to the Si-O-Zn band observed in the IR spectra. X-ray diffraction results verify the presence of amorphous SiO2 and the polycrystalline nature of the samples having particle size in nanometer scale. Furthermore, diffraction spectra show that thermal annealing removes the zinc complexes resulting to more pronounced hemispherical tips of ZnO-SiO2.

1. Introduction

Zinc oxide (ZnO) is an *n*-type semiconductor that acquired great interest due to its attractive properties such as 3.37 eV wide band gap [1], large exciton binding energy (60 meV), good piezoelectric characteristics and good optical and electrical properties [2]. It has potential applications in the field of optoelectronics for making devices like photodetectors, light emitting diodes, electroluminescence devices, solar cells, flat cathode ray tubes [3] and short wavelength devices [4]. However, the photocatalytic activity of ZnO [5] and its poor compatibility to organic compounds that makes it to agglomerate in organic solution [6] are just some problems that hinder some potential applications of ZnO. Fortunately, these disadvantages of ZnO are avoidable through the process of surface modification [7] and this can be achieved by coating ZnO particles by surface modifiers like alumina and silica. But silica is given much attention by most scientists due to its important properties such as high transparency for visible light, low chemical activity and compatibility to ZnO [5].

Various methods were used to incorporate SiO_2 to ZnO such as are double jet precipitation [8, 9], hydrolysis method [10], sonochemical synthesis [11], and spray drying [4]. However it is difficult to obtain small size and large scale production of incorporated material with these techniques due to the need of faster chemical reaction. On the other hand, the chemical solution (sol-gel) process is of great

[3] To whom any correspondence should be addressed.

advantage due to its capability to mass produce and the chemical reaction can be controlled and is a low cost method [12].

Different studies have been reported in the preparation of ZnO-SiO_2. The group of K. Han studied zinc oxide coated with silica for its application as interconnector or nanoelectronic devices and biochemical capsules [13]. Eunice S., et al. performed random laser action in SiO_2/ZnO composite films which consist of ZnO clusters placed on silica matrix [14]. The coating of SiO_2 on ZnO is also applied to protect organic material from the photocatalytic reaction of ZnO where the effect of polyelectrolyte dispersants for obtaining uniform composite particles is studied by Wang, H., et al [15]. However in their studies they used tetraorthosilicate (TEOS) as a source of their silica which is very expensive and very toxic chemical.

In this particular work, we incorporated SiO_2 extracted from rice husk ash (RHA) to ZnO using sol-gel process for the synthesis of ZnO-SiO_2 composite material. We present the effects of SiO_2 to the morphology of bare ZnO and the effects of thermal annealing to the surface morphology and structural characteristics of the composite powders.

2. Experimental methods

The ZnO-SiO_2 composite powders were synthesized through sol-gel process. A solution of 0.03 M zinc sulphate ($ZnSO_4$) and 2.0 M ammonium hydroxide (NH_4OH) together with amorphous SiO_2 powders were magnetically stirred in a beaker for 30 minutes and the pH of the solution was measured. After 30 minutes the solution was heated and constantly stirred for 5 hours with bath temperature maintained at 70°C. After 5 hours of stirring and heating, the solution with the precipitates was cooled at room temperature (25°C). The powders were filtered and washed with distilled water in order to removed unnecessary residues present in the powders. The sample powders were thermally annealed at 400°C, 500°C and 600°C in a furnace keeping as-grown (un-annealed) powder as control sample. The morphological characteristics of the samples were studied using JEOL JSM-6510LA analytical scanning electron microscope and the structural characteristics of the powders were analysed by X-ray diffraction (XRD) using CuKα1 radiation (λ=0.154 nm) at 0.010 increment per step and by Perkin Elmer Spectrum 100 infrared spectrometer for functional groups determination.

3. Results and discussions

3.1. Surface morphology

Figure 1 shows the micrographs of bare ZnO and as grown ZnO with SiO_2 from RHA. The bare ZnO rods show hexagonal tips, but upon adding SiO_2, the surface of hexagonal ZnO was modified to hemispherical tips. Since all the growth parameters were held the same for both samples, it is very apparent that SiO_2 from RHA modified the surface structure of ZnO. SiO_2 possibly capped the surface of the ZnO nanorods. Also observed in the figure are some flaky particles associated to the complexes of zinc.

As the ZnO-SiO_2 composite powders were annealed at 400°C, 500°C and 600°C, more apparent hemispherical tips were observed. Annealing the samples at 400°C and 500°C, flaky particles present in the as-grown ZnO-SiO_2 were removed and the tips of the nanorods are more pronounced hemispherical tips, which further suggest the dissociation of the weakly bonded zinc complexes (zinc sulphate and zinc hydroxide) as seen in figure 2 (a) and (b). However, annealing the sample at 600°C, a disintegration of the nanorods was observed. The hemispherical tips were modified to sharp tips with remarkable decrease of the diameter of the nanorods after thermal annealing at this elevated temperature. It can be said from the surface properties of the samples that the formation of the hemispherical tips depends on the chemical activity on the surface, which relies on the concentration of the *OH*⁻ groups and the presence of the silanol (Si-OH) and siloxane (Si-O-Si) functional groups of SiO_2 that act as the center of molecular adsorption. The functional groups of silica were formed during the synthesis through rehydroxylation process inside an aqueous solution allowing it to bind with the $Zn(OH_2)$ which is the seed nuclei for the formation of ZnO nanorods. The resulting hemispherical tips evidently show the integration of amorphous SiO_2 at the surface of ZnO allowing for the formation of ZnO-SiO_2 composite powder with SiO_2 capping the surface of ZnO.

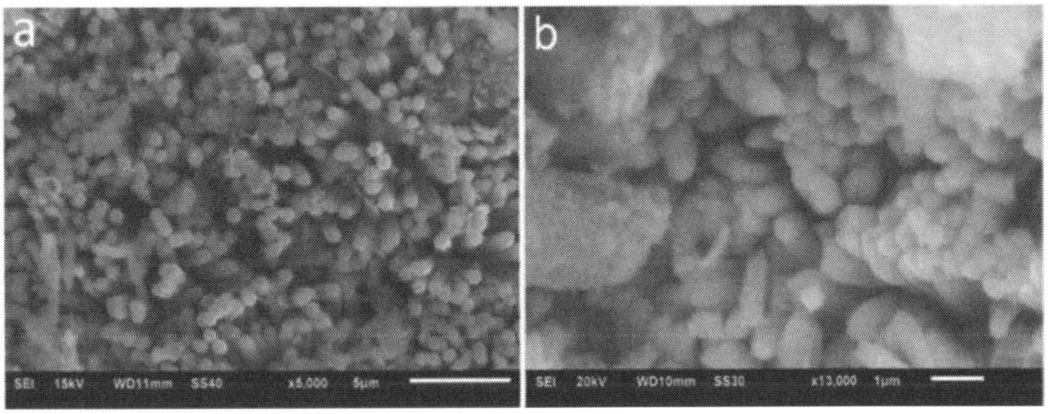

Figure 1. SEM micrographs of (a) as grown bare ZnO powder and (b) as grown ZnO-SiO_2 composite powder.

Figure 2. SEM images of ZnO-SiO_2 composite powder annealed at (a) 400°C, (b) 500°C, and (c) 600°C.

To supplement the morphological result on the capping of SiO_2 at the surface of ZnO, FTIR spectroscopy was done on the samples. As presented in figure 3, the presence of characteristic peaks of amorphous SiO_2 in as-grown and annealed samples is observed. These peaks are found at 961 cm^{-1} for Si-OH stretching, 1099 cm^{-1} for Si-O-Si asymmetric stretching 471 cm^{-1} for Si-O-Si bending and 803 cm^{-1} for Si-O-Si symmetric stretching. The broad band at 3476 cm^{-1} is attributed to the O-H stretching vibration. The appearance of absorption band at 933 cm^{-1} in the IR spectra is associated to the Si-O-Zn absorption band which confirms that SiO_2 indeed binds with ZnO. The Si-O-Zn band weakens as the annealing temperature increases from 500°C to 600°C and along with it is the appearance of Si-OH stretching and ZnO_4 (605 cm^{-1}). Noticeable also is the shifting of Si-O-Si asymmetric stretching band to a lower frequency at temperatures 500°C and 600°C. The shifting of Si-O-Si asymmetric stretching band is probably due to the bonding change around the SiO_4 tetrahedral and can be directly associated to the binding of ZnO and SiO2 causing the hemispherical tips formation. The presence of peaks corresponds to Si-O-Zn bond and ZnO_4 at 500°C and 600°C suggests a possible formation of zinc silicate (Zn_2SiO_4). This might be the reason also for the change in surface morphology of ZnO with amorphous SiO_2sample when annealed at 600°C. Further investigations are being done to support this effect.

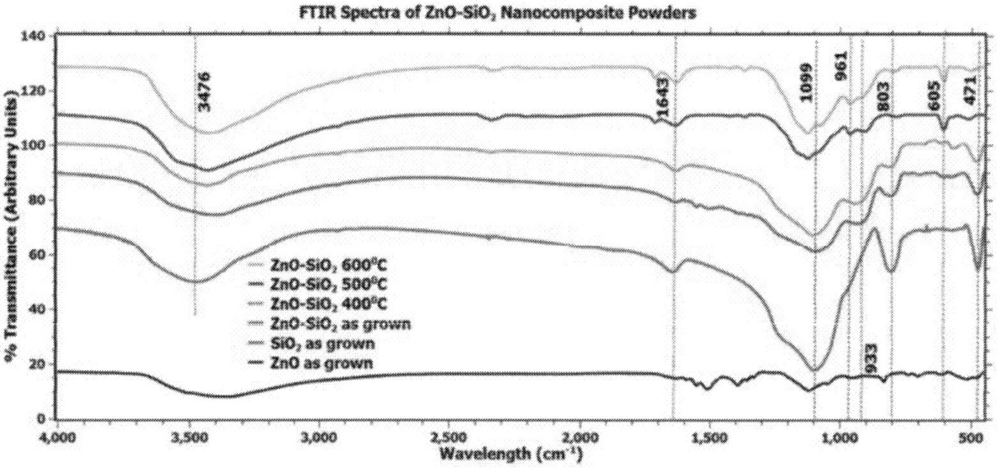

Figure 3. FTIR spectra of ZnO, SiO_2 and ZnO-SiO_2 composite powders at different annealing temperatures.

3.2. Structural characteristics

Figure 4 shows the diffraction patterns of amorphous SiO_2 and the polycrystalline nature of as grown ZnO-SiO_2 and ZnO-SiO_2 powders annealed at temperature range of 400-600°C. The ZnO-SiO_2 powders exhibit strong 2θ peak centred at 31.74 which corresponds to the (101) peak of highly c-axis oriented wurtzite ZnO. The sample also has other peaks which correspond to (100), (002), (102) planes, etc., which may be attributed to the rodlike nature of the samples as observed in the micrographs result. Furthermore, the presence of $Zn(OH)_2$, $Zn(HSO_4)_2$ and $ZnSO_3H_2O$ for the as-grown ZnO-SiO_2 powder were observed indicating that the as-grown powder is indeed composed of zinc complexes while annealing the samples at 400-600°C shows the removal of these complexes supporting our results in the SEM images. Additionally, investigation on the XRD patterns of the sample showed that as the ZnO-SiO_2 powders were annealed, the full width at half maximum (FWHM) of the (101) plane decreases which indicate increase of crystallinity of the samples. However, it is worth noting that even without annealing the as-grown ZnO-SiO_2 exhibits high crystallinity.

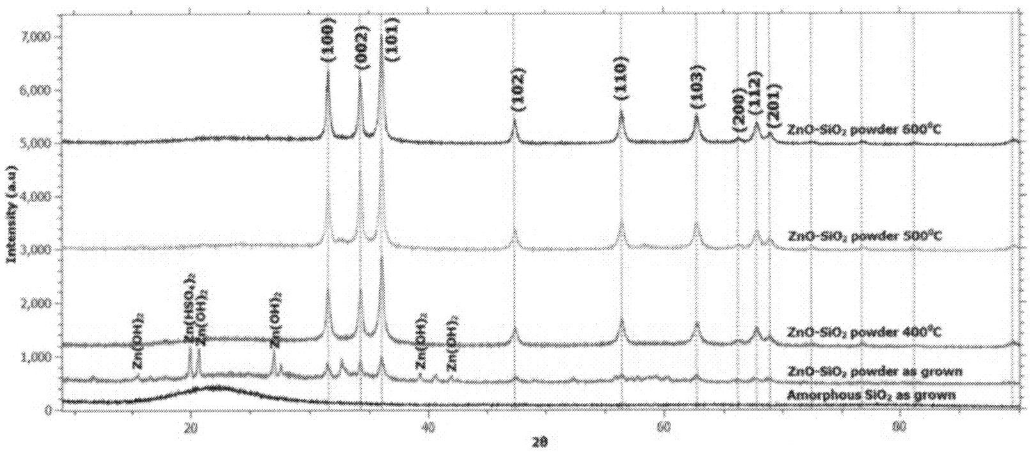

Figure 4. X-ray diffraction spectra of amorphous SiO_2 and ZnO-SiO_2 composite powder at varying temperatures.

Figure 5 (a), shows the dependence of the FWHM at the (101) plane to annealing temperature. To further determine the effect of SiO_2 and annealing temperature to the structural characteristics of ZnO, we calculated the lattice parameters of the samples. The calculated lattice parameters were slightly greater than the standard lattice parameter of wurtzite ZnO may be due to the attachment of SiO_2 to ZnO that results to the realignment of ZnO atoms onto SiO_2 atoms. Figure 5 (b) shows the plot of the difference between the standard lattice parameters of ZnO and calculated lattice parameters of the ZnO-SiO_2 powders. The calculated particle sizes ranges from 13.62 nm to 24.24 nm which clearly shows that the synthesized powder samples are in nanometer scale. These samples have promising potential applications in nanotechnology like in sensor technology.

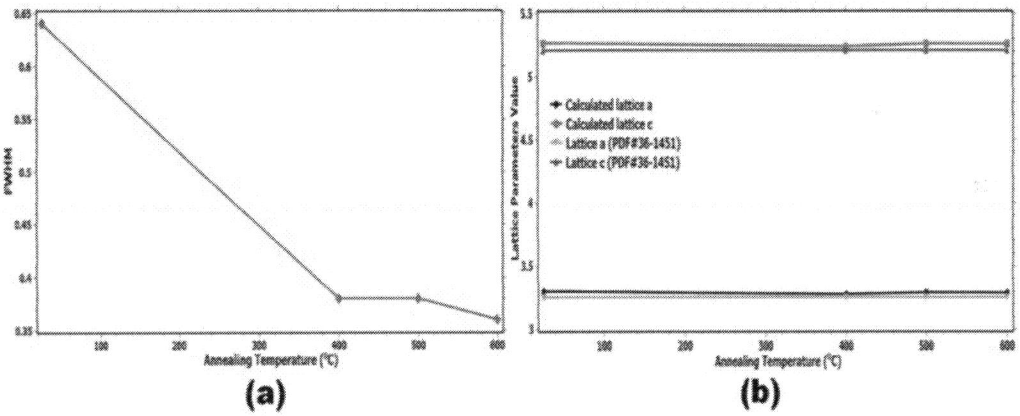

Figure 5. (a) Dependence of full width at half maximum (FWHM) of the (101) plane of ZnO to annealing temperatures and (b) plot of the calculated lattice parameters compared to the standard lattice parameters of ZnO (PDF#36-1451) at various annealing temperatures.

4. Conclusions

Nanocomposite powders of ZnO-SiO_2 were successfully synthesized through sol-gel process. The surface properties based on SEM images showed that the tips of ZnO-SiO_2 powders were hemispherical, far different from bare ZnO which is hexagonal in form. Infrared spectra show the functional groups of SiO_2 and Si-O-Zn absorption band indicating the attachment of SiO_2 to ZnO. XRD analysis affirmed that the composite powder was polycrystalline and were indexed to wurtzite nature of ZnO with strong peak at (101) plane suggesting the c-axis orientation of the nanorods with hemispherical tips due to SiO_2 capping. Annealing the samples removes the complexes of zinc and made the sample more crystalline indicative of the decrease of full width at half maximum. The particle sizes ranges from 13.62 nm to 24.24 nm which can be a promising material for nanotechnology applications.

5. References

[1] Maensiri S, Laokul P and Promarak V 2006 *Journal of Crystal Growth* **289** 102–6
[2] Suwanboon S 2008 *Science Asia* **34** 031–4
[3] Lu Y F, Ni H Q, Mai Z H and Ren Z M 2000 *J. Appl. Phys.* **88** 498–502
[4] WangY G, Yuen C, Lau S P, Yu S F and Tay B K 2003 *Chemical Physics letters* **377** 329–32
[5] Cao Z, Zhang Z, Wang F and Wang G 2009 *Colloids and Surfaces* A: *Physicochem. Eng. Aspects* **340** 161–7
[6] Hagura N, OgiT, Shirahama T, Iskandar F and Okuyama K 2010 *Journal of Luminescence* **131** 921–5
[7] Cao Z andZhang Z 2010 *Applied Surface Science* **257** 4151–8
[8] Xia H Land Tang F Q 2003 *S. Phys. Chem.* B **107** 9175–8
[9] Tang F, Ding J,Chen D, Huang Z, Xia H and Meng X 2003 *Technical Proceedings of the Nanotechnology Conference and Trade Show* **3** 262–5
[10] Abdullah M, Shibamoto S and Okuyama K 2004 *Optical Materials* **26** 95–100
[11] Sounderya N and Zhang Y 2008 *Recent Patents on Biomedical Engineering* **1** 34–42
[12] Cui H, Zayat M and Levy D 2009 *Journal of Alloys and Compounds* **474** 292–6
[13] Han K, Zhao Z,Xiang Z, Wang C, Zhang J and Yang B 2007 *Materials Letters* **61** 363–8
[14] LeongE S P, ChongM K, Yu S F and Pita K 2004 *IEEE Photonics Technology Letters* **16** 2418–20
[15] Wang H, Nakamura H, Yao K, Uehara M, Nishimura S, Maeda H and Abe E 2002 *J. Am. Ceram. Soc.* **85** 1937–40

Acknowledgment
The authors would like to acknowledge the Department of Science and Technology of the Philippines for the financially support of this work through ASHTHRD Scholarship grant.

IOP Conf. Series: Materials Science and Engineering **79** (2015) 012025 doi:10.1088/1757-899X/79/1/012025

Development of functional ZnS nanospheres as active material for acetic acid detection

A D M V Peguit, R T Candidato Jr[1] and A C Alguno

Physics Department, College of Science and Mathematics, Mindanao State University-Iligan Institute of Technology, Iligan City, 9200, Philippines

Email: r.candidatojr@gmail.com

Abstract. We have successfully synthesized zinc sulphide (ZnS) nanospheres deposited on glass and silicon on insulator substrates as an acetic acid sensor. Results show that nanospheres deposited on silicon on insulator substrate at lower $ZnCl_2$ concentration show better response and good recovery. We found out that the sensitivity of the ZnS nanospheres were dependent on the surface morphology and that the morphology is affected by the $ZnCl_2$ concentrations and the substrates used. Our results show a promising potential of ZnS nanospheres as an inexpensive alternative sensing material to the existing acetic acid detectors.

1. Introduction

Among the dangerous pollutants that can have a compounding long term health effects are the volatile organic compounds (VOCs). Acetic acid (HOAc) is one of the many VOCs widely used in many industries. Large scale industry utilization of acetic acid employs a lot of people who may be exposed to this potentially hazardous organic compound since it is classified as acutely toxic to humans because it is highly corrosive to skin and mucous linings even at 50ppm exposure. Although there are existing methods and materials developed in monitoring acetic acid, however most of these monitoring devices are very expensive and are not good for on-site monitoring. Recently, a variety of sensing materials, such as carbon fiber, graphite, carbon nanotube and boron-doped diamond are employed [1-2] but these materials are quite expensive. One promising material that can be used for acetic acid sensing is zinc sulphide *(ZnS)* due to its superior chemical and thermal stability, good sensitivity, fast-time response and high selectivity [3] and can be fabricated in a relatively inexpensive way.

Among the various techniques employed in synthesizing *ZnS*, chemical bath deposition (CBD) is attractive because of its convenient, cost effective and good for large scale application [4]. This method offers an inexpensive fabrication route because it eliminates the need for vacuum conditions and complex equipment.

In this work, the sensitivity of the synthesized ZnS nanospheres deposited on glass and silicon on insulator (SOI) substrates to acetic acid is investigated. Moreover, the effects of using different kinds of substrate and precursor concentrations to the response of ZnS nanospheres to acetic acid as well to its morphology were also explored.

[1] To whom any correspondence should be addressed.

2. Experimental methods

ZnS nanospheres are deposited on glass and SOI substrates from an aqueous solution using 0.01M and 0.03M zinc chloride ($ZnCl_2$), 0.05M urea (CH_4N_2O) and 0.3M thioacetamide (C_2H_5NS). The total volume is 200 mL by the addition of distilled water. The pH of the solution is maintained at 2.5 by drop wise addition of HCl. The solution is then magnetically stirred at 360 at 70°C temperature for 30 minutes. Subsequently, the substrates are immersed in the solution for three hours to deposit ZnS nanostructures. Current – voltage *(I-V)* characteristics and sensitivity to acetic acid gas of the ZnS nanospheres are performed via two-probe method inside an improvised gas sensing box for three cycles. Each cycle of exposure and without exposure to acetic acid gas is about three hundred sixty seconds (360 s). The surface morphology and elemental composition of synthesized ZnS samples were examined using JEOL JSM-6510LA Analytical Scanning Microscope.

3. Results and discussions

3.1. Sensitivity and recovery characteristics of ZnS nanospheres to acetic acid

Figure 1 shows the sensitivity of ZnS grown on glass substrate with different $ZnCl_2$ concentrations under exposure to HOAc gas for three cycles. The sample synthesized using 0.03M $ZnCl_2$ (sample 3G) exhibits higher response (maximum of ~20% sensitivity) compared to the sample using 0.01M $ZnCl_2$ (sample 1G) having only a maximum sensitivity of ~9%. The higher response on sample 3G might be due to the agglomeration of the ZnS nanospheres expose to HOAc. It is believed that agglomeration of nanospheres is possible when using higher $ZnCl_2$ concentration. The presence of the agglomerates might also be the reason in its difficulty in recovering to its original state after exposure to HOAc as observed in the *I-V* characteristics in Figure 2. On the other hand, the nanospheres grown using 0.01M $ZnCl_2$ show good recovery which is an essential characteristic for every gas sensors.

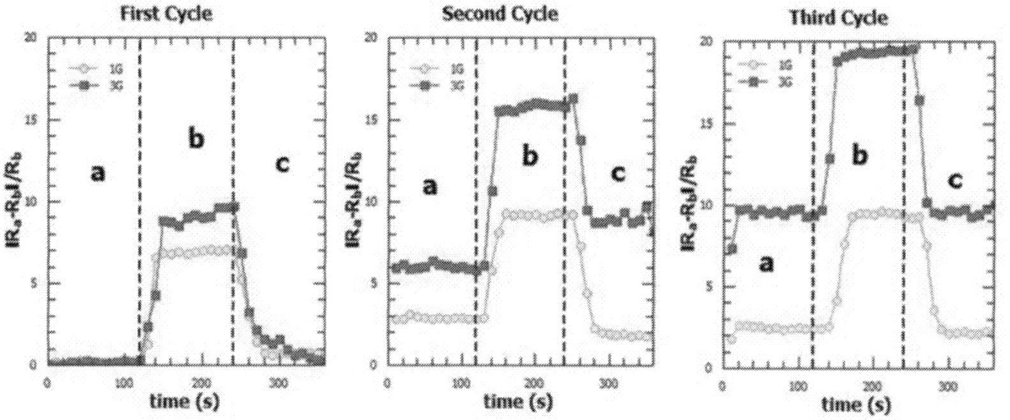

Figure 1. The sensitivity plot of ZnS samples deposited on glass substrate (a) before, (b) during and (c) after exposure to HOAc gas. (red-0.03M $ZnCl_2$; yellow-0.01M $ZnCl_2$).

Figure 3 shows the graph of the sensitivity of the fabricated ZnS sample deposited on SOI substrate. It is observed that the sensitivity decreases as the concentration of $ZnCl_2$ is increased and that it has almost the same sensitivity in each cycle. ZnS deposited using 0.01M $ZnCl_2$ (sample 1S) has higher sensitivity of ~12% compared to ZnS grown using 0.03M $ZnCl_2$ (sample 3S) having only ~5% sensitivity. Comparing the sensitivity of ZnS nanospheres grown on different substrates using 0.01 concentration of $ZnCl_2$, higher response has been observed for ZnS on SOI substrate as compared to ZnS on glass substrate attributed to the less wetting property of SOI substrate. The recovery of the samples were also determined and it was found out that the synthesized ZnS sample using 0.01 $ZnCl_2$ concentration shows better recovery compared to the grown ZnS sample using 0.03M $ZnCl_2$.

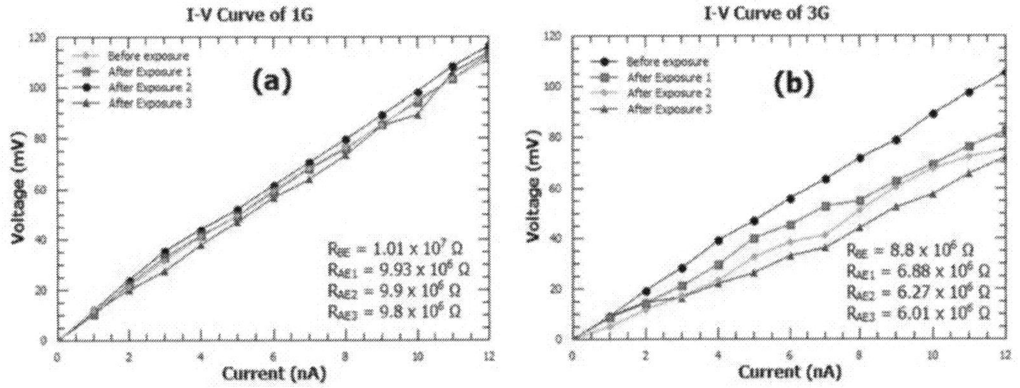

Figure 2. *I-V* curve of ZnS samples grown on glass substrate using different $ZnCl_2$ concentrations showing the good recovery of ZnS using 0.01M $ZnCl_2$.

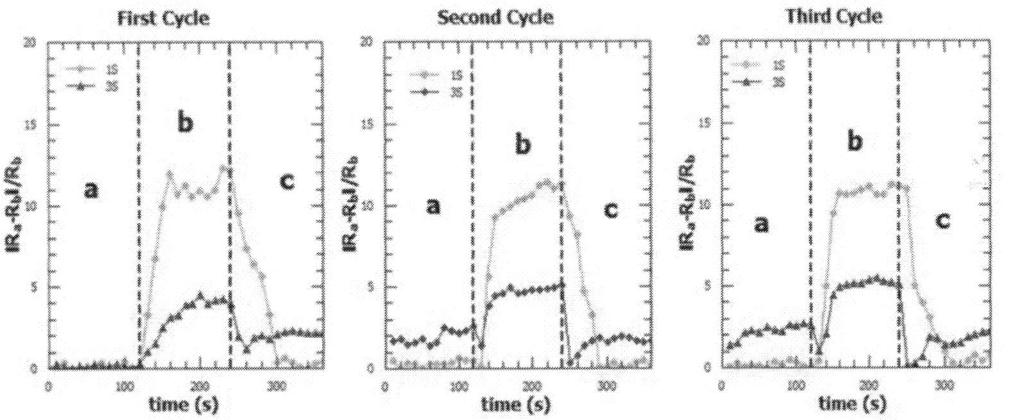

Figure 3. The sensitivity plot of ZnS samples deposited on SOI substrate (a) before, (b) during and (c) after exposure to HOAc gas. (blue-0.03M $ZnCl_2$; green-0.01M $ZnCl_2$).

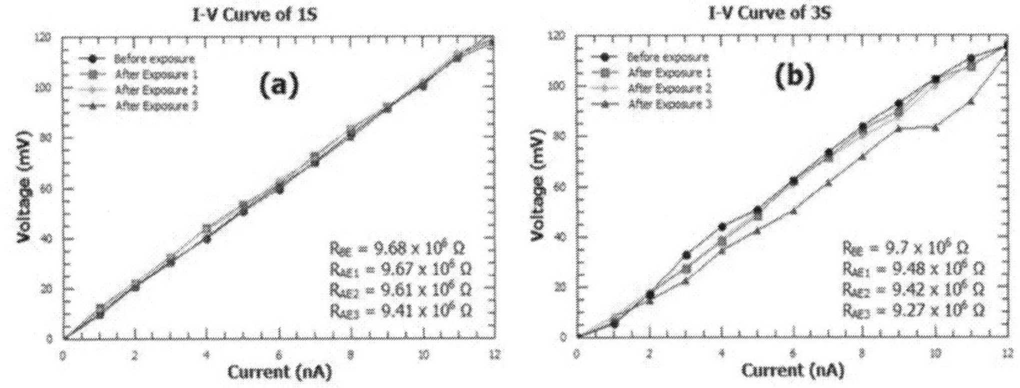

Figure 4. *I-V* curve of ZnS samples grown on SOI substrate using different $ZnCl_2$ concentrations showing the good recovery of ZnS using 0.01M $ZnCl_2$.

3.2. Surface morphology of the ZnS nanospheres

Micrographs of ZnS deposited on glass and SOI substrates are presented in Figures 5 and 6, respectively. The nanostructures are generally spherical in shape. The morphology of the samples grown on glass substrate revealed grainy structures randomly grouped together forming islands as shown in Figure 5(a). It is observed that compact and well defined nanospheres are grown as shown at higher magnification in Figure 5(b) having average diameter of ~ 195 nm.

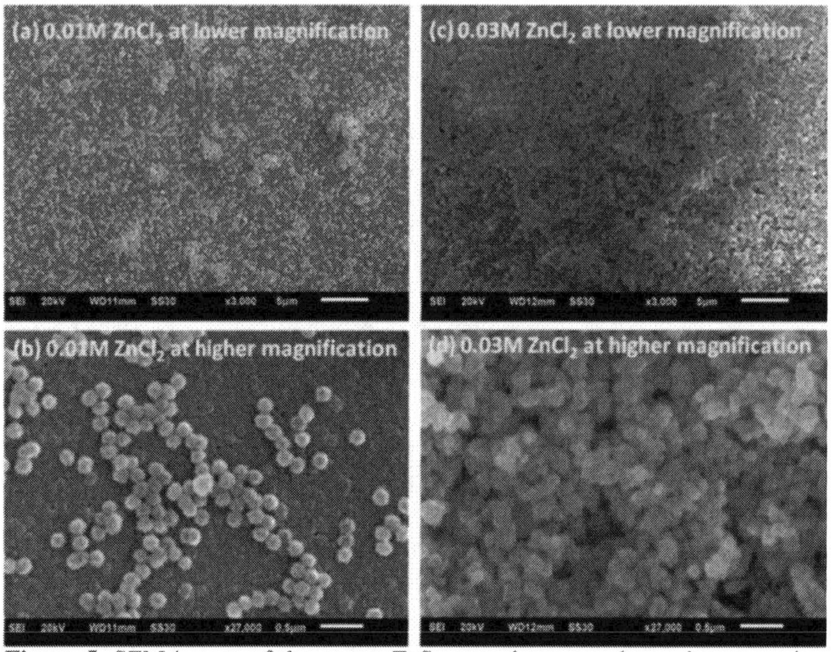

Figure 5. SEM images of the grown ZnS nanospheres on glass substrates using 0.01M $ZnCl_2$ at (a) 3000x and (b) 27000x magnification and using 0.03M $ZnCl_2$ at (c) 3000x and (d) 27000x magnification.

However, at higher concentration of $ZnCl_2$ (Figure 5(c)), dense islands composed of nanospheres that have random sizes and are less defined, randomly covered the whole surface of the substrate. Agglomeration of nanospheres is very apparent at higher magnification as shown in Figure 5(d). This agglomeration is due to the super-saturation of Zn^{+2}ions that add more nucleation sites for other atoms to attach or bind to Zn. This would increase the number of nanospheres that will be adsorbed at the surface resulting to piling up of the uneven distribution of nanospheres. This supports our claim that this is one possible reason in its difficulty in recovering to its original state after exposure to HOAc as observed in Figure 2.

On the other hand, the surface morphology of the ZnS nanostructures grown on SOI revealed uniformly distributed nanospheres as shown in Figure 6(a). At higher magnification (Figure 6(b)), compact and well-defined nanospheres are observed. Additionally, using higher concentration of $ZnCl_2$ revealed randomly agglomerated nanospheres and less-defined nanospheres forming islands as shown in Figure 6(c). As observed from the SEM images shown in Figure 5 and 6, both substrates revealed that at higher concentration of $ZnCl_2$ would produce random islands composed of less defined nanospheres that will agglomerate. Higher concentration of $ZnCl_2$ in the solution will increase the number of Zn^{2+}ions that participate in the nucleation increasing the possibility of collisions between particles as they move randomly. In contrast, ZnS nanospheres grown on SOI substrates are smaller (~152 nm) and uniformly distributed compared to those nanospheres grown in glass substrates even with same concentration of $ZnCl_2$. The less-wetting property of SOI surface than that of the glass by water of aqueous chemical bath could be the explanation of the smaller grain size of ZnS grown on SOI. The elemental composition of the grown ZnS nanostructures grown in different substrates is shown in Table 1. It is noted that the Zn to S ratio has a value close to one suggesting that the nanospheres are most purely composed of ZnS. Furthermore, it is observed that at lower concentrations, higher Zn content is present.

Table 1. Elemental composition of the ZnS nanospheres with varying concentration of $ZnCl_2$ grown on different substrates using EDX measurement.

Substrate	$ZnCl_2$	% Zn	% S	[Zn]/[S]
glass	0.01M	51.04	48.96	1.04
glass	0.03M	47.74	52.26	0.91
SOI	0.01M	50.78	49.22	1.03
SOI	0.03M	50.69	49.31	1.02

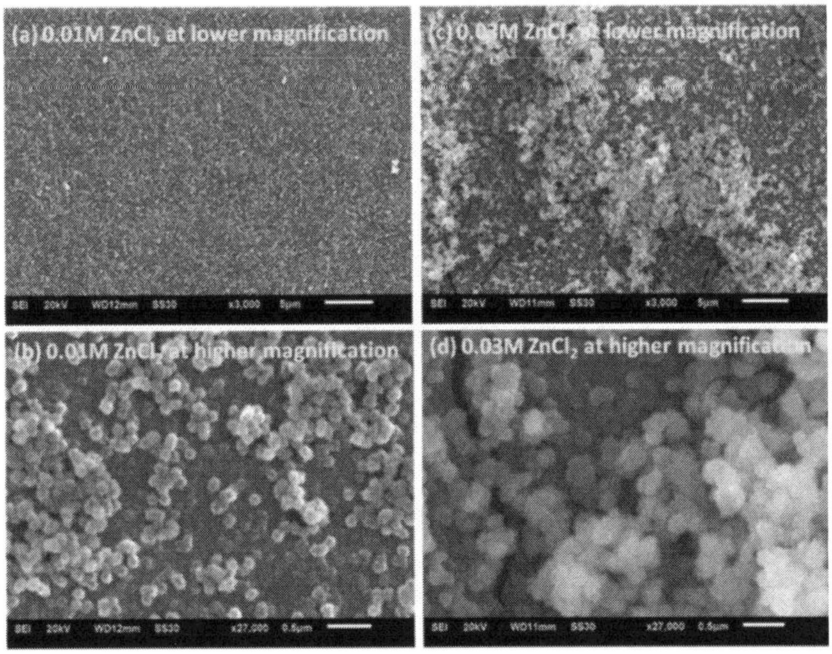

Figure 6. SEM images of the grown ZnS nanospheres on SOI substrates using 0.01M $ZnCl_2$ at (a) 3000x and (b) 27000x magnification and using 0.03M $ZnCl_2$ at (c) 3000x and (d) 27000x magnification.

4. Conclusions

ZnS nanostructures fabricated as an HOAc sensor were successfully grown on glass and SOI substrates via chemical bath deposition using acidic solution of 0.01M and 0.03M $ZnCl_2$ and is found out to be responsive to HOAc as analyte gas. The surface morphology of the grown samples is influenced by the concentration of $ZnCl_2$ and type of substrate used. Agglomerated nanospheres are observed on glass and SOI substrates at higher concentration of $ZnCl_2$.

It was also observed that the surface morphology would have an influence on the sensitivity of the fabricated ZnS-based sensor. Among the ZnS nanostructures produced, sample deposited on SOI substrate using 0.01M $ZnCl_2$ showed good response and exhibited good recovery whenever the analyte would be cut-off from the environment. ZnS nanostructures showed good potential as an inexpensive alternative material to current HOAc sensor in the market.

5. References

[1] Jiao Z, Wu M, Qin Z, Gu J and Lu M 2003 *Sensors and Transducers Magazine* **37** 24–9
[2] Gana T, Hua C, Chen Z and Hua S 2011 *Talanta* **85** 310–16
[3] Fang X, Zhai T, Gautam U K, Li L, Wu L, Bando Y and Golberg D 2011 *Progress in Materials Science* **56** 175–287
[4] PawarS M, PawarB S, Kim J H, Joo O S and Lokhande C D 2011 *Current Applied Physics* **11** 117–61

Acknowledgment

The authors would like to thanks the Office of the Vice-Chancellor for Research and Extension (OVCRE) of the Mindanao State University-Iligan Institute of Technology (MSU-IIT) for funding this research work.

Growth of chemically deposited ZnO and ZnO-SiO_2 on Pt buffered Si substrate

A D M V Peguit[1], R T Candidato Jr, F R G Bagsican, M K G Odarve, M E Jabian, B R B Sambo, R M Vequizo and A C Alguno

Department of Physics, College of Science and Mathematics,
MSU-Iligan Institute of Technology, Tibanga, Iligan City, 9200 Philippines

E-mail: deamarie.peguit@gmail.com

Abstract. Growing ZnO on Si via low-cost CBD is difficult owing to the large lattice mismatch between ZnO and Si and the intricate control of nanoparticle aggregation. In this work, a Pt buffer layer and addition of SiO_2 on the chemical solution were introduced. The effect of these parameters on the resulting morphology and composition were investigated using SEM-EDX and FTIR. Pt-coated Si showed higher density of ZnO nanostructure growth than bare Si due to the additional nucleation sites provided by Pt. Moreover, SiO_2 addition resulted to a different ZnO nanostructure.

1. Introduction

The drive for nanosized semiconductor led to a large number of researches on different semiconductor for various applications. One of the most promising materials available is zinc oxide (ZnO) which has a direct wide bandgap of 3.37 eV, a large exciton binding energy of 60meV at room temperature and a stable wurtzite structure [1]. Aside from those mentioned, ZnO is easily acquired and is a relatively low cost material making it a popular choice as semiconductor. Also, it can be grown on various substrates using chemical bath deposition (CBD), a low cost simple deposition method for synthesizing a wide variety of ZnO nanostructures [2]. But some substrates such as the extensively used sapphire are very expensive and are electrically insulating materials that introduce difficulty in device fabrication. Alternatively, silicon (Si) substrates have higher carrier mobility and can be easily integrated with on-chip circuitry and Si-based devices. However, the growth of ZnO is particularly challenging due to the large lattice mismatch between ZnO and Si.

This problem can be solved by the introduction of a buffer layer to facilitate ZnO growth. Platinum (Pt) is a good candidate due to its smaller ZnO plane misfit [3]. In addition, there is a tendency of ZnO to adsorb strongly on the noble metals such as Pt since they are known to be good catalyst for deposition. Also, Pt displayed good stability under harsh conditions due to its resistance to oxidation [4] and Pt-coating on substrates showed enhancement of electrical contact during growth. Though there is inadequate works done focusing on its effect, it is believed that the metal coating of the substrate has an influence on the morphology, size and shape of ZnO [5]. In spite of this, ZnO

[1] To whom any correspondence should be addressed.

synthesis in the desired dimension specifically as nanosized structure has limitations due to difficulties in controlling the nanoparticle aggregation [6].

Researches focusing on the surface modification of ZnO found out that the addition of surface cappers such as the chemically stable silica (SiO_2) will inhibit particle growth and aggregation [7] leading to better control on the shape and size of the ZnO structures in the particle growth solutions resulting to higher surface to volume ratio. Furthermore, the addition of SiO_2 which acts as an embedding matrix modifies defect structures and transition mechanisms [8]. But most studies done concerning silica-added ZnO (ZnO-SiO_2) are synthesized using sol-gel technique that yielded powders instead of depositing these nanostructures on substrates.

Despite it being a fascinating area of research, there are limited amount of literature available that tackles it. This problem leaves a dent in research advancement such as the morphology, growth behaviour and unique properties of ZnO-SiO_2 deposited on commonly used Si substrates which could possibly lead to interesting discoveries. This study therefore examines the effect of Pt buffer layer to the growth of chemically deposited ZnO and ZnO-SiO_2 by comparing their growths on bare and Pt-coated Si substrate. This paper would also explore the possibility of enhancing the ZnO structures grown on Si by the addition of silica powders. The resulting morphology and composition were examined.

2. Experimental details

In this work, ZnO and ZnO-SiO_2 nanostructures were both synthesized using ammonium zincate solutions from 0.03M zinc sulphate and 1.0M ammonium hydroxide. The solutions were stirred at 360 rpm for thirty minutes to allow proper mixing of the precursors. But in preparing ZnO-SiO_2, silica was added after mixing the precursors. Both solutions were then immersed in a water bath and heated to 70°C while stirring them at 360 rpm. Pre-cleaned silicon (Si) substrates, with and without Pt-coating, were then immersed in the solution baths for five hours allowing ZnO and ZnO-SiO_2 nanostructures to deposit on the substrates. The substrates were then taken out of the bath and rinsed with distilled water preheated to 70°C to remove loosely adhering particles. The fabricated samples were scanned using JEOL JSM-6510LA Analytical Scanning Electron Microscope (SEM) coupled with Energy Dispersive X-ray Spectroscopy (EDS) employing 20kV accelerating voltage to obtain micrographs of the morphology of the deposited structure and its composition. Also, the samples were scanned in the range of 650-4000cm^{-1} via Perkin Elmer Spectrum 100 FT-IR spectrometer to identify the functional groups present.

3. Results and Discussions

SEM micrographs of ZnO deposited on bare and Pt-coated Si substrates are obtained and displayed in Figure 1. Figure 1(a) reveals that there are very few structures grown directly on the substrate. A larger magnification, shown as Figure 1(b), indicates that these round structures are urchin-like clusters made up of radially adhering hexagonal nanorods with flat tips averaging at 478 nm wide. EDS data of these structures, listed in Table 1, indicate very high Zn and O content which confirms that these are actually ZnO structures. Figure 1(c), on the other hand, shows that the Pt-coated Si surface have numerous grown structures. It is obvious that this surface has higher growth density of structures as compared to the bare Si substrate. Greater magnification, shown as Figure 1(d), reveals that clusters of flat-tipped hexagonal nanorods are grown on the surface. Growth aggregation can also be observed as thinner rods grow along the structure of thicker rods. These thinner rods having an average diameter of 122 nm accumulate in the spacings between thicker rods having an average diameter of 325 nm. Both the thin and thick rod structures are believed to be ZnO as indicated by their EDS data.

These results show that ZnO deposited on bare Si have lower growth density of structures compared to those deposited on Pt-coated substrate. The ZnO growth density difference between these substrates can be attributed to the larger lattice mismatch on ZnO-Si than on Pt-coated ZnO. This eventually led to the difficulty in growing ZnO on bare Si and also explains the higher growth density on Pt-coated Si substrate. Aside from that, the presence of Pt makes ZnO easier to nucleate. The Pt layer had acted as a nucleant layer promoting the growth of additional nucleation sites for ZnO to adhere. This is also supported by a more stoichometric [Zn]/[O] ratio of 1.75 for Pt-coated Si compared to 2.24 ratio for bare Si. In addition, during the deposition of ZnO, a secondary growth of ZnO was observed. The cluster of thicker nanorods of ZnO served as spherical seed for the growth of thinner ZnO nanorods which were preferentially formed on these clusters. These primary clusters served as building blocks for the dense growth of more ZnO nanorods.

Table 1. EDS data of ZnO deposited on bare and Pt-coated Si substrate.

Substrate	%O	%Zn	[Zn]/[O]
bare Si	30.86	69.14	2.24
Pt-coated Si	36.38	63.62	1.75

Figure 1. SEM micrographs of ZnO deposited on bare Si substrates at (a)1000x and (b)10,000x magnification and on Pt-coated Si substrates at (c)1000x and (d)10,000x magnification.

These results show that ZnO deposited on bare Si have lower growth density of structures compared to those deposited on Pt-coated substrate. The ZnO growth density difference between these substrates can be attributed to the larger lattice mismatch on ZnO-Si than on Pt-coated ZnO. This eventually led to the difficulty in growing ZnO on bare Si and also explains the higher growth density on Pt-coated Si substrate. Aside from that, the presence of Pt makes ZnO easier to nucleate. The Pt layer had acted as a nucleant layer promoting the growth of additional nucleation sites for ZnO to adhere. This is also supported by a more stoichometric [Zn]/[O] ratio of 1.75 for Pt-coated Si compared to 2.24 ratio for bare Si. In addition, during the deposition of ZnO, a secondary growth of ZnO was observed. The cluster of thicker nanorods of ZnO served as spherical seed for the growth of thinner ZnO nanorods which were preferentially formed on these clusters. These primary clusters served as building blocks for the dense growth of more ZnO nanorods.

In contrast, ZnO-SiO_2 deposited on bare Si, shown in Figure 2(a), exhibited a fibrous-like layer with cracks that evenly covered the whole substrate. A larger magnification, shown in Figure 2(b), revealed that the layer wrapping the surface has very high Si content. On the other hand, ZnO-SiO_2 deposited on Pt-coated silicon, shown in Figure 2(c), had formed structures on the Si surface. The surface has low density of growth but is better compared to bare Si which showed no rod-like structure at all. A higher magnification, indicated as Figure 2(d), shows that these structures are clustered together and appeared to form flower-like structures. These structures have radially adhering rods that narrows towards its tips. These rods are not smooth but are irregularly shaped. Its EDS data, also, revealed very high Zn content. These results further support the belief stated earlier about the role of Pt on the growth. The presence of Pt layer provided nucleation sites that promote the growth of ZnO structures on the surface.

Table 2.EDS data of ZnO-SiO_2 deposited on bare and Pt-coated Si substrate.

Substrate	%Si	%O	%Zn	[Zn]/[O]
bare Si	50.11	39.89	10.00	0.25
Pt-coated Si	10.58	18.91	70.51	3.73

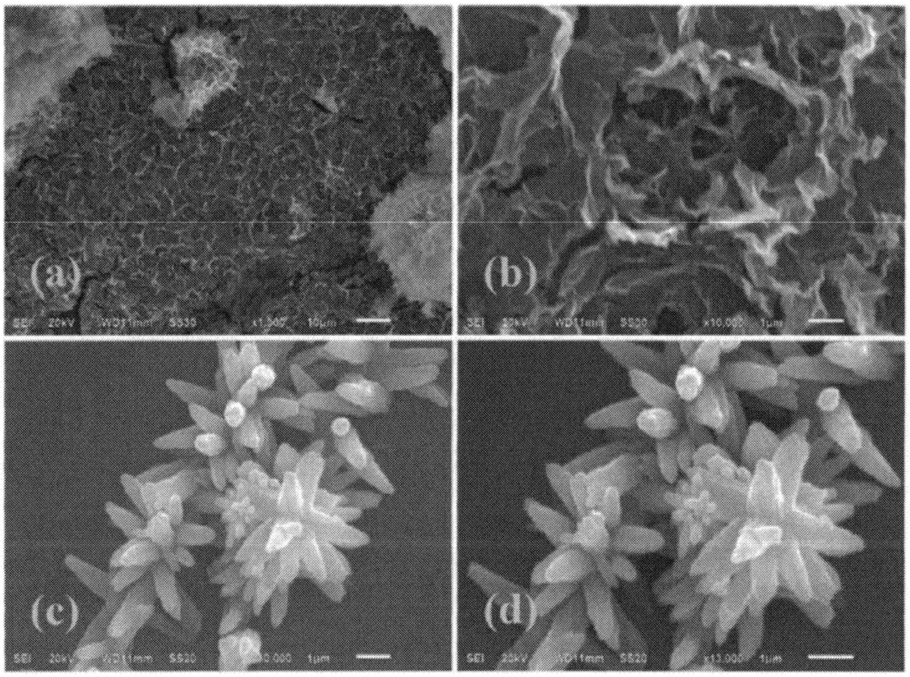

Figure 2. SEM micrographs of ZnO deposited on (a) bare and (b) Pt-coated Si substrates and ZnO-SiO_2 deposited on (c) bare and (d) Pt-coated Si substrates.

Comparison of Figures 1 and 2 gives insight to the effect of SiO_2 addition to the structure. Without adding SiO_2, the structures formed were urchin-like clusters made up of hexagonal rods with flat tips. However, the addition of SiO_2 has significantly changed the morphology on the surface. For bare Si, a SiO_2 layer was observed while for Pt-coated Si, a flower-like structure with less aggregation was

IOP Conf. Series: Materials Science and Engineering **79** (2015) 012026 doi:10.1088/1757-899X/79/1/012026

perceived. This morphological modification was clearly due to the addition of SiO_2 into ZnO. Since SiO_2 is a nonpolar network, it attaches itself to the non-polar facets of ZnO and leaves the polar face exposed for crystal growth. That attachment to the non-polar sides inhibits the would-be-hexagonal structure into something different such as that seen in Figure 2(d). The adherence of SiO_2 on the non-polar sides leaves ZnO growth only to its polar facets thus significantly decreasing the aggregation of structures.

The bonding interactions are also investigated using FTIR spectroscopy with results shown in Figure 3. The peaks detected in 706-789 cm^{-1}range is due to Zn-O bonding. These peaks are visible to all samples indicating the presence of ZnO deposition for all samples. Also, the points in the range of 3220-3240 cm^{-1}, 1391-1508 $cm^{-1}$and 1030-1090 $cm^{-1}$are attributed to Si-OH vibrations, Si-O-Si bending and Si-O-Si stretching vibrational modes, respectively. These peaks are present to ZnO on bare Si and to ZnO-SiO_2 on both substrates and are absent to ZnO on Pt-coated substrate further supporting the SEM-EDX results. ZnO was directly deposited on bare Si thus Si-O bonds are expected unlike ZnO deposited on Pt-coated Si where Si-O bonds are deficient due to the presence of Pt between ZnO and the Si substrate. This result implies that the Pt acted as a buffer that discouraged Zn or O interaction to Si substrate.

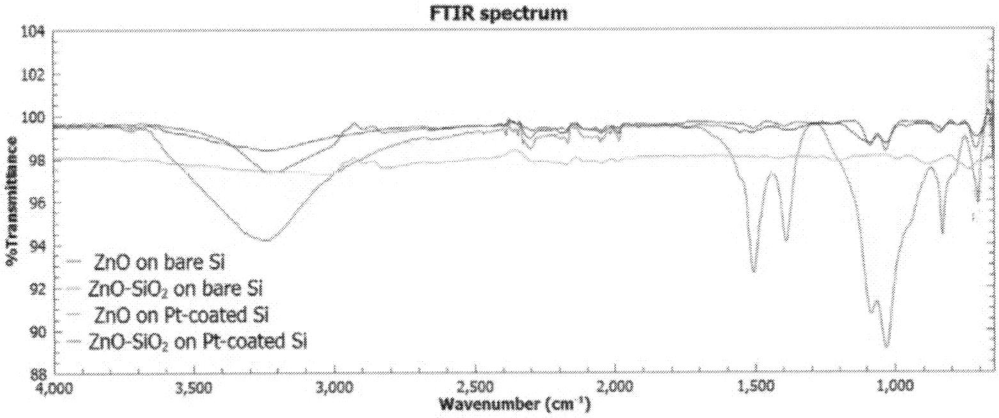

Figure 3. FTIR spectrum of ZnO on bare Si (black line) and Pt-coated Si substrates (green line) and ZnO-SiO_2 on bare Si (red line) and Pt-coated Si substrates (blue line).

The same can be said with ZnO-SiO_2 on bare Si, however the addition of SiO_2 means that there will be more Si-O bonds than on bare Si which is obvious with its FTIR spectrum (red line) results. This further suggests that the layer seen on Figure 2(a) and 2(b) is composed mainly by SiO_2 as verified by its EDS data. Also, ZnO-SiO_2 on Pt-coated Si still registered Si-O bonds due to the addition of SiO_2 thus validating the absorption peaks observed in its FTIR spectra (blue line). Since Pt-coated the substrate, Si-O bonds detected in its IR spectra suggests that SiO_2 may have adhered with the non-polar sides of the grown ZnO nanostructure. This results to a different nanostructure and the reason why hexagonal rods are not detected in its morphology.

4. Conclusion

ZnO and ZnO-SiO_2 were grown on bare and Pt-coated Si substrates. Few hexagonal ZnO nanostructures adhered to bare Si while numerous hexagonal ZnO nanostructures grow on Pt-coated Si substrate. The catalytic Pt buffer layer promotes growth of ZnO by providing additional nucleation sites on the surface. In contrast, ZnO-SiO_2 deposited on bare Si revealed only a porous layer with the absence of any ZnO structure however ZnO-SiO_2 grown on Pt-coated Si showed low density of irregularly shaped rods. The adhesion of SiO_2 network discourages the formation of ZnO hexagonal

IOP Conf. Series: Materials Science and Engineering **79** (2015) 012026 doi:10.1088/1757-899X/79/1/012026

structures. This result is further supported by the FT-IR where numerous Si-O-Si functional groups were detected particularly to SiO_2 added ZnO samples.

5. References

[1] Hassan N K, Hashim M R, Mahdi M A and Allam N K 2012 *ECS Journal of Solid State Science and Technology* **1** 86–9

[2] Lokhande C D, Gondkar P M, Mane R, Shinde V R and Han S H 2009 *Journal of Alloys and Compounds* **475** 304–11

[3] Yamada H, Ushimi Y, Takeuchi M, Yoshino Y, Makino T and Arai S 2004 *Vacuum* **74** 689–92

[4] Duclère J R, Mc Loughlin C, Fryar J, O'Haire R, Guilloux-Viry M, Meaney A, Perrin A, McGlynn E, Henry M O and Mosnier J P 2006 *Thin Solid Films* **500** 78–83

[5] Bakar M, Hamid M and Shamsudin R 2013 *Journal of Physics: Conference Series* **431**

[6] Lihitkar P B, Violet S, Shirolkar M, Singh J, Srivastava O N, Naik R H and Kulkarni S K 2012 *Materials Chemistry and Physics* **133** 850–6

[7] Yang Y, Li Y Q, Shi H Q, Li W N, Xiao H M, Zhu L P, Luo Y S, Fu S Y and Liu T 2011 *Composites: Part* B **42** 2105–10

IOP Conf. Series: Materials Science and Engineering **79** (2015) 012027 doi:10.1088/1757-899X/79/1/012027

Surface geometry based hydrophobicity of the PDMS for microfluidic devices

J C Pelayo[1,3] , R A Badiola[1], J Castañares[2], U Pili[1], R Violanda[1] and R Bacabac[1]

[1]Department of Physics, University of San Carlos, Nasipit, Talamban, Cebu City, 6000, Philippines

[2]Department of Chemistry, University of San Carlos, Nasipit, Talamban, Cebu City, 6000, Philippines

E-mail: carlospelayo38@gmail.com

Abstract. In this report, the surface hydrophobicity of PDMS was investigated using two methods of preparations. The first method was performed by changing the surface roughness through the use of different molds. The second method was performed by varying the reconstitution ratio (volume of elastomer base to volume of elastomer curing) of the PDMS. Variation in the hydrophobicity of the PDMS was characterized by measuring the contact angle of a liquid droplet against the surface of the PDMS. The results showed that both the surface roughness and the reconstitution ratio of the PDMS positively correlated with the contact angle measured regardless of the liquid used. The maximum and minimum contact angle obtained were $\theta_r = (138 \pm 3)°$ and $\theta_r = (99 \pm 3)°$, respectively. The results demonstrate a straightforward way of fabricating microfluidic devices using PDMS with controlled hydrophobicity.

1. Introduction

Microfluidic devices have driven considerable progress in the development of microsystems, predominantly for chemical, biological and medical applications. In the past years microflucidic devices has evolved from silicon and glass structures towards polymers. The commonly used polymer in this field is PDMS (polydimethylsiloxane) which generated interest due to its low cost and ease of fabrication and its properties which includes: thermal stability, transparency and its surface properties. However PDMS surface properties needs to be modified in order to achieve desirable functionality [1]. This modification leads to the degree of wetting of the PDMS, particularly its hydrophobic or hydrophilic nature. This property of the material can be altered by many ways ranging from different etching methods to chemical alterations. This can be characterized through the contact angle of a droplet with the surface of PDMS, these variations in the hydrophobicty of PDMS is essential in different applications, such as in cell culture, self-cleaning of the device and others [2].

[3] To whom any correspondence should be addressed.

2. Theory and methodology

2.1. Fabrication

The main material used in this experiment was PDMS. This was reconstituted by mixing the elastomer base and the elastomer curing agent with a 10:1 ratio. In order to remove the unnecessary bubbles, the PDMS was degassed using a bell jar for a time interval of 4 hrs. Mold designs for the surfaces were made afterwards. The designs that were chosen were the molds for paper, vinyl sticker, and aluminum foil tape. These were made by cutting a part of each of the materials and pasting it on a glass slide. The elastomer was then poured into the design (which is attached to the slide) and was cured in an oven with a temperature of 80 ℃ for an hour. When the curing was done, the PDMS was peeled off the slide and the resulting surfaces for the three materials were obtained.

2.2. Characterization

Surface roughness was characterized by taking an image of the surface of the PDMS under a bright field microscope and a histogram of the gray level image was then obtained. By fitting the histogram to a Gaussian function and obtaining the FWHM (Full Width at Half Maximum) the distribution of the pixel value would then indicate the roughness of the surface.

Surface interaction between a solid medium and a liquid is often characterized by the degree of wetting of the surface, hence the contact angle of a liquid to the surface of a solid. The Cassie-Baxter equation for a heterogeneous surface is shown below [3],

$$\cos\theta_r = rf\cos\theta - (1-f) \tag{1}$$

where θ_r is the contact angle between the surface and the droplet, r is the roughness ratio, $\boldsymbol{\theta}$ is the contact angle of an ideal surface and f is the fraction of solid surface area wet by the liquid. If we decrease f we can see that θ_r should increase, the decrease in f would imply greater roughness hence, as surface roughness increases the contact angle increases. A contact angle of $\boldsymbol{\theta_r}$<90° is a hydrophilic surface and contact angle of 90°≤$\boldsymbol{\theta_r}$<150° is a hydrophobic surface and a contact angle of $\boldsymbol{\theta_r}$≥150° is a super hydrophobic surface.

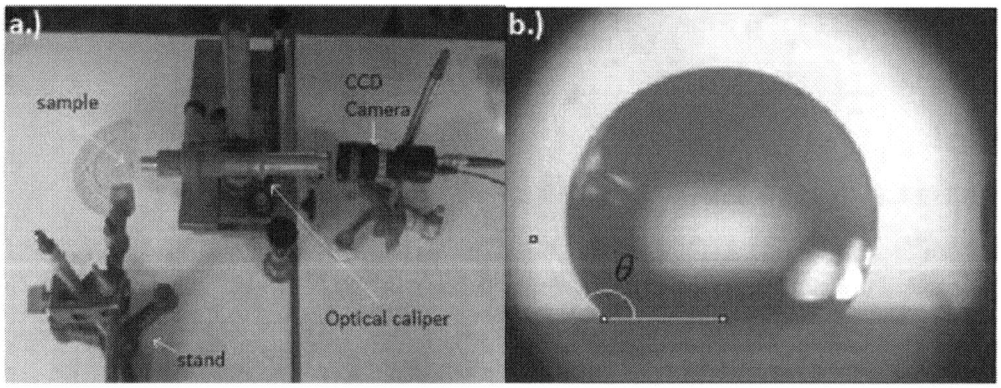

Figure 1. (a)The experimental set-up consists of an optical caliper, an iron stand, a CCD camera connected to a computer installed with National Instrument Vision Assistant software, and a sample. (b) The contact angle of the sample droplet was obtained using the software ImageJ.

After fabricating the surfaces, the next step was obtaining the contact angle of a droplet for each of the sample. The experimental set-up for the contact angle measurement is shown in figure 1a. The set-up consists of an optical caliper, a CCD camera connected to a computer with National Instrument Vision Assistant software, and a stand for the sample. A droplet of water with a volume of >5$\boldsymbol{\mu L}$ was placed on its surface using a pipette and an image of the droplet on the surface was taken. Using the software ImageJ, the contact angle of the droplet was obtained as shown in figure 1b. This was repeated for different liquids namely, Hydrochloric acid (1M), methyl orange, Sodium Hydroxide (1M),solution of HCl and methyl orange and distilled water. The PDMS was also reconstituted using a different elastomer base to elastomer curing ratios of 10:2, 10:1, 10:0.5 and 10:0.25. The surface of the paper and sticker was used as the mold for this part of the experiment and the only liquid use was water. The contact angle was also measured for the varied reconstitution ratio.

3. Results and discussions

From the FWHM of the different surfaces it can be inferred that the PDMS molded in paper proved to be the roughest surface, followed by the Al foil tape and sticker. The greater the FWHM the rougher should be the surface since the tendency of a rough surface is to scatter light in many directions hence the distribution of grey pixel value widens.

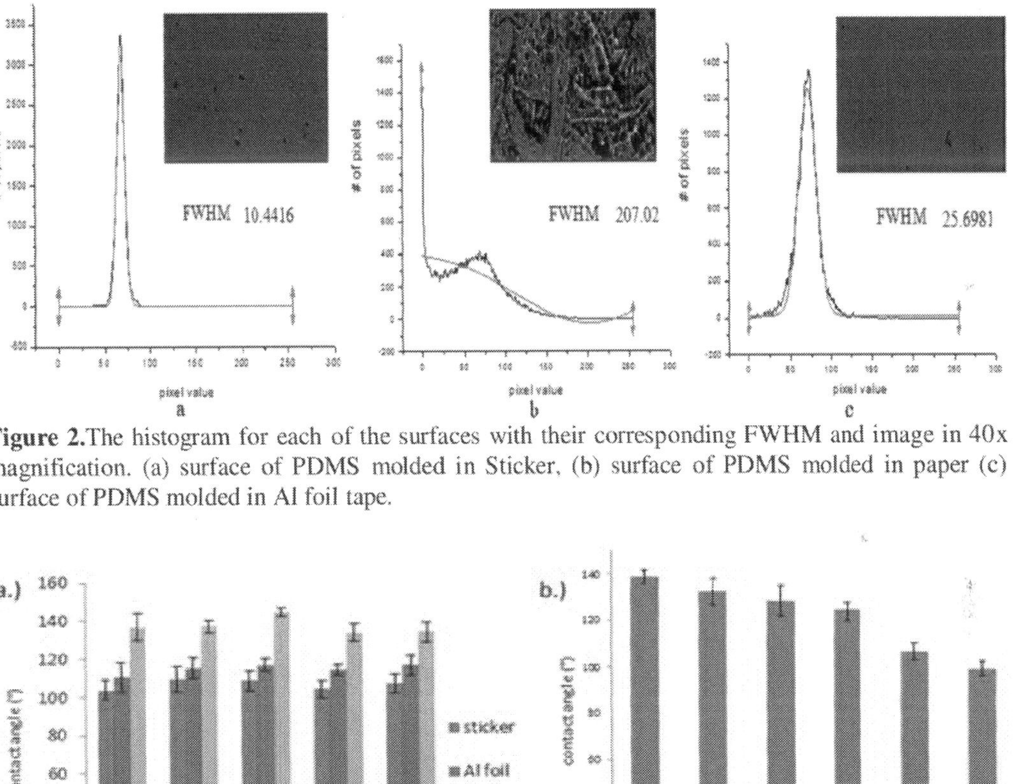

Figure 2. The histogram for each of the surfaces with their corresponding FWHM and image in 40x magnification. (a) surface of PDMS molded in Sticker, (b) surface of PDMS molded in paper (c) surface of PDMS molded in Al foil tape.

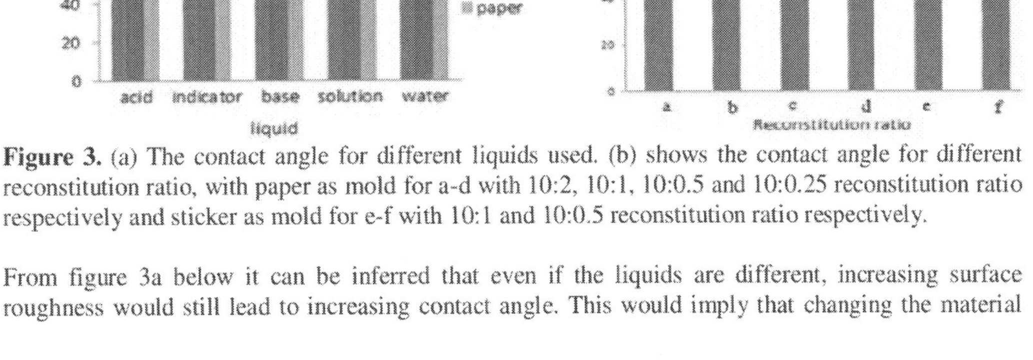

Figure 3. (a) The contact angle for different liquids used. (b) shows the contact angle for different reconstitution ratio, with paper as mold for a-d with 10:2, 10:1, 10:0.5 and 10:0.25 reconstitution ratio respectively and sticker as mold for e-f with 10:1 and 10:0.5 reconstitution ratio respectively.

From figure 3a below it can be inferred that even if the liquids are different, increasing surface roughness would still lead to increasing contact angle. This would imply that changing the material used as a mold would directly correlate to a change in the hydrophobicity of the PDMS and holds true for a large number of liquid. However the effect in the interaction of different liquids with the surface of PDMS cannot be determined given the data, since the differences in the contact angles are inconclusive. From figure 3b below it can be inferred that by increasing the reconstitution ratio between the elastomer base and the elastomer curing, yields a positive correlation with the contact angle of the liquid and implies that the reconstitution ratio affects the hydrophobicity of the PDMS. This effect is attributed to the change in adhesion of the PDMS as the reconstitution ratio is changed.

4. Conclusion

The hydrophobicity of the PDMS was investigated using two methods, first by using different molds that would act as the surface of the PDMS, and the other was achieved by changing the reconstitution ratio of the PDMS. By combining the two methods, the maximum and minimum contact angle obtained are $\boldsymbol{\theta_r} = (\mathbf{138 \pm 3})°$and $\boldsymbol{\theta_r} = (\mathbf{99 \pm 3})°$ respectively. The maximum contact angle was about 10° less, to consider the surface as super hydrophobic while the minimum contact angle was about 10° greater, to consider the surface as hydrophilic. These results demonstrate a straightforward way of fabricating microfluidic devices using PDMS with controlled hydrophobicity.

5. References

[1] Fuard D *et al.* 2008 *Microelectronic Engineering* **85** 1289–93
[2] Zimmerman W B 2011 *Chemical Engineering Science* **66** 1412–25
[3] Ingram P *et al.* 2011 *15th International Conference on Miniaturized Systems for Chemistry and Life Sciences* 1539–41
[4] Jin M *et al.* 2005 *Macromol.Rapid Commun.* **26** 1805–9

IOP Conf. Series: Materials Science and Engineering **79** (2015) 012028 doi:10.1088/1757-899X/79/1/012028

X-ray diffraction and infrared spectroscopy analyses on the crystallinity of engineered biological hydroxyapatite for medical application

G M Poralan Jr[1,3], **J E Gambe**[1],**E M Alcantara**[2] **andR M Vequizo**[1]

[1]Materials Science, Laboratory, Physics Department, MSU - Iligan Institute of Technology, A. Bonifacio Avenue, Tibanga, 9200 Iligan City, Philippines

[2]College of Arts and Sciences, Capitol University, Corrales Avenue Osmeña St., 9000 Cagayan de Oro City, Philippines

Email: gilbertporalan@gmail.com, vequizorey2@gmail.com, diyes2003@gmail.com, eric_gordon19@yahoo.com

Abstract. Biological hydroxyapatite (BHAp) derived from thermally-treated fish bones was successfully produced. However, the obtained biological HAp was amorphous and thus making it unfavorable for medical application. Consequently, this research exploits and engineers the crystallinity of BHAp powders by addition of $CaCO_3$ and investigates its degree of crystallinity using XRD and IR spectroscopy. On XRD, the HAp powders with [Ca]/[P] ratios 1.42, 1.46, 1.61 and 1.93 have degree of crystallinity equal to 58.08, 72.13, 85.79, 75.85% and crystal size equal to 0.67, 0.74, 0.75, 0.72 nm, respectively. The degree of crystallinity and crystal size of the obtained calcium deficient biological HAp powders increase as their [Ca]/[P] ratio approaches the stoichiometric ratio by addition of $CaCO_3$ as source of Ca^{2+} ions. These results show the possibility of engineering the crystallinity and crystal size of biological HAp by addition of $CaCO_3$. Moreover, the splitting factor of PO_4 vibration matches the result with % crystallinity on XRD. Also, the area of phosphate-substitution site of PO_4 vibration shows linear relationship (R^2=0.994) with crystal size calculated from XRD. It is worth noting that the crystallinity of the biological HAp with [Ca]/[P] ratios 1.42 and 1.48 fall near the range 60-70% for highly resorbable HAp used in the medical application.

1. Introduction

Hydroxyapatite (HAp) with the chemical formula $Ca_{10}(PO_4)_6(OH)_2$ is one of the most important bioceramic material that has been extensively used in orthopedics and dentistry for implant fabrication and drug delivery system owing to its biocompatibility and bioactivity [1]. Numerous types of natural or synthetic HAp are commercially available not only for use in the biomedical field but also for chromatographic and catalytic applications [2]. In addition, it has been an excellent material for ion exchange material [2], adsorbent [2], ionic conductor [2], chemical gas sensor [3], fuel cells [4], physical sunscreen and sunscreen firming [5].

Hydroxyapatite derived from thermally-treated fish bones was successfully produced. However, the obtained biological HAp is calcium deficient and amorphous. In medical application, amorphous HAp

[3] To whom any correspondence should be addressed.

is unfavorable when used for bone graft because of its very fast dissolution rate. Recent studies found out that the resorbability of HAp can be developed by controlling its degree of crystallinity to 60–70% and reducing its grain size to nanolevel [6,7].This study aims to investigate the HAp powders with crystallinity useful in the medical application by engineering the produced amorphous HAp from thermally-treated fish bones by addition of $CaCO_3$. X-ray diffraction (XRD) is used to characterize the phase composition and degree of crystallinity of the produced biological HAp. Additionally, infrared spectroscopy is used to elucidate the degree of crystallinity of HAp by establishing an index of crystallinity.

Considering the crystallography of HAp, it occurs as a hexagonally packed crystal belonging to $P6_3/m$ space group. The unit cell consists of Ca^{2+}, PO_4^{2-}, and OH^- groups closely packed together in a hexagonal arrangement. The OH^- group serves as the backbone for HAp. The six phosphates, PO_4, are in a helical arrangement around the *c*-axis. These phosphate groups form a skeletal frame structural network which provides the stability of HAp. On X-ray diffraction, the crystallite size can be described by the diffracting plane (002) of HAp at 2θ=25.50° for the reason that this miller index correspond to the *c*-axis length. The full width at half maximum (FWHM) of the 002 reflection is inversely proportional to the crystallite length along the *c* unit cell direction (*c*-axis length) [8]. Thus, as HAp becomes more crystalline, the *c*-axis length increases which can be verified by the diffraction peak for (002) becoming narrower. Furthermore, the 100% crystalline HAp has diffraction peaks for (112) and (300) at equal intensities. Thus, the change in the ratio of the intensity of the trough between the (112) and (300) peaks to the intensity of (300) diffraction peak of HAp is a strong indication of change in the degree of crystallinity.

On infrared spectroscopy, information about the molecular components and structures are presented as a spectrum of absorption bands. At 500-620 cm^{-1} spectral region, the bending mode of PO_4 vibration consists of two peaks corresponding to transverse and longitudinal optical frequency. However, amorphous calcium phosphate gives a broad single absorption band in this region because of structural distortion on the lattice. As HAp becomes more crystalline, the hexagonal structure becomes evident. Consequently, the PO_4 vibration at this region becomes well define and a doublet peak is observed. Thus, this splitting of the bending mode of PO_4 vibration on the transformation of HAp from amorphous to crystalline phase can be an index of crystallinity. Furthermore, at 900-1200 cm^{-1} spectral region, a broadening of the spectra is observed between the anti-symmetric stretch modes of PO_4 vibration at 1040 and 1090 cm^{-1}. Incorporation of different types of anions in the HAp unit cell which affects the vibration of PO_4 and induces local structural distortions around these ions might be the reason of the broadening on this region [9]. Thus, the area of phosphate-substitution site might give as information on the crystal structure of HAp. These relationships between the planes (002), (112) and (300) for HAp as observed in XRD and the phosphate absorption bands of HAp on FTIR could be the bases on the correlation of the degree of crystallinity of Hap from the two characterization techniques.

This study opens a door for the modification of HAp by addition of $CaCO_3$ to tailor HAp phase which is good for biomedical applications. Additionally, the study on the crystallinity of this obtained HAp powders would greatly help in the production of quality HAp powders since crystallinity primarily influences the properties and performance of HAp.

2. Methodology

2.1. HAp powder preparation

The bones are boiled and washed then crushed and blended. Four samples are prepared and added with 0.2, 0.3, 0.4 and 0.5g $CaCO_3$. The samples are then dissolved in a HNO_3 solution. The solution is heated at 150 °C until only the powders are left. The dried samples are contained in a crucible then calcined for 5 hrs. at 900 °C. The samples are subjected to FTIR spectroscopy for chemical purity, XRD for crystallographic properties and SEM-EDX for elemental composition. The crystallinity of the powder samples is elucidated from the obtained XRD and FTIR spectra.

2.2. Crystallinity and crystal size calculation using XRD characterization

The degree of crystallinity was calculated using diffraction peaks (112) and (300) of HAp with the equation [10] given as:

$$X_c = \frac{I_{300} - V_{112/300}}{I_{300}} \cdot 100\ [\%] \qquad (1)$$

Where X_c is the fraction of crystalline phase, I_{300} is the intensity of (300) diffraction peak and $V_{112/300}$ is the intensity of the trough between (112) and (300) diffraction peaks of HAp. The range of 2θ where the peaks of interest fall is graphed separately. These peaks are then fitted using Gussian/Lorentzian function in Qtiplot.

The average particle size is calculated from XRD data using the Debye–Scherrer approximation:

$$\tau = \frac{k\lambda}{\beta_\tau \cos\theta} \qquad (2)$$

where τ is the particle size in nanometer, as calculated for the *hkl* reflection, k is the wavelength of $Cu_{K\alpha 1}$ radiation (1.5406 Å), β_τ is the full width at half maximum for the diffraction peak under consideration (in radian), θ is the diffraction angle and k is the broadening constant chosen as 0.9.

2.3. Index of crystallinity using FTIR spectroscopy

The splitting factor is taken as the area between peaks 570 and 602cm^{-1}. The area is integrated using trapezoidal rule in Qtiplot. Afterwards, the selected region is fitted using Gaussian/Lorentzian function to obtain the FWHM of 602cm^{-1} vibration.

2.4. Correlation of area of phosphate-substitution site and crystal size

The 900-1200 cm^{-1} spectral region is graphed using Qtiplot software. Lorentzian/Gaussian function is used for curve-fitting. For each spectrum, the regression coefficient is always better than 0.998 in order to have a consistent result. Initial peak positions are obtained from second-derivative spectra of the raw data. The area of the subband at approximately 1060cm^{-1} is determined from the area given by the software from the fitting result. The calculated area of the subband is reported as a percentage of the total area of all the components. This method was presented by Pleshko *et al.* [8]. This area is labelled as %B and is graphed with respect to the crystal size calculated from XRD.

2.5. Characterizations

2.5.1. XRD characterization

X-ray diffraction (XRD) patterns are obtained in the range of 2θ = 10-70° using a step size of 0.02°/min with a Shimadzu XRD-700 x-ray diffractometer operating at 40kV and 30mA producing $Cu_{K\alpha}$ with λ= 0.154 nm. The obtained raw data are plotted using Qtiplot© free software for analysis of the samples followed by indexing the corresponding peaks using standard diffraction files of apatite (JCPDS# 09-0432), α-TCP (JCPDS# 09-0348), β-TCP (JCPDS# 09-0169), CaO (JCPDS# 37-1497), $Ca(OH)_2$ (JCPDS# 04-0733) and $CaCO_3$ (JCPDS# 29-0305).

2.5.2. FTIR characterization

FTIR transmittance spectra are obtained in the 4000-450 cm^{-1} infrared region. The obtained spectra is compared to the standard KBr and reference materials and compared to correlation charts for more precise assignments of the observed vibrational peaks.

3. Results and discussions

3.1. Crystallinityanalyses using X-ray diffractometry

The XRD patterns of the samples displayed in Figure 1 show the characteristic peaks of HAp and TCP. Although there are TCPs present, however,majority of the peaks belong to HAp. The presence of α-TCP affects the degree of crystallinity of the obtained biological HAp powders in view of the fact that TCP competes with HAp formation and causes distortions on the HAp crystal.

However, it is worth noting that as the amount of $CaCO_3$ is increased, the presence of α-TCP and β-TCP peaks become noticeably reduced with increasing growth of HAp peaks. The reason why the TCP peak weakens might be that TCP undergoes transition to HAp phase upon addition of $CaCO_3$.This implies the possibility of tailoring HAp phase by subsequent addition of $CaCO_3$.

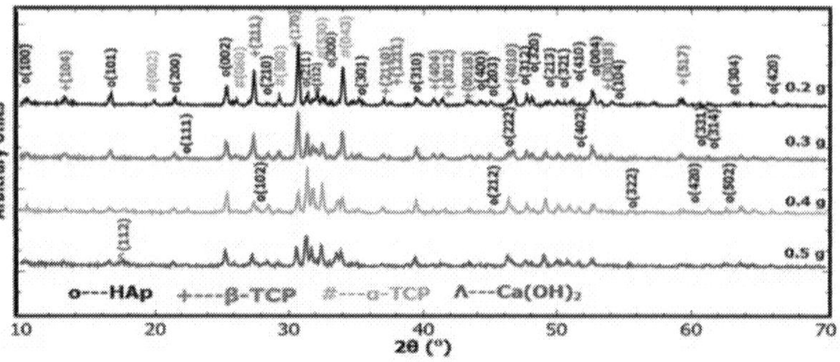

Figure 1. X-ray diffractogram of biological HAp powders with various amounts of $CaCO_3$calcined at 900°C.

The HAp powders with [Ca]/[P] ratios 1.42, 1.46, 1.61 and 1.93 have degree of crystallinity equal to 58.08, 72.13, 85.79, 75.85% and crystal size equal to 0.67, 0.74, 0.75, 0.72 nm, respectively. From these results, it can be observed that there is an increasing degree of crystallinity as the amount of $CaCO_3$ is increased except for the apatite powders with 0.50g $CaCO_3$ which has degree of crystallinity at 75.85% only.Taking into account the [Ca]/[P] ratio of the obtained HAp powders, it appears that it is the reason why the degree of crystallinity of the biological HAp sample with 0.50 g $CaCO_3$ decreases. As the [Ca]/[P] ratio approaches the value of 1.67, for stoichiometric HAp, the degree of crystallinity increases as seen from Figure 2a. The HAp sample with [Ca]/[P] ratio closest to 1.67 has the highest degree of crystallinity. However, as [Ca]/[P] ratio goes beyond 1.67, a decrease on the degree of crystallinity is observed.Although the TCP phase is the least in this sample, another impurity which is $Ca(OH)_2$ is formed which might be from the excess calcium present in the sample.

Considering the crystal size of the HAp samples, the result shows that as the HAp sample becomes stoichiometric, the crystal size increases as seen from Figure 2b. However, crystal size decreases as it goes beyond the stoichiometric ratio. Hence, as the HAp sample becomes crystalline, the crystal size increases. This is due to the increase of the crystallinity which is probably cause by the formation of the crystal structure. The results show that the degree of crystallinity and crystal size are associated with the amount of impurity and the stoichiometry of the obtained biological HAp powders.

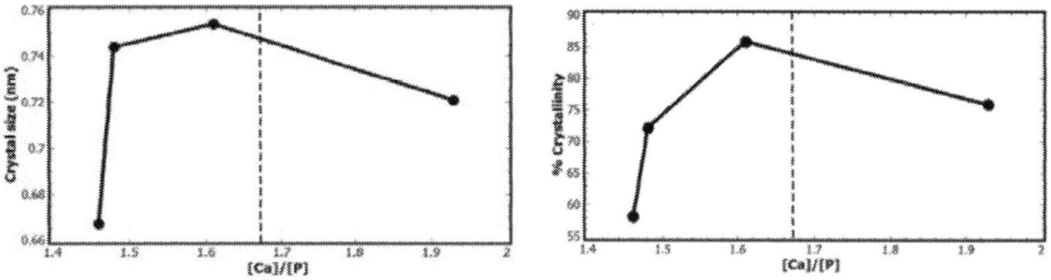

Figure 2. Relationship between (a) Xc and [Ca]/[P] ratio and (b) crystal size with respect to its [Ca]/[P] ratio.

It is worth noting that the degree of crystallinity of the obtained biological HAp samples with [Ca]/[P] ratios 1.42 and 1.48 fall on the range of the desired crystallinity for medical application. Also, the crystal sizes of these samples are at nanolevel and thus have a high surface area for good

IOP Conf. Series: Materials Science and Engineering **79** (2015) 012028 doi:10.1088/1757-899X/79/1/012028

resorption. Additionally, resorption can be regulated by combination of several calcium phosphate phases. TCPs are very soluble while HAp is stable. Combination of these two phases makes the obtained biological HAp samples added with $CaCO_3$ possess the solubility of TCPs and the stability of HAp.

3.2. Crystallinity analyses using FTIR spectroscopy

The powder IR absorption spectra of the investigated samples displayed in Figure 3 show the usual absorption bands of HAp. The intense bands observed at 900-1200 cm^{-1} and 500-620 cm^{-1} are related to internal vibrations of the PO_4 tetrahedra. They are referred to as symmetric stretch, bending mode, and anti-symmetric stretch. Another bands related to the stretching of OH groups are observed in the IR spectrum of the HAp samples at 3572 and 3643 cm^{-1}. The OH vibration seen at 3643 cm^{-1} corresponds to $Ca(OH)_2$. This peak is strongest at the biological HAp sample with 0.5g $CaCO_3$. This may be due to the excess Ca^{2+} ions on the sample that bind with OH^- ion from water or due to the CaO produced from the decomposition of $CaCO_3$. Bands related to the adsorbed H_2O are observed at 1634 and 3433 cm^{-1}. Lastly, vibrations corresponding to carbonate group is seen as a singlet peak at 875 cm^{-1} and doublet peak around 1420 and 1461 cm^{-1}. These peaks are contributed from the added $CaCO_3$ to the biological HAp samples. The spectra of the HAp samples display a systematic splitting of the anti-symmetric stretch and bending mode of PO_4 bands.

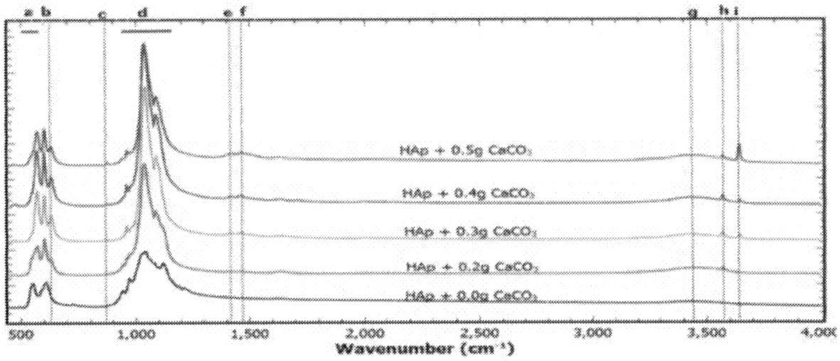

Figure 3. FTIR spectra of biological HAp powders with various amounts of $CaCO_3$ calcined at 900°C with peaks at (a) 570 and 602 cm^{-1}, (b) 630 cm^{-1}, (c) 875 cm^{-1}, (d) 963, 1094 and 1089 cm^{-1}, (e) 1420 cm^{-1}, (f) 1461 cm^{-1} (g) 3433 cm^{-1}, (h)3572 cm^{-1} and (i) 3643 cm^{-1}.

The SF for the apatite samples with [Ca]/[P] ratios 1.46, 1.48, 1.61 and 1.93 are 14.85, 16.99, 19.70 and 15.71, respectively. The results show that as [Ca]/[P] ratio approaches the stoichiometric ratio, the splitting factor increases as seen from Figure 4a. The HAp sample with [Ca]/[P] ratio closest to the stoichiometric ratio has the highest splitting factor. However, the splitting factor decreases as the [Ca]/[P] ratio exceeds 1.67. It is worth noting that these results obtain from infared spectroscopy match the results with % crystallinity obtain from XRD.

At 900 and 1200 cm^{-1} spectral region, a broadening of the spectra is observed between the anti-symmetric stretch modes at 1040 and 1090 cm^{-1}. Incorporation of different types of anions in the HAp unit cell which affects the vibration of PO_4 and induces local structural distortions around these ions might be the reason for the broadening on this region. Thus the area of phosphate-substitution site might give information on the crystal structure of HAp. This research uses the method done by Pleshko [8] on measuring the area of phosphate-substitution site named as band B. The HAp powders added with 0.2, 0.3, 0.4, 0.5 g $CaCO_3$ has %B equal to 20.88, 12.42, 12.03, 14.85%, respectively. From the results it can be seen from Figure 4b that %B and crystal size has a linear relationship with correlation coefficient (R^2) = 0.994. This suggests that the decrease of the area of phosphate-substitution site indicates increase in the crystal size. Hence the area of phosphate-substitution site might correspond to the *c*-axis length of HAp.

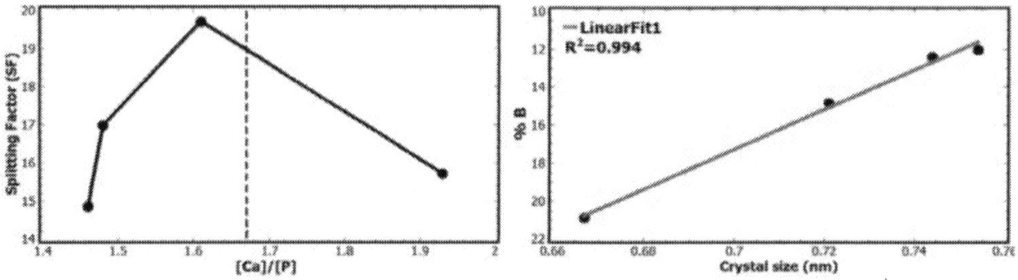

Figure 4. Graph of (a) the splitting factor (taken as the area between peaks 570 and 602 cm^{-1}) as an index of crystallinity with respect to [Ca]/[P] ratio and (b) %B with respect to crystal size showing linear relationship.

4. Conclusions

Engineering of the crystallinity of biological HAp powders from thermally-treated fish bones by addition of $CaCO_3$ was successfully done. The degree of crystallinity and crystal size of the obtained biological HAp powders were successfully calculated using diffraction planes (300), (112) and (002) in XRD. The indices of crystallinity for biological HAp were established using the splitting factor of the bending mode of PO_4 vibration at 500-620 cm^{-1} and area of phosphate-substitution site of the symmetric-antisymmetric stretching modes of PO_4 vibrations at 900-1200 cm^{-1} in FTIR.

Results show that the degree of crystallinity and crystal size of the obtained biological HAp are influenced by its stoichiometry. By addition of $CaCO_3$ as source of Ca^{2+} ions, the [Ca]/[P] ratio of the obtained calcium deficient biological HAp powders approaches the stoichiometric ratio. As a result, the degree of crystallinity and crystal size increase. However, crystallinity and crystal size decrease with [Ca]/[P] > 1.67. Hence, engineering the crystallinity and crystal size of the thermally-treated biological HAp can be possible by addition of $CaCO_3$. Significantly, degree of cystallinity of the HAp powders with [Ca]/[P] ratio equal to 1.46 and 1.48 fall near the range 60-70% for highly resorbable HAp used for medical application.

The splitting factor of the bending mode of phoshate vibration at 500-620 cm^{-1} matches the result with % crystallinity on XRD. Hence, the splitting factor of phoshate vibration is a good index of crystallinity. Moreover, the area of phosphate-substitution site of PO_4 vibration at 900-1200cm^{-1} infrared spectral region corresponds to the length of the *c*-axis of HAp. IR spectroscopy and X-ray diffraction are thus both suited to determine crystal size and crystallinity of HAp and should be used in conjunction when possible to obtain the maximum amount of structural information.

References

[1] Kumar GS, Girjija EK, Thamizhavel A, Yokogawa Y and Kalkura SN 2010 *J. Colloid and Interface Science* **349** 56–62
[2] Mahabole MP, Aiyer RC, Ramakrishna CV, Sreedhar B and Khairnar RS 2005 *Bull. Mater. Sci.* **28** 535–45
[3] Mene RU, Mahabole M P and Khairnar R S 2011 *Radiation Physics and Chemistry* **80** 682–7
[4] Zaragoza DL, Guzman ETR and Gutierrez LRR 2007 *J. Minerals and Materials Characterization and Engineering* **8** 591–609
[5] Rigano L *et al.* 2009 *SOFW-Journal* **135** 20–7
[6] Prabakaran K and Rajeswari S 2009 *Spectrochimica Acta Part* A **74** 1127–34
[7] Fathi MH, Hanifi A and Mortazavi V 2008 *Journal of Materials Processing Technology* **202** 536–42
[8] Pleshko N, Boskey A and Mendelsohn R 1991 *Biophys. J.* **60** 786–93
[9] Abraham JA, Sanchez HJ, Marcelli CA, Grenon M, Guidi MC and Piccinini M 2011 *Anal. Bioanal. Chem.* **399** 1699–704
[10] Landi E, Tampieri A, Celotti Gand Sprio S2000 *J. Eur. Ceram. Soc.* **20**, 2377–87

Characteristics of zinc sulphide nanostructures grown onsilica modified-polyaniline with polymerization time-dependent

M Potestas[1], A Alguno, R Vequizo, B R Sambo and M K Odarve

Materials Science Laboratory, Department of Physics, MSU-Iligan Institute of Technology, Andres Bonifacio Avenue, Tibanga, Iligan City, 9200 Philippines

E-mail: melchorjpotestas@gmail.com

Abstract. Growth of zinc sulphide (ZnS) nanostructures on silica modified-polyaniline (SM-PAni) with polymerization time-dependent was prepared using chemical bath deposition (CBD) technique. The grown samples were characterized by scanning electron microscopy (SEM) equipped with energy dispersive x-ray spectroscopy (EDS) and fourier transform infrared spectroscopy (FTIR). SEM images revealed that polyaniline rod-like nanostructures and ZnS nanospheres were successfully grown. The average diameter of the grown ZnS nanospheres did not significantly change by changing the growth time of the polyaniline. However, ZnS nanospheres grown with longer polymerization time of PAni is less dense and loosely bound as compared to shorter polymerization time of PAni. The less density of ZnS nanostructures with longer polymerization time of PAni may be due to the presence of PAni agglomerates that hinders the growth of ZnS nanospheres. Furthermore, FTIR spectra confirmed that the grown polyaniline is of emeraldine salt oxidation state which is the most conductive state of PAni.

1. Introduction

Polyaniline (PAni) is one of the most studied forms of conducting polymers [1-5]. PAni has three distinct oxidation states, namely; the leucoemeraldiene, emeraldine base, and pernigraniline [6]. However, emeraldine salt oxidation state of PAni which is derived from the emeraldine base is the most investigated form due its conductivity, environmental stability, and simple fabrication method [7-8]. The synthesis of this polymer can produce a variety of "one-dimensional" morphologies, like rectangular structures [9], nanotubes [10], nanofibers [1, 11], micromats [11], or even nanodisks [11], which made them unique materials for specific device application. One way to tune and improve the morphology of PAni nanostructures is by the introduction of a suitable steric stabilizer [12]. Smoother and more uniform surface will be achieved due to the non-formation of PAni agglomerates upon the introduction of steric stabilizer [13]. It is also believed that aside from the influence of steric stabilizer, polymerization time of PAni may also affect the growth of the grown nanostructures [14].

Another promising material that has been widely studied in the field of optoelectronics application is zinc sulphide (ZnS) [15-17]. ZnS belongs to II-VI group compound semiconductors and has a wide energy band gap [18, 19]. Several studies present different techniques in its fabrication which leads to its different morphologies [19-21]. One noble way in preparing ZnS nanostructures is by chemical

[1] To whom any correspondence should be addressed.

bath deposition (CBD). In this method, nanostructures are deposited on substrate immersed in a solution containing metal ions source, hydroxides, and sulphide ions [22]. The morphology of the ZnS nanostructures is also affected by the chemical precursors used in CBD [23, 24].

In this work, we report on the growth of zinc sulphide nanostructures on the surface of the time-dependent silica modified-polyaniline films on glass substrate. Elucidation on the surface morphology, elemental composition, and the vibrational modes of the functional groups of the grown SM-PAni/ZnS nanostructures were thoroughly investigated.

2. Experimental details

2.1. Preparation of substrate

Rectangular glass substrates with a dimension of 26mm x 76mm x 1mm are cleaned with acetone, ethanol, and hydrochloric acid, and distilled water. This method is necessary to remove the impurities and unnecessary compounds present on the glass slides which may affect the growth of the nanostructures. The purpose of acetone in substrate cleaning is to remove oils and organic substances while ethanol and HCl were employed to remove the residue of acetone and oxides in the substrate respectively. The cleaned substrates were then air dried at room temperature.

2.2. Synthesis of silica modified-polyaniline and zinc sulphide

Silica modified-polyaniline (SM-PAni) nanostructures were produced on glass surfaces immersed in the reaction mixture during the oxidative polymerization of aniline. Liquid aniline was dissolved in hydrochloric acid to form 50mL aniline hydrochloride solution. Steric stabilizer in the form of silica (SiO_2) extracted from rice husk ash was introduced forming the silica-aniline hydrochloride (SiO_2 – AHCl) solution and oxidized with ammonium peroxydisulfate (APS) solution for the synthesis of SM-PAni. The progress of polymerization was monitored by the change in temperature of the solution recorded by a thermometer. Two SM-PAni nanostructures, with 16 minutes and 25 hours polymerization time were produced. These nanostructures grown on the glass substrates were washed with HCl to remove the residual monomer, oxidants and its decomposition products; and ethanol to remove low-molecular-weight organic intermediates and oligomers; and then air dried at room temperature.

The air dried samples were then subjected to chemical bath deposition (CBD) for the synthesis of zinc sulphide. ZnS nanostructures were deposited from the stirred aqueous solution containing zinc (II) chloride ($ZnCl_2$), urea ($(NH_2)_2CO$), and thioacetamide (CH_3CSNH_2). The total volume of the solution is increased to 200mL by the addition of distilled water. The pH of the solution was monitored by OAKTON Acorn pH meter and adjusted to the desired value which is 2.0 to 4.5 by dropwise addition of HCl. The solution was stirred using a magnetic stirrer and heated on the hot plate while monitoring its temperature with a maintained pH. The SM-PAni thin films were then dipped into the solution for 90 minutes. After the deposition time, samples were removed and washed with distilled water then air dried at room temperature.

2.3. Characterization of the nanostructures

The fabricated SM-PAni/ZnS nanostructures were characterized with JEOL Model JSM-6510LA scanning electron microscope (SEM) equipped with energy dispersive x-ray spectroscopy (EDS) for its surface morphology and elemental analysis respectively. Samples were scanned at a working distance of 12mm and accelerating voltage of 20kV to generate its SEM micrographs. Furthermore, Fourier Transform Infrared Spectroscopy (FTIR) with Perkin Elmer Spectrum 100 Spectrometer was also used for the determination of the vibrational modes and to find other functional groups present on the surface of the grown nanostructures. The spectrum has been recorded using the technique of Universal Attenuated Total Reflection (UATR) at 20 scans per sample in the region of 650 cm^{-1} to 4000 cm^{-1}. Data were carbon dioxide corrected and were fitted using Gaussian fitting for the determination of the peaks.

IOP Conf. Series: Materials Science and Engineering **79** (2015) 012029 doi:10.1088/1757-899X/79/1/012029

3. Results and discussion

3.1. Morphology and elemental analysis of the grown nanostructures

Zinc sulphide nanostructures in the form of nanospheres are successfully grown on polymerization time-dependent silica modified-polyaniline as shown in Figure 1. Figure 1(a) shows the SEM micrograph of SM-PAni/ZnS nanostructures with 16 minutes polymerization time of SM-PAni. It is observed that the rod-like nanostructures polyaniline (white arrow) and ZnS nanospheres which are uniform in size were formed on the surface of the film. The diameters of the ZnS nanospheres are about 100nm and the PAni rod-like nanostructures are about 800nm.The length of the PAni rod-like nanostructures extends from 20μm to 30μm. On the other hand, Figure 1(b) shows the SEM image of the 25 hours polymerization time of SM-PAni/ZnS nanostructures. It is observed that more PAni agglomerates on the surface are present compared with the 16-minute polymerization time. The PAni rod-like (white arrow) nanostructures of the 25hrs polymerization time have larger diameter when compared with the 16 min polymerization time. Furthermore, the length of the PAni rod-like nanostructures present on the 25hrs polymerization time is longer when compared to 16min polymerization time. This result suggests that the formations of bigger and longer PAni rod-like structures might be due to the longer submersion of the substrate on the reaction mixture. Formation of the rod-like nanostructures might be attributed to the low acidity (pH > 4) of the reaction mixture at the start of the polymerization. Low-molecular weight aniline oligomers are produced during the first phase of oxidation. These aniline oligomers serve as a soft template for the growth of the PAni rod-like nanostructures [25, 26]. On the other hand, both samples shown in Figures 1(a) and 1(b) have agglomerates. These agglomerates are believed to be PAni precipitates which are produced during bulk polymerization [27, 28]. It can be observed that 25-hour polymerization time of SM-PAni/ZnS has more PAni agglomerates. These may be due to the neutral aniline molecules that continually oxidize after the exothermic reaction during oxidative polymerization. The oxidized neutral aniline molecules lead to the formation of short aniline oligomers at the trimer level which serve as soft templates for the formation of PAni agglomerates [29].

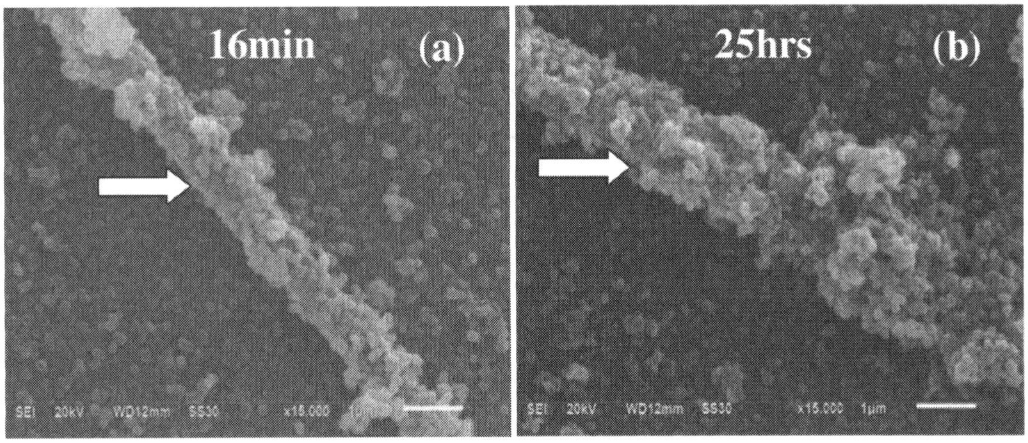

Figure 1. SEM micrograph images of the grown ZnS nanostructure on SM-PAni with different polymerization time (a) 16 minutes and (b) 25 hours.

ZnS nanospheres are present on the film and on the top of PAni precipitates as observed from the SEM images. These nanospheres have the same average diameter for both samples. However, ZnS nanospheres on 16min polymerization time SM-PAni is denser and more compact compared with the 25hrs SM-PAni. This result can be attributed to the agglomeration of PAni nanostructures that may hinder the attachment of ZnS nanospheres onto the SM-PAni film. Wei *et al.* [30] proposed that formation of ZnS nanostructures during CBD is dominated by cluster-by-cluster process instead of ion-by-ion process. In cluster-by-cluster process, ZnS nanoparticles preformed in the solution and then adsorbed onto the surface. Grown ZnS nanostructures by this process possess poor adherence on the film. For the two samples, 16 min polymerization time SM-PAni has minimal agglomeration and more uniform compared with the 25 hours polymerization time.

The morphology of SM-PAni and ZnS nanostructures grown on glass substrate are shown in Figures 2 (a) and 2 (b) respectively. SEM micrograph of SM-PAni reveals the formation of PAni rod-like nanostructures (block arrow) and PAni agglomerates (line arrow). It is also observed that PAni agglomerates also grow on top of the PAni rod-like nanostructures. Furthermore, Figure 2 (b) shows that the grown ZnS nanostructures are nanospheres. The morphology of ZnS nanospheres grown on glass substrate agrees with the nanospheres structure from Figures 1(a) and 1(b).

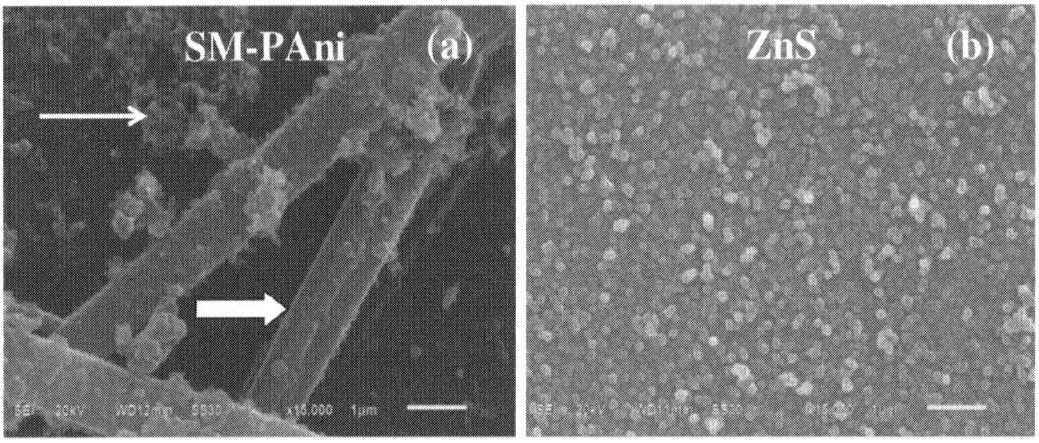

Figure 2. SEM micrograph images of the grown SM-PAni and zinc sulphide nanostructures on glass substrate.

The elements present on the samples that correspond to Figures 1(a) and 1(b) are shown in Table 1 including the bare glass substrate. The presence of other elements such as oxygen (O), Sodium (Na), magnesium (Mg), aluminum (Al), silicon (Si), and calcium (Ca) is attributed to the components of the glass substrate being used. One way to determine the quality of the grown ZnS nanostructures for both samples is to calculate the ratio between the zinc atoms and sulfur atoms (i.e. [Zn]/[S]) from the results of EDS. The ideal ratio for zinc versus sulfur atoms is 1. Film sample with 16min polymerization time of SM-PAni/ZnS obtains a ratio of 0.97 compared with 25hrs polymerization time of SM-PAni/ZnS which obtains 0.91 for zinc atoms versus sulfur atoms. From this result, it is believed that more ZnS nanostructures have grown for 16min polymerization time of SM-PAni compared with the other sample as supported also by the SEM images shown in Figure 1(a) and 1(b). Excess atoms of sulfur can be attributed to the presence of sulfuric acid as the by-product during oxidative polymerization of polyaniline. Large percentage of carbon (C) atoms is detected on the sample which attributes to the quinone and benzene ring forming the polymer chains [31]. Nitrogen (N) atoms are also present in the samples which constitute the imine and amine nitrogen groups in the polymer chains [26]. Traces of Chlorine (Cl) are detected for both samples. The presence chlorine elements can be traced from the HCl that was being used in oxidative polymerization of SM-PAni.

Table 1. Elemental analysis of ZnS nanospheres grown on polymerization time dependent SM-PAni.

Sample	Elements in Percent (%) Atom										
	C	N	O	Na	Mg	Al	Si	S	Cl	Ca	Zn
Glass Substrate	-	-	43.07	8.58	2.29	1.00	40.69	-	-	4.37	-
SM-PAni(16min)/ZnS	26.55	3.32	38.17	5.07	1.35	0.46	20.49	1.20	0.28	1.95	1.16
SM-PAni(24hrs)/ZnS	44.20	7.14	28.84	3.42	0.71	0.24	10.93	1.62	0.39	1.03	1.48

Table 2 shows the elemental analysis corresponding to Figure 2 (a) and 2 (b). Traces of chlorine and sulphur atoms are detected for SM-PAni sample which are by-products during oxidative polymerization. Furthermore, elemental analysis for ZnS nanostructures shows that the ratio of zinc atoms to sulfur atoms is 0.96 which is very close to the theoretical value. Chlorine atoms are not detected in the spectrum which is probably removed during the rinsing of distilled water.

Table 2 Elemental analysis of the grown SM-PAni and zinc sulphide nanostructures corresponding to the SEM images in Figure 2.

Sample	Elements in Percent (%) Atom										
	C	N	O	Na	Mg	Al	Si	S	Cl	Ca	Zn
SM-PAni	32.55	3.23	41.82	5.16	1.10	0.32	14.30	0.12	0.08	1.32	-
ZnS	-	-	56.61	8.71	1.93	0.60	26.88	1.42	-	2.48	1.37

3.2. FTIR spectra of the grown nanostructures

FTIR spectra of SM-PAni, as-grown ZnS, and ZnS nanostructures on polymerization dependent SM-PAni are shown on Figure 3. Since the FTIR spectrum has been taken in the specified range, the presence of the chemical constituents of zinc sulphide cannot be determined because the metal sulphides bond occurs within 200 cm^{-1} to 400 cm^{-1}[32].

Absorption peaks of all the functional groups of polyaniline are present for both samples. The peak at around 1651 cm^{-1} corresponds to the O-H bending vibration which indicates the presence of water moisture in the sample [32]. This peak is not observed for SM-PAni on glass substrate (Figure 3 (c)) which confirms that water moisture are produced during the CBD of ZnS nanospheres since the reaction between hydronium ions and hydroxides are favorable in the acidic reaction bath. The main peaks at around 1585 cm^{-1} and 1497 cm^{-1} corresponds to quinine and benzene stretching ring-deformation of polyaniline [33]. Strong absorption band at around 1306 cm^{-1} is attributed to the C–N stretching of secondary aromatic amine strengthened during the protonation of SM-PAni [34]. The signature peak at around 1245 cm^{-1} confirms that the PAni films being produced are of the emeraldine salt oxidation state. This peak is interpreted as C–N+• stretching vibration in the polaron structure [33, 34]. Absorption band at around 1380 cm^{-1} is due to Zn-O (HCO_2) which indicates the possible formation of ZnO during CBD [32]. Another absorption peak is observed at around 1144 cm^{-1} which is assigned to the vibration mode of –NH+= structure produced during protonation [35]. Strong absorption band at around896 cm^{-1} is assigned Si-H bending from the glass substrate used [36].

4. Conclusion

ZnS nanospheres were successfully grown on polymerization time-dependent polyaniline. SEM images show that the ZnS nanospheres are denser and more compact for 16 minutes polymerization time of SM-PAni compared to 25 hours polymerization time. Tailoring the size of PAni nanostructures and density of ZnS nanostructure can be done by changing the polymerization time. These results proved that polymerization time of SM-PAni nanostructures would influence the growth of ZnS nanostructures. The absorption peak at around 1245 cm^{-1} from FTIR spectra confirmed that the grown polyaniline nanostructures in the glass substrate are of the emeraldine salt oxidation state.

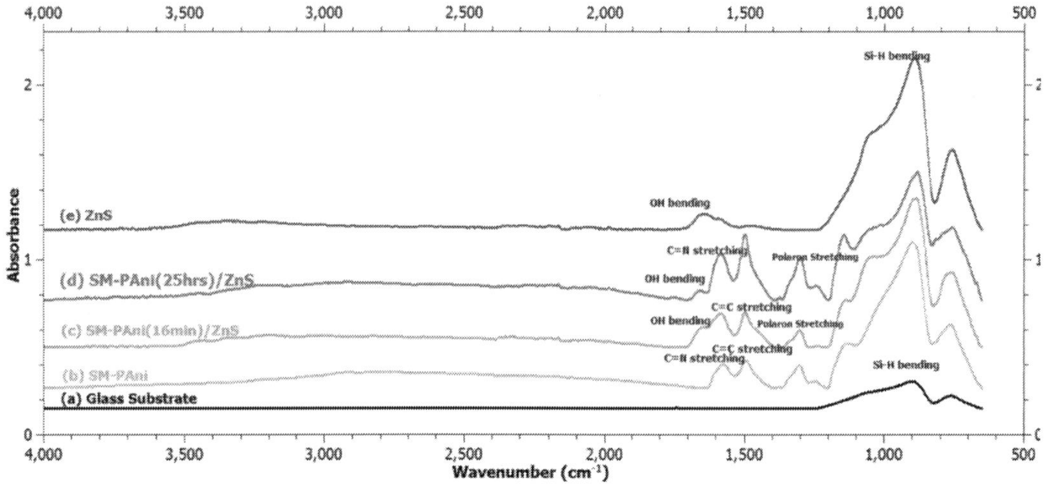

Figure 3. FTIR Spectra of the grown nanostructures on glass substrate: (a) bare glass (b) SM-PAni (c) 16min SM-PAni/ZnS (d) 25 hours SM-PAni/ZnS and (e) ZnS.

5. References

[1] Sadek A Z, Wlodarski W, Kalantar-Zadeh K, Baker C and Kaner R B 2007 *Sensors and Actuators* A **139** 53–7
[2] Deshpande N G *et al.* 2009 *Smart Mater. Struct.* **18** 095010
[3] Xie D *et al.* 2002 *Sensors and Actuators* B **81** 158–64
[4] Xing S, Zhao C, Jing S, WuY and Wang Z 2006 *European Polymer Journal* **42** 2730–5
[5] Nicolas-Debarnot D and Poncin-Epaillard F 2002 *Analytica Chimica Acta* **475** 1–15
[6] Skotheim T A and Reynolds J R2007 *Conjugated Polymers Processing and Applications 3rd Edition* (New York: Taylor & Francis Group, LLC) pp 2-4–2-6
[7] Stejskal J, Sapurina I, Prokes J and Zemek J 1999 *Synthetic Metals* **105** 195–202
[8] Riede A, Helmstedt M, Riede V and Stejskal J 1997 *Colloid Polym. Sci.* **275** 814–20
[9] Lee B, Kim J, Fang F, Choi H and Feller JF 2011 *Macromol. Chem. Phys.* **212** 2300–7
[10] Blinova N, Stejskal J, Trchova M, Sapurina I and Ciric-Marjanovic G 2008 *Polymer* **50** 50–6
[11] Wang X, Li Y, ZhaoY, Liu J, Tang S and Feng W 2010 *Synthetic Metals* **160** 2008–14
[12] Stejskal J and Sapurina I 2005 *Pure Appl. Chem.* **77** 815–26
[13] Riede A, Helmstedt M and Riede V 2000 *American Chemical Society, Langmuir* **16** 6240–4
[14] Sapurina I, Riede A and Stejskal J 2001 *Synthetic Metals* **123** 503–7
[15] Oladeji I and Chow L 2004 *Thin Solid Films* **474** 77–83
[16] Soetedjo H and Prabowo GS 2010 *Sensors & Transducers Journal* **123** 100–5
[17] Stupca M *et al.* 2012 *Journal of Applied Physics* **112** 074313
[18] Nagamani K, Prathap P, Lingappa Y, Miles R W and Reddy K T R 2012 *Physics Procedia* **25** 137–42
[19] Meshram R S and Thombre R M 2012 *Proc. 2012 ICBEST Conference* 22–5
[20] Wang Z *et al.* 2005 *Nature Materials* **4** 922–7
[21] Lee C, Nakamura Y, Nakamura H, Uehara M and Maeda H 2011 *Anyagtudomàny Materials Science* 52–6
[22] Goudarzi A, Aval G, Sahraei R and Ahmadpoor H 2007 *Thin Solid Films* **516** 4953–7
[23] Shinde M, Ahirrao P and Patil R 2011 *Archives of Applied Science Research* **3** 311–7
[24] Ernits E *et al.* 2010 *Advances in Materials Science and Engineering* **2009**
[25] Sedenkova I, Trchova M, Stejskal J and Bok J 2007 *Applied Spectroscopy* **61** 1153–62
[26] Stejskal J, Sapurina I, Trchova M, Konyushenko E and Holler P 2006 *Polymer* **47** 8253–62
[27] Stejskal J, Sapurina I, Prokes J and Zemek J 1999 *Synthetic Metals* **105** 195–202

[28] Xing S, Zhao C, Jing S, Wu Y and Wang Z 2006 *European Polymer Journal* **42** 2730–5
[29] Stejskal J, Sapurina I and Trchova M 2010 *Progress in Polymer Science* **35** 1420–81
[30] Wei A, Liu J, Zhuang M and Zhao Y 2013 *Materials Science in Semiconductor Processing* **16**1478–84
[31] Skotheim T A1986 *Handbook of Conducting Polymers* vol 1 (New York: Marcel Dekker, Inc.) chapter 1pp 2–38
[32] Dhanam M and Kavitha B 2009 *Chalcogenide Letters* **6** 299–307
[33] Sapurina I *et al.* 2002 *Synthetic Metals* **129** 29–37
[34] Trchova M, Sedenkova I and Stejskal J 2005 *Synthetic Metals* **154** 1–4
[35] Trchova M and Stejskal J 2011 *Pure Appl. Chem.* **83** 1803–17
[36] Stuart B 1996 *Modern Infrared Spectroscopy Analytical Chemistry by Open Learning1st Edition*(New York: John Wiley & Sons) chapter 1pp 109–11

Acknowledgment
One of the authors (MJP) would like to acknowledge Commission on Higher Education (CHED) for the scholarship grant. The Department of Science and Technology (DOST) and CHED-PHERNet are also acknowledged for the research grant. The authors would like to thank Ms. Piagola for technical help in the experiment.

IOP Conf. Series: Materials Science and Engineering **79** (2015) 012030 doi:10.1088/1757-899X/79/1/012030

Electrical properties of an individual ZnO micro/nanorod

A B Santos-Putungan[1,2,3], **L M Bambao**[2] **and R V Sarmago**[2]

[1] Physics Division, Institute of Mathematical Sciences and Physics, University of the Philippines, Los Baños, Laguna, Philippines

[2] National Institute of Physics, University of the Philippines, Diliman, Philippines

Email: absantos4@up.edu.ph

Abstract. Free standing, highly crystalline ZnO micro/nanorods have been successfully fabricated using low temperature hydrothermal synthesis. Current-voltage characteristic curves show that ZnO micro/nanorods' resistances are proportional to their geometrical ratios, satisfying classical Ohm's law. Temperature-dependent resistance measurements reveal exponential decay of current with temperature, implying good semiconducting properties. Finally, three different thermally-activated impurity levels were identified from the measurements, and these are attributed to Zn interstitials in the bulk. The results are important in supplementing research on nanoelectronics and nanocircuitries.

1. Introduction

Nanotechnology has influenced the way humans live, ranging from biology, chemistry, physics, and medicine. In addition, nanotechnology also paved the way for the miniaturization of materials and complex electronic devices that are found to be useful in many practical and scientific applications. In this respect, extensive researches are being conducted to improve these technologies, especially in the semiconductor field where nanotechnology is said to have the most varied applicability. These materials have electrical properties that are in between of a conductor and an insulator. Commonly utilized semiconductors include the elemental ones (Si, Ge) and the compound ones (GaAs, GaN, InP).

Another group known as wide gap semiconductors consists of materials that have energy band gap of more than ~ 2 eV, which include ZnO. It is a II-VI semiconductor having a hexagonal wurtzite crystal structure (lattice parameters of a = 0.325 nm and c = 0.521 nm). Due to its unique properties such as its wide gap of 3.34 eV and high excitonic binding energy of 60 meV, it is being utilized in different applications as sensors, UV emitters, LEDs and piezoelectric devices [1-6]. These make ZnO a prospect material in nanotechnology applications, especially the micro/nanorod structures. This entails that these structures must possess good electrical properties, and that techniques of measuring and analyzing these must be studied thoroughly as well. Several studies have been done [7-9] on the assessment of ZnO micro and nanostructures' electrical properties grown using different fabrication techniques. In this work, ZnO micro and nanostructures grown via hydrothermal growth method are characterized with focus on their electrical properties. A single ZnO micro/nanorod is mounted on

[3] To whom any correspondence should be addressed.

patterned substrate through an Ohmic contact connection, and subsequently its I-V curve characteristics and RT measurements are obtained and analyzed.

2. Methodology

The synthesis of ZnO structures was done using the hydrothermal method, in which zinc acetate dihydrate $Zn(CH_3COO)_2 \cdot 2H_2O)$ and hexamethylenetetramine (HMTA : $C_6H_{12}N_4$) as primary precursors. Equimolar amounts of these precursors were used in the synthesis, ranging from 0.008 M, 0.009 M and 0.010 M. The chemicals were diluted in fixed amount of water and were sonicated for 20 minutes, and subsequently were mixed and sonicated further for another 20 minutes. The proposed reaction for ZnO growth is

$$Zn(CH_3COO)_2 \cdot 2H_2O \leftrightarrow Zn^{2+} + 2CH_3COO^- + 2H_2O$$
$$C_6H_{12}N_4 + 6H_2O \leftrightarrow 4NH_3 + 6HCHOO$$
$$NH_3 + H_2O \leftrightarrow NH_4^+ + OH^-$$
$$Zn^{2+} + 4OH^- \leftrightarrow Zn(OH)_4^{2-}$$
$$Zn(OH)_4^{2-} + heat \leftrightarrow ZnO + H_2O + 2OH^-$$

Silicon substrate was used to collect the synthesized ZnO structures, which was then brought into heat treatment at 95°C for 120 minutes using a hot plate. The system consisting of collected ZnO structures in Si substrate was washed with water to remove residual salts, air dried and annealed at 500° C for 20 minutes.

Electrical characterization of the ZnO samples were divided into two parts: the current-voltage measurements (I-V measurement) and the resistivity versus temperature (RT) measurements. Preliminary sample preparation was done for the electrical characterizations. ZnO rods were sonicated in ethanol and was dispersed using a micropipette on prefabricated chips with electrodes. Platinum was deposited using Focused Ion Beam (FEI NOVA 600) to form contacts with the ZnO rods and the electrodes. The I-V characteristic curves were measured using four-point probe, while RT measurements were done using the two-point probe method. A10nAcurrent source was used to supply current and the voltage output is measured afterwards for the RT measurements.

3. Results and discussions

The hydrothermally grown ZnO structures were determined using x-ray diffraction, and shown in figure 1 are the x-ray diffractograms of the samples in different molarity concentrations. All the samples showed a prominent peak at 31.70° which corresponds to $(10\bar{1}0)$ growth direction. It can also be noticed that the reaction indeed produced ZnO in different growth directions as the other peaks show. Furthermore, it is noteworthy to mention that no significant and additional peaks other than ZnO peaks were observed. Thus, ZnO has been synthesized from the reaction and the peaks are in agreement with the JCPDS database with card no. 36-1451 for hexagonal wurtzite ZnO.

For the electrical characterization, the samples grown using 0.009 M growth parameter were used. Figure 2 shows the SEM micrographs of the individual ZnO micro/nanorod with their corresponding I-V characteristic curves. The plots show perfect linear trend indicating that the samples have Ohmic electrical properties, and that the contacts between the micro/nanorods are Ohmic in nature, as oppose to Schottky type that results to potential drops and thus deteriorates the electrical properties. This finding is of importance in the utilization of ZnO micro/nanorods as key materials in the fabrications of nanoscale electronics. The slope of each I-V curve denotes the resistance of a single ZnO micro/nanorod, and the resistances measured from the samples are of the order of 10^3 Ohms, which implies that the samples are highly resistive.

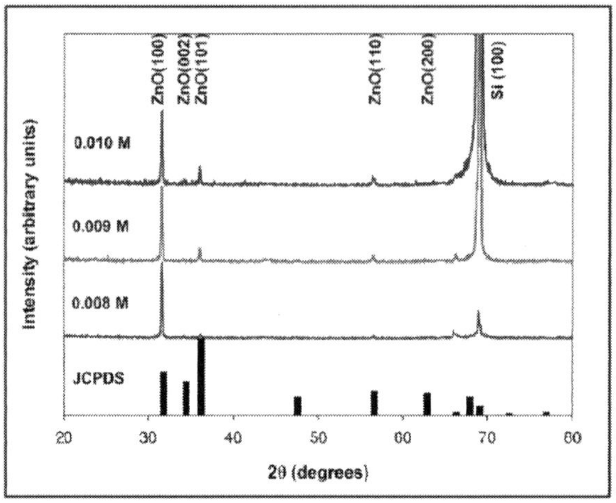

Figure 1.XRD spectra of ZnO grown hydrothermally using different equimolar concentrations of $Zn(CH_3COO)_2 \cdot 2H_2O$ and HMTA.

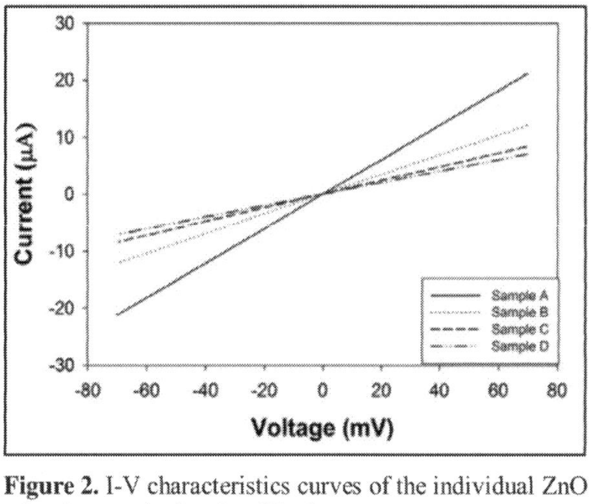

Figure 2. I-V characteristics curves of the individual ZnO micro/nanorods showing Ohmic property.

The geometrical ratio of each sample were obtained and summarized in Table 1. For sample A, the geometrical ratio is 3.35 ± 0.78 μm^{-1}, and the calculated resistance is 3.3 kΩ. On the other hand, for sample B having geometrical ratio of 7.98 ± 0.45 μm^{-1}, a higher value for the resistance was obtained, which is 5.8 kΩ. Furthermore, for the remaining samples C and D with geometrical ratios of 10.72 ± 0.85 μm^{-1} and 17.23 ± 4.78 μm^{-1}, the computed resistances are 8.3 kΩ and 9.9 kΩ, respectively. Thus, it was illustrated that higher geometrical ratio amounts to higher resistance, which is a signature property of classical Ohmic device.

The resistivity of the samples, ρ, is obtained statistically following Ohm's law behaviour relating the resistance R to its aspect ratio, given by

$$R = \rho \frac{L}{A} \qquad (1)$$

Table 1. Summary of the dimensions, geometrical ratio and resistance obtained for each ZnO micro/nanorod.

Sample	Length (μm)	Area (μm^2)	Ratio (L/A) (μm^{-1})	Resistance (kΩ)
A	2.194 ± 0.032	0.502 ± 0.019	3.35 ± 0.78	3.3
B	3.59 ± 0.016	0.502 ± 0.0037	7.98 ± 0.45	5.8
C	5.04 ± 0.014	0.426 ± 0.0058	10.72 ± 0.85	8.3
D	2.46 ± 0.084	0.235 ± 0.00969	17.23 ± 4.78	9.9

From the samples considered above, the resistivity of the ZnO micro/nanorods is calculated to be equal to 4.83×10^{-2} Ω·m.

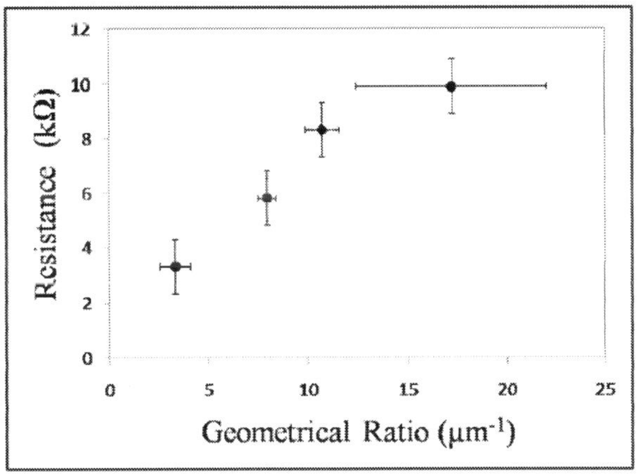

Figure 3. Plot of resistance versus geometrical ratio of ZnO micro/nanorods.

Figure 4 shows the resistance versus temperature (RT) plot of one of the four samples used in the I-V curve analysis. The exponential decay nature of the plot illustrates the semiconducting property of ZnO micro/nanorod sample. This is because for typical semiconductors, the resistance decreases exponentially with increasing temperature. This is due to the fact that electrons are free to move at higher temperatures as a result of thermal energy and this contributes to the lowering of the resistance. At a lower temperature however, electrons are frozen due to deficiency of thermal energy, making the resistance rise significantly.

Shown in figure 5 is the plot of the conductance G as a function of1/T using the ZnO micro/nanorod grown from the 0.009 M sample. It can be observed that the plot can be divided into three regions or transitions. The slopes of the lines give information on the nature of the electronic transitions, whether it is from the valence band or from impurity levels within the bandgap. It is worthwhile to mention that the RT plot's temperature range is limited only up to room temperature, thus eliminating the possibility of a direct valence to conduction band transition, in which case the slope would give the band gap energy. ZnO is an n-type semiconductor intrinsically, and thus the majority carriers are electrons. The slopes mentioned above would then give us the idea that these transitions are actually due to electrons trapped in impurity levels, which may be shallow or deep as measured from the conduction band. The three different slopes obtained point to the energy positions of these impurity levels.

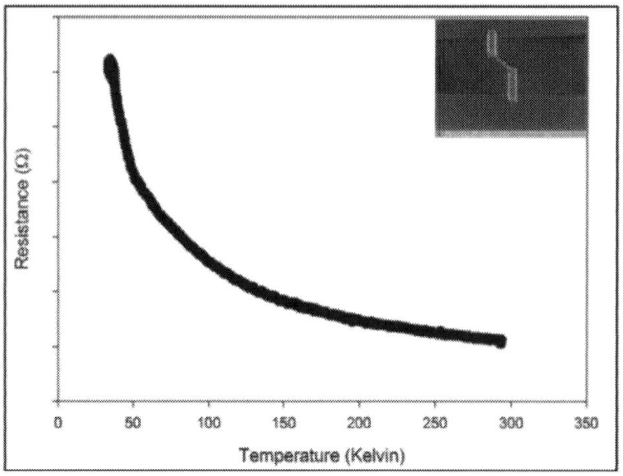

Figure 4. Resistance versus temperature curve of an individual ZnO micro/nanorod using Sample D (inset).

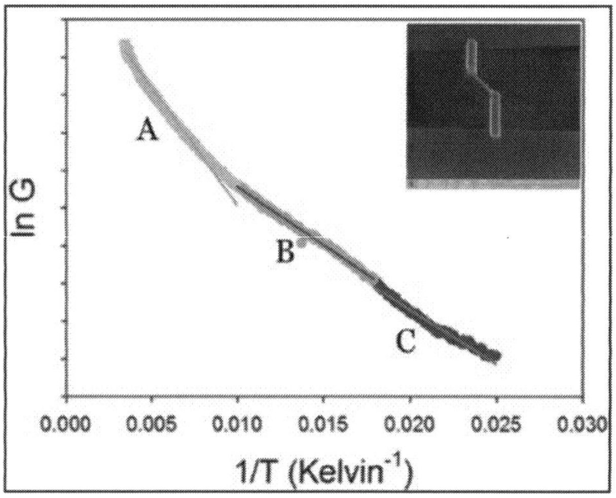

Figure 5. Conductance plot with temperature of an individual ZnO using Sample D (inset).

Focusing on the transitions of the sample, the activation energy of the electrons can be known using equation (2)

$$slope = \frac{-(E_c - E_d)}{2.3k} = \frac{-E_{impurity}}{2.3k} \tag{2}$$

where k is the Boltzmann's constant. From these data, the energy for the impurity level from the conduction band is computed to be 24.6 - 24.9meV for transition A, 12.1 - 12.3 and 11.1 - 11.8 meV for transitions Band C, respectively. These activation energies from the conduction band have the same order of magnitude for a Zn impurity [10, 11]. There has-been a consensus that Zn impurities form shallow levels measured from the conduction band, although there is no general agreement as to what are the exact positions of these levels. Look and Remsky (1999), and Reddyet. al (2008) reported values that form a range of plausible positions of the levels, in which this work's calculated values fit in [10, 11].

Table 2. Summary of activation energy obtained from the sample grown using 0.009 M concentration

Region	E_a (meV)
A	24.6 – 24.9
B	12.1 – 12.3
C	11.1 – 11.8

4. Conclusion

Synthesis of ZnO micro/nanorods was successfully done, and the grown ZnO structures were investigated. Free-standing ZnO micro/nanorods were effectively produced using a low temperature method of fabrication, the hydrothermal method. XRD results show that the method produces highly crystalline ZnO with no signs of other chemical impurities from the solution. The I-V characteristic curves of the grown ZnO micro/nanorods were obtained, and it can be concluded that the resistance of the ZnO micro/nanorods is proportional to the aspect ratio of the samples. Also, the grown ZnO micro/nanorods electrical properties follow an Ohmic behavior. Furthermore, the RT curve of the samples using two-point probe method showed the semiconducting behavior of the ZnO micro/nanorods.The slopes represent the activation energies for thermal-assisted conduction persisting in the samples.

5. References

[1] Bill J *et al.* 2005 *Mater. Res. Soc. Symp. Proc.* **873** EK 5.3.1–7
[2] Kim J B *et al.* 2008 *Semicond. Sci. Technol.* **23** 095004
[3] Mahmoodi S R *et al.* 2009 *Procedia Chemistry* **1** 947–50
[4] Liu W J *et al.* 2008 *Sensors and Actuators* B **133** 664–70
[5] Wu W Y *et al.* 2009 *Nanoscale Res. Lett.* **4** 513–7
[6] Choi J *et al.* 2010 *Journal of Electrical Engineering & Technology* **5** 640–5
[7] Schlenker E *et al.* 2007 *Phys. Stat. Sol.*B **244** 1473–7
[8] Yoon S W *et al.* 2009 *Thin Solid Films* **517** 4003–6
[9] Sakurai M *et al.* 2009 *Nanotechnology* **20** 155203
[10] Look D and Hemsky J 1999 *Physical Review Letters* **82**
[11] Reddy N K, Ahsanulhaq Q, Kim J and Hahn Y B 2008 *Europhysics Letters* **81**

Acknowledgement
We would like to acknowledge the Emergent Science and Technology grant from the University of the Philippines Office of the Vice President for Academic Affairs (OVPAA).

Growth mechanism of chemically prepared *ZnO-SiO_2* nanostructures grown on glass and silicon substrates

K T Simfroso[1]*, A C Alguno[1], F R Bagsican[1], M E Jabian[2], M K G Odarve[1], G G Paylaga[1], B R B Sambo[1], R M Vequizo[1], and R T Candidato, Jr[1].

[1] Department of Physics, College of Science and Mathematics, MSU-Iligan Institute of Technology, A. Bonifacio Avenue, Tibanga, 9200, Iligan City, Philippines.

[2] Department of Electrical, Electronics & Communication & Computer Engineering, College of Engineering, MSU-Iligan Institute of Technology, A. Bonifacio Avenue, Tibanga, 9200, Iligan City, Philippines.

*E-mail: simfrosokey@gmail.com

Abstract. Different surface morphology of ZnO-SiO_2 has been successfully grown on glass and silicon substrates using chemical bath deposition method. The dependence on the substrates used on the morphology of ZnO-SiO_2 was investigated. The morphology of ZnO-SiO_2 on glass substrate was flake-like in form and changes to protruding hemispherical structures when grown on silicon substrate. Elemental composition analysis verified the presence of ZnO and SiO_2. Infrared characteristics showed an absorption band for the binding of ZnO and SiO_2 and revealed the presence of zinc complexes. We proposed a mechanism on the growth of ZnO-SiO_2 on glass and silicon substrate using low temperature deposition technique.

1. Introduction

Among inorganic semiconductors, zinc oxide (ZnO), having a wide band gap of 3.37 eV at room temperature and a large exciton binding energy of 60 meV, attracted much attention and candidate for many applications such as gas sensors [1], optical wave guides [2], solar cells [3] and SAW ultraviolet photodetectors [4]. Because of these applications, many growth techniques have been used for deposition of ZnO nanostructures and these include sputtering [5], chemical vapor deposition [6], pulsed laser deposition [7], chemical bath deposition [8], atomic layer deposition [9], spray pyrolysis [10] and molecular beam epitaxy [11]. Although ZnO has many optoelectronic applications, it has also its limitations such as its high catalytic activity that may degrade its functions. To overcome its limitations, a concept of composite material is done. Among ZnO-based composite materials, ZnO-SiO_2 is gaining much attention due their compatibility and stability. There are existing works in the synthesis of ZnO-SiO_2, however, most of the processes involve sophisticated and expensive method like double jet precipitation [12-13], hydrolysis method [14], sonochemical synthesis [15], and spray drying [16]. Studies conducted by references [17-19] were able to synthesized ZnO-SiO_2 in where they used tetraorthosilicate (TEOS) as a source of SiO_2 which is toxic chemical and very expensive. However, a technique called chemical bath deposition is extremely attractive for the synthesis of ZnO-SiO_2 due to its advantageous features such as it is a low temperature and low cost technique, easy coating of large surfaces and the growth conditions can be easily controlled. Aside from this, there is no current work on the synthesis of ZnO-SiO_2 through chemical bath deposition method deposited on glass and silicon substrates using silica obtained from rice husk ash (RHA). RHA is considered as a waste material in the locality.

In this work, low temperature chemical bath deposition method is used for the synthesis of ZnO-SiO_2. The growth mechanism on flake-like and hemispherical formation of ZnO-SiO_2 on glass and silicon substrates is proposed. Detailed analysis on surface morphology and infrared characteristics of ZnO-SiO_2 is done and effect of OH^- concentration to the growth of ZnO-SiO_2 is also discussed.

2. Experimental

2.1 Preparation of the substrates

The substrates are cleaned first before the deposition process. The silicon substrates with dimension of 1.5 cm x 1.5 cm were first washed with distilled water. These substrates were soaked with ethanol then rinsed with distilled water. After that, they were immersed in ethanol and washed again by distilled water. Each step is done for every 5 minutes for three times. After cleaning, silicon substrates were dried in air. The amorphous glass substrates (25.4 mm x 76.2 mm) on the other hand were thoroughly cleaned using acetone, ethanol, hydrochloric acid and distilled water. The glass substrates were first immersed in acetone swaying the substrates 24 times. It is then followed by degreasing the substrates using distilled water and then immersed in ethanol and again rinsed with distilled water. The amorphous substrates were washed by 1.0 M hydrochloric acid and completely degreased with distilled water. These methods were performed three times. The glass substrates were dried in ambient air condition.

2.2 Synthesis and characterization of ZnO-SiO_2

ZnO-SiO_2 composite was deposited on silicon and glass substrates through chemical bath deposition method using laboratory grade chemical reagents. A 3.0 M ammonium hydroxide (NH_4OH) and 0.03 M zinc sulphate heptahydrate ($ZnSO_4 \cdot 7H_2O$) were used to form the aqueous bath solution. Silica (SiO_2) powder from RHA was added to the aqueous solution and the resulting mixture was magnetically stirred at 360 rpm for 30 minutes and heated up to 70°C using magnetic stirrer with hot plate apparatus. After 30 minute of stirring, the substrates were immersed in the prepared solution and again magnetically stirred for five hours. After the deposition, the substrate was rinsed with heated deionized water and dried at room temperature.

The surface morphology and elemental composition of the ZnO-SiO_2 composite was observed using scanning electron microscopy-energy dispersive X-ray (SEM-EDX) using JEOL M-6510LA analytical scanning microscope. The functional groups (chemical bonding) and impurities present in the samples were investigated by Fourier transform infrared (FTIR) spectroscopy using Perkin Elmer Spectrum 100 FTIR spectrometer with scanning range of 600-4000 cm^{-1}.

3. Results and Discussion

3.1. Surface Morphology

Figure 1(a) shows the SEM image of ZnO deposited on glass substrate. It is apparent in the figure that flake-like formations were deposited on glass substrates. This formation of flake-like structures is attributed to the excess OH^- ions which are responsible for the convolution of Zn^{2+} ions forming flake-like structure than sea urchin like ZnO.

Figure 1(b) is the surface morphology of ZnO-SiO_2 which also shows flake-like structures that is larger in size compared to the flake-like formation of ZnO alone. This formation is due to the amorphous nature of the glass substrate used and to the excess OH^- ions that facilitates faster nucleation of the $Zn(OH)_2$ seed nuclei. In an amorphous glass substrate, many phase boundaries act as nucleation sites. At the nucleation sites, the effective surface energy is lower, thus diminishing the free energy barrier and facilitating faster nucleation. The $Zn(OH)_2$ formed must have attached quickly to the glass substrate and the excess OH^- ions attacked the Zn^{2+} ions decreasing Zn^{2+} ions that combines to O^{2-} to form urchin like ZnO structure. When SiO_2 from RHA was added, the silanol and siloxane bridges of SiO_2 attached to the OH^- ions which competes with $Zn(OH)_2$, leaving many free Zn^{2+} ions, thus forming ZnO-SiO_2 with zinc complexes (sulfates and hydroxides of zinc).

Figure 2 shows the morphological structure of ZnO and ZnO-SiO_2 deposited on silicon substrate. It can be observed from Figure 2(a) that a sea-urchin like structure of ZnO was formed far different from the flake-like formation using glass as a substrate. The nanostructures exhibits sharpened tips due to the reactive basic solution resulting to a crystal etching process. The formation of sea-urchin like nanostructures is attributed to substrate used, since silicon has higher surface energy compared to the glass substrate, then nucleation is not fast since high energy is required for the seed nuclei to adhere on the surface. In the process, the $Zn(OH)_2$ seed nuclei dissociates providing Zn^{2+} and O^{2-} that will allow the gradual formation of the sea-urchin like nanostructure. When SiO_2 from RHA was added to the bath solution, hemispherical protruding structures were observed having diameter of about ~4 μm comparable to the diameter of the ZnO sea-urchin like structure. It is possible that the functional groups of SiO_2 from RHA bonded with the OH^- ions of $Zn(OH)_2$. As the competition of ZnO and SiO_2 in attracting the OH^- ions of the $Zn(OH)_2$ continues, it is possible that more SiO_2 particles are attached to the OH^- group of the seed nuclei. The extra OH^- ions in the solution have attached to the SiO_2 particles and will cover the growing sea urchin-like ZnO. The SiO_2 has a lower surface energy than ZnO thus SiO_2 will adhere to the surface on top of ZnO. In these cases, formation of zinc complexes would likely to

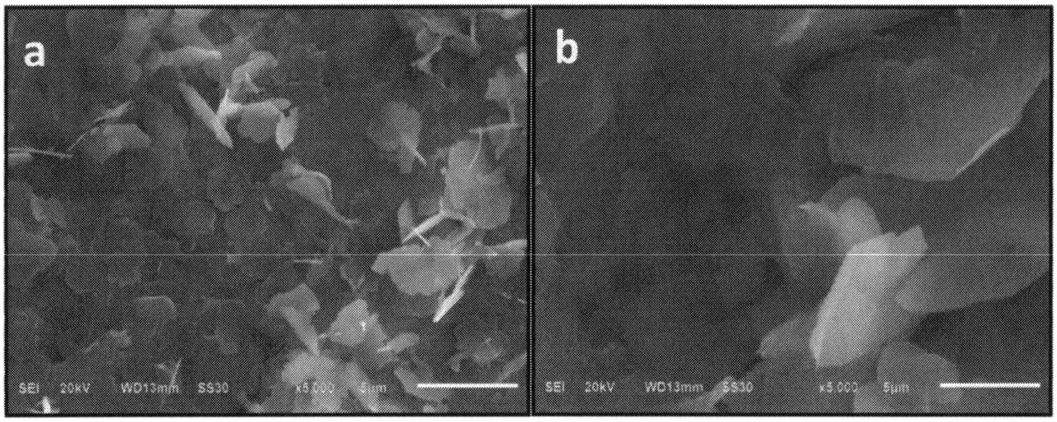

Figure 1. Scanning electron micrograph of (a) ZnO and (b) ZnO-SiO_2 deposited on glass substrate.

Figure 2. Scanning electron micrograph of (a) ZnO and (b) ZnO-SiO_2 deposited on silicon substrate.

happen and these may affect the formation of the hemispherical structures. Elemental composition analysis results verified the presence of ZnO and ZnO-SiO_2, however it showed the non-stoichiometric ratio of Zn, O and Si suggesting that Zn-complexes are indeed present in the samples.

Table 1 shows the elemental composition of the grown ZnO and ZnO-SiO_2 nanostructures deposited on glass and silicon substrates. For ZnO deposited on glass, the [Zn] to [O] ratio is not equal to unity. It is believed that the ZnO structures are formed with some zinc complexes such as $Zn(HSO)_4$ and $(NH_4)_2ZnO_2$. For ZnO-SiO_2 also on glass, large atom composition of Zn and O atoms were detected. The solution is oxygen rich due to the precursors and the substrate. Silicon atom has significant percentage observed after the incorporation of SiO_2 powder. The ratio of Zn to O is not significant, it decreases with respect to [Zn]/[O] ratio of the bare ZnO, showing high percentage of oxygen atom. The number of oxygen atoms also increases when added with amorphous SiO_2 powder. It further suggests that the formations onto the substrate are not pure ZnO-SiO_2, but there is also zinc complexes incorporated during the formations.

For the ZnO on silicon substrate, the [Zn] to [O] ratio is equivalent to 4.21 depicting that an excess of zinc is present on the grown ZnO nanostructures. For the ZnO-SiO_2 on silicon substrate, the hemispherical structures are composed rarely of Zn with a trace amount of SiO_2 due to SiO_2 powder at the surface which indicates that the SiO_2 are possibly covers the ZnO since large percentage are attributed for Si and O atoms.

Table 1. Quantitative EDS data (% atom) of ZnO and ZnO-SiO_2 nanostructures deposited on glass and silicon substrates

	Glass substrate		**Silicon substrate**	
Element	ZnO	ZnO-SiO_2	ZnO	ZnO-SiO_2
Zn	38.25	22.05	77.81	1.01
O	60.20	56.35	18.48	55.83
Si	1.55	21.60	3.71	43.16

3.2. Infrared Characteristics

The infrared characteristics of ZnO-SiO_2 nanostructures deposited on glass and silicon substrates are shown in Figure 3. The characteristic peaks at 738 cm^{-1} and 642 cm^{-1} are observed indicating the presence of ZnO. The peaks at 1042 cm^{-1}, and 830 cm^{-1} are the characteristic peaks for Si-O-Si asymmetric and symmetric stretching, which indicate the presence of SiO_2 on the sample. The two distinct peaks for Si-O-Si bending at 1502 cm^{-1} and Si-OH at 1386 cm^{-1} are attributed to the SiO_2 particles from RHA that attached to ZnO. The characteristic peaks for O-H stretching and bending vibration absorption at 3329 cm^{-1}, 1642 cm^{-1}, and 704 cm^{-1} are shown also in the FTIR spectra which correspond to the adsorbed water and for the Zn-complexes like zinc hydroxides and zinc sulfates. The presence of numerous O-H vibrations suggests that there is significant volume of Zn-complexes deposited for both substrates. This supports the EDX results that show a ratio of Zn and O which is near to the stoichiometric ratio of Zn to O for $Zn(OH)_2$.

The absorption peak at 940 cm^{-1} that corresponds to Si-O-Zn stretching is also observed which confirms that the SiO_2 from RHA has bonded with ZnO nanoparticles. A span of peaks is observable between 2400 cm^{-1} and 1900 cm^{-1} and this may be due to the dissolved volatile species like CO_2 which is acquired when exposed to air. The spectrum of ZnO-SiO_2 deposited on glass has more pronounced absorption peaks than the spectrum of the nanostructures deposited on silicon because of the SiO_2 composition of the amorphous glass substrates.

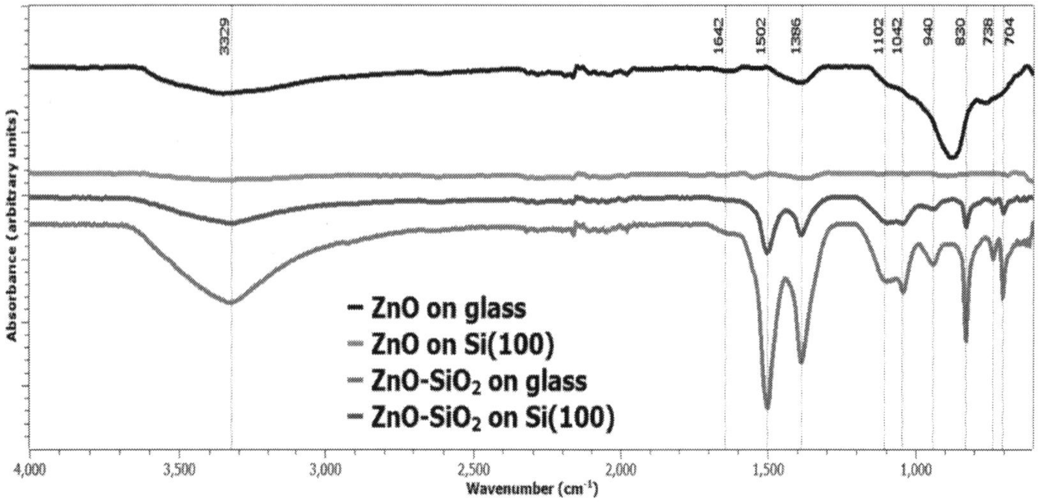

Figure 3. Infrared spectra of ZnO and ZnO-SiO_2 deposited on glass and silicon substrates.

4. Conclusions

ZnO-SiO_2 nanostructures have been deposited on glass and silicon substrates via chemical bath deposition method. SEM-EDX and FTIR reveals the incorporation of SiO_2 on ZnO surface. Flake-like ZnO-SiO_2 is deposited on glass substrate while protruding hemispherical structures are deposited on silicon substrate. Flake-like ZnO-SiO_2 nanostructures are due to a high nucleation rate and a high growth rate in a high OH^- concentration. Precipitates quickly formed at a high concentration and random sites acted as nucleation sites for the nanostructures. As a result, random growth directions of ZnO-SiO_2 causing flake-like structures. The high growth rate does not allow enough time for ions to move into proper sites. The surface morphology and the infrared characteristics of the ZnO-SiO_2 nanostructures depend on the concentration and substrate being used. Results also showed that the synthesized nanostructures are not pure ZnO-SiO_2 but there are also Zn-complexes incorporated into the formations.

References

[1] Suchea M, Christoulakis S, Moschovis K, Katsarakis N, and Kiriakidis G 2006 *ZnO Thin Solid Films* **515** 551-554

[2] Struk P, Pustelny T, Gut K, Golaszewska K, Kaminska E, Ekielski M, Pasternak I, Lusakowska E and Piotrowska A 2006 *Acta Physica Polonica A* **116** 3

[3] Sekine N, Chou C, Kwan W L, and Yang Y, 2009 *Organic Electron.* **10** 169392 1473–1477

[4] Emanetoglu N W, Zhu J, Ying Chen, Zhong J, Yimin Chen, and Y. 2004 *Appl. Phys. Lett.* **85** 17 3702-3704

[5] Van de Pol F C M, Blom F R, and Popma T J A *1991 Thin Solid Films* **204** 349-364

[6] Choi J H, Seo J S, Cha S N, Kim H J, Kim S M, Park Y J, Kim S, Yoo J, and Kim J M 2011 *Japan. J. Appl. Phys.* **50** 015001 1-4

[7] Mosnier J, O'Haire R J, McGlynn E, M. Henry O, McDonnell S J, Boyle M A, and McGuigan K A 2009 *Sci. and Technol. of Advanced Materials* **10** 045003 1-10

[8] Jia G, Wang Y and Yao J 2010 *J. Ovonic Research* **6** 6 303-307

[9] Guziewicz E, Godlewski M, Krajewski T, Wachnicki L, Szczepanik A et al. 2009 *J. Appl. Phys.* **105** 122413 1-5

[10] Mazon C, Muci J, Sa-Neto A, Ortiz-Conde A and Garcia F J 1991 *Photovoltaic Specialists Conference 1991 Conference Record of the Twenty Second IEEE* **2** 1156-1161

[11] Xiu F, Yang Z, Zhaoa D, Liu J, Alim K A, Balandin A A, Itkis M E and Haddon R C 2006

J. Cryst. Growth **286** 61-65
[12] Xia H L and Tang F Q 2003 *S. Phys. Chem. B* **107** 9175-9178
[13] Tang F, et al. 2003 *Technical Proceedings of the 2003 Nanotechnology Conference and Trade Show* **3** 262-265
[14] Abdullah M, Shibamoto S and Okuyama K 2004 *Optical Materials* **26** 95-100
[15] Sounderya N and Zhang Y 2008 *Recent Patents on Biomedical Engg.* **1** 34-42
[16] Cui H, Zayat M and Levy D 2009 *J. of Alloys and Compounds* **474** 292-296
[17] Han K, et al. 2007 *Materials Lett.* **61** 363-368
[18] Tanigaki T, et al. 2003 *J. Cryst. Growth* **256** 317-323
[19] El Mir L, Amlouk A, Barthou C and Alaya S 2007 *Physica B* **388** 412-417

Acknowledgment

The authors would like to thank the Department of Science and Technology (DOST) of the Philippines for the equipment used for the synthesis and characterization of the samples. To the Commission on Higher Education - Philippine Higher Education Research Network (CHED-PHERNet) for funding this research work.

IOP Conf. Series: Materials Science and Engineering **79** (2015) 012032 doi:10.1088/1757-899X/79/1/012032

Morphological and voltammetric characterization of different concentrations of spin coated Nafion-$Ru(bpy)_3^{2+}$ thin films

B SB Tuason[1], S Tiong-Palisoc and M Tiamzon-Natividad

Materials Science Laboratory, Physics Department, De La Salle University, 2401 Taft Ave., Manila, Philippines

Email: benjamin.tuason@dlsu.edu.ph

Abstract. Different concentrations of $Ru(bpy)_3^{2+}$ were successfully deposited on ITO coated glass substrates employing a simple method using a spin coater. The surface morphology was determined by scanning electron microscopy (SEM). The transport mechanism and the diffusion coefficient of the redox mediators within the films were characterized using cyclic voltammetry (CV). The concentrations were varied by dissolving different amounts of the redox mediator in methanol and 5% Nafion. SEM micrographs showed that the roughness of the surface increased with concentration of redox mediator. CV showed successful incorporation of $Ru(bpy)_3^{2+}$. The order of the magnitude of the diffusion coefficients confirmed that the redox mediators were immobilized within the Nafion thin film. The resulting amount of redox mediator immobilized can be manipulated by simply varying the concentration of the casting solution.

1. Introduction

Nafion has received a considerable interest in technological and scientific researches due to its stability in a wide range of applications. Nafion is biologically inert and is mechanically, chemically and electrochemically stable [1].These properties make Nafion to be the primary material in numerous applications such as fuel cells, water electrolysers, chloralkali cells, chemically modified electrodes, ion-selective electrodes and especially, sensors.

In sensor applications, electrodes are coated with mediator doped Nafion films since the electrode surface is prone to fouling effect which is caused by the organic components' adsorption[2]. Redox mediators are often used in reactions of organic compounds which react very slowly with the bare electrode surface. It functions as an intermediate oxidizing/reducing agent, which facilitates the transport of electrons between the organic species in the bulk solution and the electrode. Nafion, as the cation exchange polymer is a suitable coating because it prevents anionic compounds from scavenging the cationic mediators as it would if the mediators are situated in the bulk solution as well.

In electroanalytical studies, Nafion coated modified electrodes are commonly used for determining trace concentrations of electroactivecations[3]. Voltammetric signals relevant to the preconcentrated analytes are dependent on the physical feature of the Nafion film. This is due to the fact that the ionic conductivity of Nafion is strongly dependent on the morphological characteristics of the film [3]. It is

[1] To whom any correspondence should be addressed.

this reason why exploring different methods of depositing Nafion films and studying the resulting electrochemical behavior is relevant.

Nafion, when doped with the redox mediator, ruthenium(II) bipyridylRu(bpy)$_3^{2+}$,allows the flow of electrons to the ITO coated electrode surface. This conducting material is widely used because of its high transmittance and low electrical resistivity.

This study focuses on the variation of the concentration of the thin films of Nafion integrated with Ru(bpy)$_3^{2+}$ using the spin coating technique. The surface morphology was determined by scanning electron microscopy. The transport mechanism, and the diffusion coefficient of the redox mediators within the films were characterized using cyclic voltammetry.

2. Experimental

Modified electrodes were fabricated from different concentrations of casting solutions using the spin coating technique. The concentrations were based on those used in the one step procedure by Bertoncello *et al.* [4]. Films with different thicknesses were produced by varying the speed at which the films were thinned out. The spin coated substrates were baked in a furnace at around 70-80°C for 30 minutes.

2.1. Preparation of substrates

Indium tin oxide (ITO) coated glass substrates (sheet resistance 15-25 Ω/sq. slide) with dimensions of 7.5 cm x 1.0 cm were cut using a diamond tip glass cutter on the uncoated part the glass. The substrates were cut into 2.5 cm x 1.0 cm strips for use in fabricating the electrodes and ~1.25 cm x 1.0 cm for the SEM and characterization. The substrates were sonicated in acetone and isopropanol for 5 minutes, respectively. The substrates were then placed in a petri dish until it dried and was immediately used for the film depositions.

2.2. Preparation of Nafion-Ru(bpy)$_3^{2+}$stock solutions

A protocol based on the one-step procedure of Bertoncello *et al.*[4] was used to incorporate the cationic redox-active Ru(bpy)$_3^{2+}$in the Nafion film. Casting solutions of Nafion-Ru(bpy)$_3^{2+}$ were prepared by dissolving 2 mg, 1 mg and 0.2 mg of the redox mediator in 20 mL of reagent grade methanol. 0.2 mL of 5% Nafion solution was subsequently added using a Transferpette® micropipette. For simplicity, concentrations of redox mediator dissolved in the casting solution is expressed in terms of the ratio of mass(mg) of Ru(bpy)$_3^{2+}$ to amount(mL) of 5% Nafion. Thus, the three casting solutions will be referred to as 10mg/mL, 5mg/mL and 1mg/mL respectively.

The amount of Tris(bipyridyl)ruthenium(II) dichloride was carefully weighed using a BOSCH SAE200 electronic balance by differential weighing, using a laboratory grade wax paper as container. The stock solutions were sonicated to ensure that the ruthenium complex was fully dissolved. The concentrations of tris(bipyridyl)ruthenium(II) dichloride were chosen to be less than or equal to that used by Bertoncello *et al.*[4]. Greater concentrations of the mediator resulted in saturation and precipitation of the ruthenium complex in agreement with Bertoncello *et al.*[4].

2.3. Fabrication of spin coated modified electrodes

The Nafion - redox mediator solution was casted onto the previously cut ITO coated glass substrates using the spin coating technique. The SPINCOAT G3P-8 spin coating unit requires a constant supply of nitrogen gas to purge the coating chamber to minimize the effects of factors such as airborne contaminants and varying humidity.

The substrates underwent two stages of coating—the deposition and the thinning stages. During the deposition stage, the substrates were spun at 750 rpm for about 6 seconds during which, the casting solutions were sprayed onto the substrates. The substrates were then accelerated until it reached the thinning stage where it was thinned out at 1500rpm for 30 seconds.

The coated substrates were baked in a Thermolyne model 48000 furnace between 70-80°C for 30 minutes to evaporate the excess methanol solvent and make sure that the film was deposited on the substrate surface.

2.4. SEM characterization
The ~1.25 cm x 1 cm substrates were brought to the surface physics laboratory for morphological characterization. Since glass is non-metallic, the substrates were coated with gold to be able to see clear SEM images and avoid it from charging up. The samples were coated for 40 seconds on each side using JEOL Model JFC-1200 gold fine coater. The samples were then analyzed for their surface morphology and thickness using the Scanning Electron Microscopy (Model Jeol 5310).

2.5. Cyclic voltammetric analysis
Cyclic voltammetry measurements were performed using a BST8-Stat Potentiostat/Galvanostat. A standard three electrode setup was utilized. The working electrode was an ITO coated glass electrode on which the Nafion-$Ru(bpy)_3^{2+}$ films were deposited. A platinum coil was used as the counter electrode and a saturated calomel electrode (KCl saturated) was used as the reference electrode.

100 mL of 0.1 M NaCl served as the supporting electrolyte for the electroanalysis. The supporting electrolyte was purged with nitrogen for about 5 minutes before performing voltammetry measurements to remove dissolved oxygen which may react and interfere with or mask the data for the redox reaction of interest.

The area of the working electrode dipped was kept constant at ~1.0 cm^2 for all measurements. From the voltammograms obtained, the diffusion coefficients were determined.

3. Results and discussion
The discussion is based on the physical and electrochemical data obtained from the procedure stated. The tests were conducted promptly after the fabrication of modified electrodes to avoid the degradation.

3.1. Fabrication of spin coated modified electrodes
Films of Nafion-$Ru(bpy)_3^{2+}$ were fabricated from 10 mg/mL, 5 mg/mL, and 1 mg/mL casting solutions. The presence of the thin film, with a very faint yellow color, could not be easily seen visually. It must be duly noted that the thin films could be mechanically removed or damaged, thus, extra care was required in handling the electrodes

3.2. Surface morphology (SEM micrographs)
Even films were deposited on the substrates for all three concentrations of the casting solutions. Aside from the edges of the substrates where there were voids, the spin coating technique was able to successfully cover the whole surface of the electrode with the Nafion solutions. This proved that the method employed was successful in depositing films on modified electrodes. The thin films deposited on the glass electrodes only had minimal irregularities and/or contamination. To check for the smoothness of the films, samples were investigated at high SEM magnification. Only at much higher magnification (7500x and 20,000x) was the roughness of the film clearly seen.

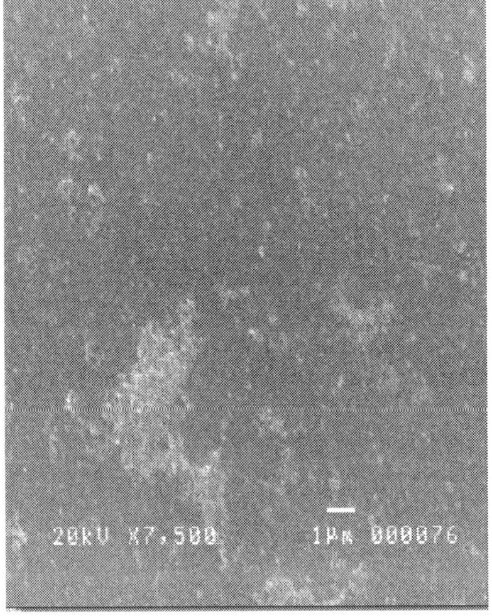

Figure 1. Surface morphology of 10mg/mL thin films at7500x magnification.

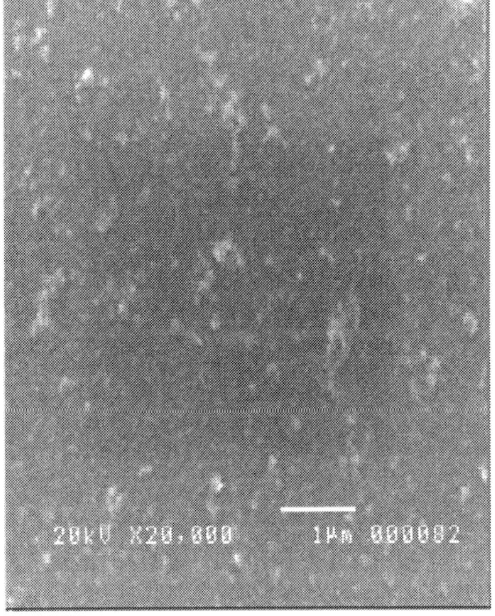

Figure 2. Surface morphology of 10mg/mL thin films at 20000x magnification.

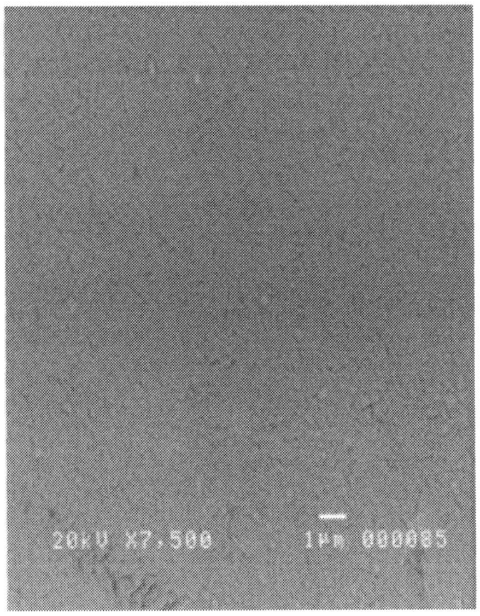

Figure 3. Surface morphology of 5mg/mL thin films at 7500x magnification.

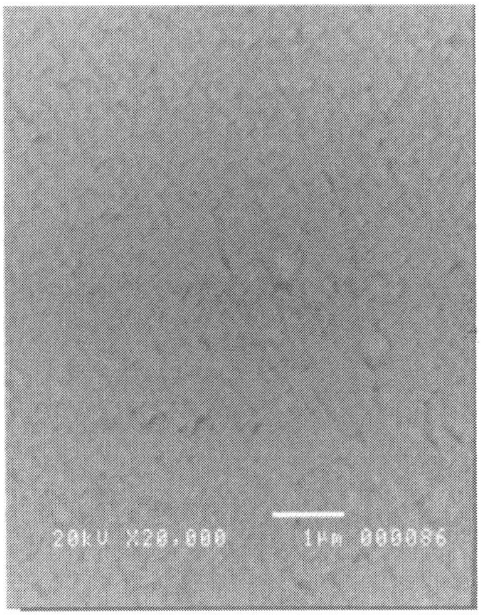

Figure 4. Surface morphology of 5mg/mL thin films at 20000x magnification.

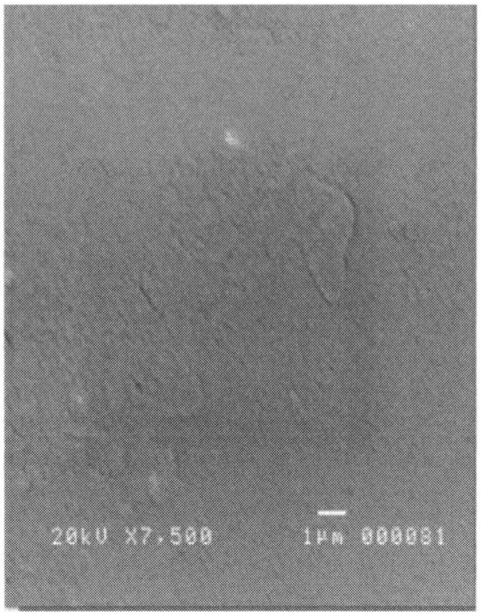

Figure 5. Surface morphology of 1mg/mL thin films at 7500x magnification.

Figure 6. Surface morphology of 1mg/mL thin films at 20000x magnification.

From the surface SEM micrographs obtained, qualitative analysis suggested a trend of decreasing roughness with increasing concentration of the casting solution.

3.3. *Film thicknesses*

The samples were viewed at an angle allowing the SEM machine to measure the thickness of the films. The technique yields measurements which were used to calculate the volume of the film which was important in determining the electrochemical properties of the thin films such as the diffusion coefficient.

Table 1. Thicknesses of three concentrations of films.

10mg/mL	5mg/ml	1mg/ml
294.00 nm	252.00 nm	248.25 nm

Obtained thickness measurements confirmed that the spin coating technique is capable of fabricating thin films with a few hundred nanometers of thickness. Increasing the concentration of the redox mediator in the casting solution increases the thickness of the film. This may be attributed to the increase in the viscosity and density of the solution as these two are factors which affect the thickness dynamics of the spin coating technique.

3.4. *Electrochemical measurements*

As a control sample, a voltammogram at 100mV/s scan rate of a bare electrode is given in the figure 7.After numerous attempts at getting voltammetric signals relevant to the immobilized redox mediator, it is concluded that the signals from electrodes fabricated with 1mg/mL Nafion–$Ru(bpy)_3^{2+}$ were too weak to be detected. Discussion, comparisons and computations will focus on two concentrations of casting solutions of the Nafion-redox mediator (10 mg/mL and 5 mg/mL) which yielded positive results. The symmetry of the voltammogram, shows that the reaction is reversible.

IOP Conf. Series: Materials Science and Engineering 79 (2015) 012032 doi:10.1088/1757-899X/79/1/012032

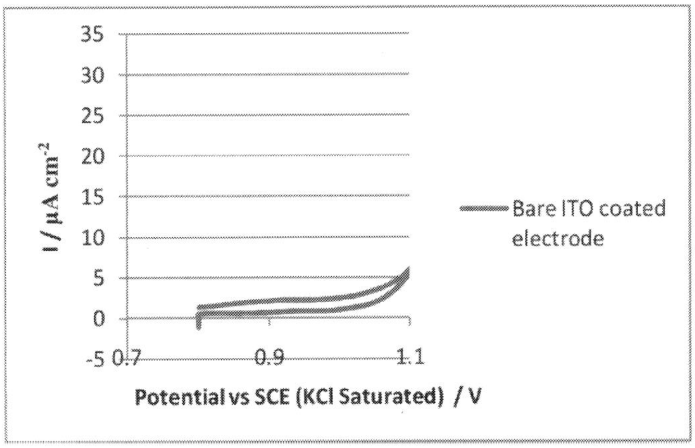

Figure 7.CV at 100mv/s of a bare ITO coated electrode.

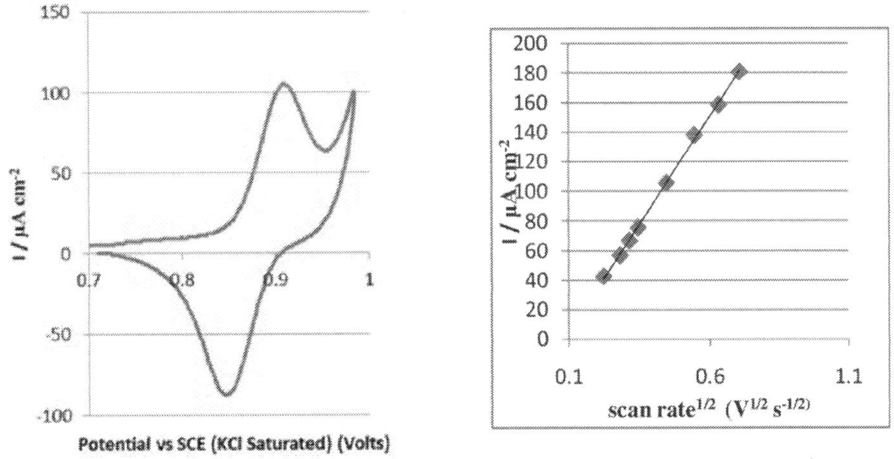

(a) (b)

Figure 8. (a) 200mV/s CV of 10mg/ml films; (b) peak current density plotted against the square root of scan rate.

The increase in peak current at increasing scan rates confirms that the diffusion layer is prevented from moving too far from the electrode surface which results in a larger current flux [5]. Also, the graph of the peak current vs. potential shows the mechanism at which the reaction proceeds. The increase in peak current with thickness corresponds to an increase in the total charge which reacted. The amount of electrons which participated in the redox reaction within the specific potential window can be accounted to the ruthenium complex incorporated within the film. This can be computed through the total charge.

The peak current densities increase monotonically with the square root of the scan rate indicating that for both 10mg/mL and 5mg/mL films, the reaction of the system is diffusion-controlled. This mechanism is operative when the thickness of the diffusion layer is thinner than the polymeric thickness [4]. Diffusion controlled reactions happen when the reaction between the electrode and species of interest occur so quickly that rate of diffusion controls the rate of the chemical reaction [6].

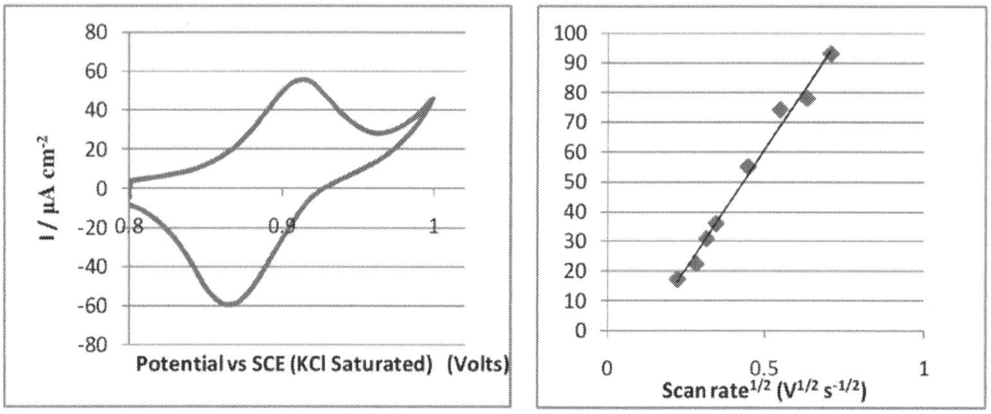

(a) (b)

Figure 9. (a) 200mV/s CVs of 5mg/ml films; (b) peak current density plotted against the square root of scan rate.

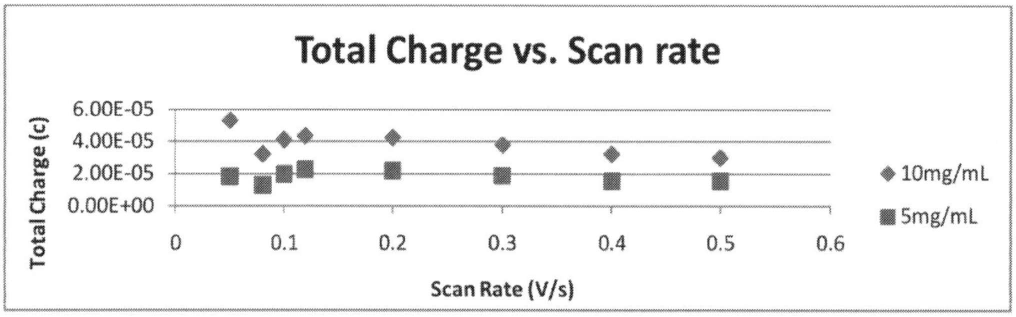

Figure 10. Total charge at varying scan rates of 10mg/mL(red), and 5mg/mL(blue) films.

The total charge (Q) was computed and it can be seen from the figure 10 that the total charge, which is proportional to the amount of redox mediator in the film, approximately remains constant even at varying scan rates. In the spin coating method, the amount of redox mediator immobilized within the Nafion film is controlled by varying the concentration of redox mediator in the casting solution.

The value for total charge obtained was used to estimate the surface coverage values, Γ, since the amount of electrons correspond to the amount of mediator which reacted during the voltammetric scan. The concentration of mediator within the film, C_p, was also computed and consequently the diffusion coefficients for the mediator within the films casted from 10mg/mL and 5mg/mL.

From the total charge Q on the cathodic(or anodic) region, the surface coverage values, Γ, were computed using:

$$\Gamma = Q / nFA \tag{3.1}$$

where n is the moles electrons involved in the reaction (equal to 1 in all cases of this study), F is the Faraday constant (96486 C / mol) and A is the geometric area of the electrode(1cm^2 in all measurements as discussed in the methodology). Dividing the surface coverage by the film thickness obtained from the SEM measurements gives the concentrations, C_p, of the redox mediators incorporated within the Nafion thin films.

$$I_p = k\, n^{3/2} A\, C\, D^{1/2} v^{1/2} \tag{3.2}$$

The Randles-Sevcik[7] (eq. 3.2) was used to calculate for the diffusion coefficient, D (cm^2s^{-1}), of the immobilized redox mediators within the films. I_p is the peak current (in amperes), and v is the scan rate. The 2.686 x 10^5 constant has units of C $mol^{-1}V^{-1/2}$. The values computed are within the typical range of values for diffusion coefficients of immobilized mediators which is around 10^{-11} to 10^{-8} cm^2s^{-1} in contrast with diffusional mediators which are in the range of 10^{-7} to 10^{-5} cm^2s^{-1} [8].

Table 2 summarizes the average values for the parameters which are relevant in characterizing the deposited films. The values computed at different scan rates were averaged to come up with one average value of diffusion coefficient per sample. The scan rate does not affect the total charge, corresponding to the amount of species which reacted, as well as the diffusion coefficient which should remain constant as it is an intrinsic property of the fabricated film.

Table 2. Average of parameters extracted using cyclic voltammetry on different thicknesses of 10mg/mL and 5mg/mL Nafion–$Ru(bpy)_3^{2+}$films.

Concentration	Thickness(cm)	Γ (mol/cm^2)	C_p (mol/cm^3)	D_{app} (cm^2/s)
10mg/mL	0.0000294	4.052E-10	1.378E-05	4.32783E-09
5mg/mL	0.0000252	1.899E-10	7.536E-06	3.19292E-09

Finally, to investigate if the modified electrodes are capable of retaining the redox mediators within the film, a 5 mg/mL electrode was subjected to continuous cycling and the total current for each cycle was measured. The results evidenced an approximate 11.2% loss in mediator after 50 cycles. The efficiency of spin coated films in retaining the mediators immobilized within the film is similar to the LS films fabricated by Bertoncello *et al.* [4] which evidenced a 10% loss of mediators after 50 cycles.

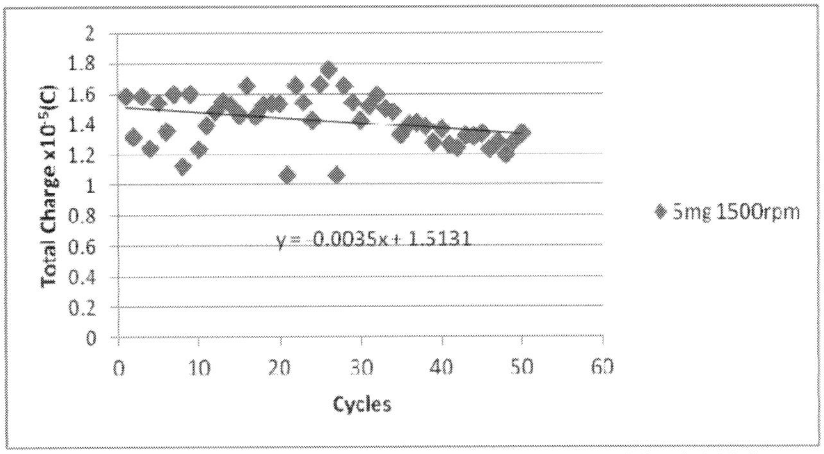

Figure 11. Total charge through continuous cycling(50 cycles) of 5mg/mL Nafion-$Ru(bpy)_3^{2+}$ electrodes.

4. Conclusions

The spin coating method was used to successfully produce redox-active thin films of Nafion. The redox mediator $Ru(bpy)_3^{2+}$ used in the study was effectively incorporated and immobilized within the Nafion modified electrodes. As seen from the results, the casting solution concentration could be manipulated to control the properties of the fabricated thin films such as the film thickness and concentration of redox mediator immobilized within the films. Thickness measurements showed that the thickness was proportional to the concentration of the casting solution, the density and/or viscosity were also factors which affected the film thickness.

IOP Conf. Series: Materials Science and Engineering **79** (2015) 012032 doi:10.1088/1757-899X/79/1/012032

SEM micrographs showed that the surface of the deposited thin films was generally smooth. However, roughness was observed for the highest concentration (10mg/mL) and reviewed further at higher magnification.

Voltammetric analyses showed that the redox-active thin film was successfully fabricated by modifying and employing the one-step method in spin coating technique. The order of the magnitude of the diffusion coefficients confirmed that the redox mediators were immobilized within the Nafion film.

5. References

[1] Bertoncello P, Ram M K, Notargiacomo A, Ugo P and Nicolini C 2000 *Physical Chemistry Chemical Physics* **4** 4036–43

[2] Neto M *et al.* 2001 *Portugaliae Electrochimica Acta* **19** 57–71

[3] Bertoncello P and Ugo P 2003 *J. Braz. Chem. Soc.* **14** 517–22

[4] Bertoncello P, Wilson N R and Unwin P R 2007 *Soft Matter* **3** 1300–7

[5] University of Cambridge*Linear Sweep* Retrieved from http://www.cheng.cam.ac.uk/research/groups/electrochem/JAVA/electrochemistry/ELEC/l4 html/cv.html

[6] Univeristy of Athens *Diffusion Controlled Electrode Processes* Retrieved from http://www.chem.uoa.gr/applets/AppletDiffus/Appl_Diffus2.html

[7] University of Adelaide *Department of Chemistry* Retrieved from http://www.chemistry.adelaide.edu.au/external/soc-rel/content/cv.htm

[8] Vielstich W, Gasteiger H A and Yokokawa H 2009 *Handbook of Fuel Cells: Fundamentals Technology and Applications*(United Kingdom: Wiley and Sons)